013759195

WITHDRAWN
FROM STOCK

Graphical Models

Graphical Models

Methods for Data Analysis and Mining

Christian Borgelt and Rudolf Kruse
Otto-von-Guericke-University of Magdeburg, Germany

JOHN WILEY & SONS, LTD

Baffins Lane, Chichester,
West Sussex, PO19 1UD, England

National 01243 779777
International (+44) 1243 779777

e-mail (for orders and customer service enquiries): cs-books@wiley.co.uk

Visit our Home Page on http://www.wiley.co.uk
or http://www.wiley.com

Other Wiley Editorial Offices

John Wiley & Sons, Inc., 605 Third Avenue,
New York, NY 10158-0012, USA

WILEY-VCH Verlag GmbH
Pappelallee 3, D-69469 Weinheim, Germany

John Wiley & Sons Australia, Ltd
33 Park Road, Milton, Queensland 4064, Australia

John Wiley & Sons (Canada) Ltd, 22 Worcester Road
Rexdale, Ontario, M9W 1L1, Canada

John Wiley & Sons (Asia) Pte Ltd, 2 Clementi Loop #02-01,
Jin Xing Distripark, Singapore 129809

Library of Congress Cataloging-in-Publication Data

Borgelt, Christian.
Graphical models : methods for data analysis and mining / Christian Borgelt, Rudolf Kruse.
p. cm.
Includes bibliographical references and index.
ISBN 0-470-84337-3 (alk. paper)
1. Data mining 2. Mathematical statistics – Graphic methods. I. Kruse, Rudolf. II. Title.

QA76.9.D43 B67 2001
006.3 – dc21
2001046761

British Library Cataloguing in Publication Data

A catalogue record for this book is available from the British Library

ISBN 0-470-84337-3

Set in LaTeX, printed from pdf files supplied by the author.
Printed and bound in Great Britain by Antony Rowe, Chippenham, Wiltshire.
This book is printed on acid-free paper responsibly manufactured from sustainable forestry, in which at least two trees are planted for each one used for paper production.

Contents

Preface

Although the origins of graphical models can be traced back to the beginning of the century, they have become truly popular only since the mid-eighties, when several researchers started to use Bayesian networks in expert systems. But as soon as this start was made, the interest in graphical models grew rapidly and is still growing to this day. The reason is that graphical models, due to their explicit treatment of (conditional) dependences and independences, proved to be clearly superior to naive approaches like certainty factors attached to if-then-rules, which had been tried earlier.

Data Mining, also called Knowledge Discovery in Databases, is a another relatively young area of research, which has emerged in response to the flood of data we are faced with nowadays. It has taken up the challenge to develop techniques that can help humans discover useful patterns in their data. In industrial applications patterns found with these methods can often be exploited to improve products and to increase turnover.

This book is positioned at the boundary between these two highly important research areas, because it focuses on learning graphical models from data, thus exploiting the recognized advantages of graphical models for data analysis and mining. Its special feature is that it is not restricted to probabilistic models like Bayesian and Markov networks. It also explores relational graphical models, which provide excellent didactical means to explain the ideas underlying graphical models. In addition, possibilistic graphical models are studied, which are worth considering if the data to analyze contains imprecise information in the form of sets of alternatives instead of unique values.

Looking back, this book has become longer than originally intended. However, although it is true that, as C.F. von Weizsäcker remarked in a lecture, anything ultimately understood can be said briefly, it is also evident that anything said too briefly is likely to be incomprehensible to anyone who has not yet understood completely. Since our main aim was comprehensibility, we hope that a reader is remunerated for the length of this book by an exposition that is clear and thus easy to read.

Christian Borgelt, Rudolf Kruse

Magdeburg, March 2002

Chapter 1

Introduction

Due to modern information technology, which produces ever more powerful computers every year, it is possible today to collect, transfer, combine, and store huge amounts of data at very low costs. Thus an ever-increasing number of companies and scientific and governmental institutions can afford to build up huge archives of tables, documents, images, and sounds in electronic form. The thought is compelling that if you only have enough data, you can solve any problem—at least in principle.

A closer inspection reveals, though, that data alone, however voluminous, are not sufficient. We may say that in large databases we cannot see the wood for the trees. Although any single bit of information can be retrieved and simple aggregations can be computed (for example, the average monthly sales in the Frankfurt area), general patterns, structures, and regularities usually go undetected. However, often these patterns are especially valuable, for example, because they can easily be exploited to increase turnover. For instance, if a supermarket discovers that certain products are frequently bought together, the number of items sold can sometimes be increased by appropriately arranging these products in the shelves of the market (they may, for example, be placed adjacent to each other in order to invite even more customers to buy them together, or they may be offered as a bundle).

However, to find these patterns and thus to exploit more of the information contained in the available data turns out to be fairly difficult. In contrast to the abundance of data there is a lack of tools to transform these data into useful knowledge. As John Naisbett remarked [Fayyad *et al.* 1996]:

> We are drowning in information, but starving for knowledge.

As a consequence a new area of research has emerged, which has been named *Knowledge Discovery in Databases (KDD)* or *Data Mining (DM)* and which has taken up the challenge to develop techniques that can help humans to discover useful patterns in their data.

In this introductory chapter we provide a brief overview on knowledge discovery in databases and data mining, which is intended to show the context of this book. In a first step, we try to capture the difference between "data" and "knowledge" in order to attain precise notions by which it can be made clear why it does not suffice just to gather data and why we must strive to turn them into knowledge. As an illustration we are going to discuss an example from the history of science. Secondly, we explain the process of discovering knowledge in databases (the *KDD process*), of which data mining is just one, though very important, step. We characterize the standard data mining tasks and position the work of this book by pointing out for which tasks the discussed methods are well suited.

1.1 Data and Knowledge

In this book we distinguish between *data* and *knowledge*. Statements like "Columbus discovered America in 1492" or "Mrs Jones owns a VW Golf" are *data*. For these statements to qualify as data, we consider it to be irrelevant whether we already know them, whether we need these specific pieces of information at this moment etc. The essential property of these statements is that they refer to single events, cases, objects, persons etc., in general, to single instances. Therefore, even if they are true, their range of validity is very restricted and thus is their usefulness.

In contrast to the above, *knowledge* consists of statements like "All masses attract each other" or "Every day at 17:04 there runs an InterRegio (a specific train) from Magdeburg to Braunschweig". Again we neglect the relevance of the statement for our current situation and whether we already know it. The essential point is that these statements do not refer to single instances, but are general laws or rules. Therefore, provided they are true, they have a wide range of validity, and, above all else, they allow us to make predictions and thus they are very useful.

It has to be admitted, though, that in daily life statements like "Columbus discovered America in 1492" are also called knowledge. However, we disregard this way of using the term "knowledge", regretting that full consistency of terminology with daily life cannot be achieved. Collections of statements about single instances do not qualify as knowledge.

Summarizing, data and knowledge can be characterized as follows:

Data

- refer to single instances
(single objects, persons, events, points in time etc.)
- describe individual properties
- are often available in huge amounts
(databases, archives)

- are usually easy to collect or to obtain
  (for example cash registers with scanners in supermarkets, Internet)
- do not allow us to make predictions

Knowledge

- refers to *classes* of instances
  (*sets* of objects, persons, events, points in time etc.)
- describes general patterns, structures, laws, principles etc.
- consists of as few statements as possible
  (this is an objective, see below)
- is usually hard to find or to obtain
  (for example natural laws, education)
- allows us to make predictions

From these characterizations we can clearly see that usually knowledge is much more valuable than (raw) data. It is mainly the generality of the statements and the possibility to make predictions about the behavior and the properties of new cases that constitute its superiority.

However, not just any kind of knowledge is as valuable as any other. Not all general statements are equally important, equally substantial, equally useful. Therefore knowledge must be assessed. The following list, which we do not claim to be complete, names some important criteria:

Criteria to Assess Knowledge

- correctness (probability, success in tests)
- generality (range of validity, conditions for validity)
- usefulness (relevance, predictive power)
- comprehensibility (simplicity, clarity, parsimony)
- novelty (previously unknown, unexpected)

In science correctness, generality, and simplicity (parsimony) are at the focus of attention: One way to characterize science is to say that it is the search for a minimal correct description of the world. In business and industry higher emphasis is placed on usefulness, comprehensibility, and novelty: the main goal is to get a competitive edge and thus to achieve higher profit. Nevertheless, none of the two areas can afford to neglect the other criteria.

Tycho Brahe and Johannes Kepler

Tycho Brahe (1546–1601) was a Danish nobleman and astronomer, who in 1576 and in 1584, with the financial support of King Frederic II, built two observatories on the island of Ven, about 32 km to the north-east of Copenhagen. Using the best equipment of his time (telescopes were unavailable then—they were used only later by Galileo Galilei (1564–1642) and Johannes Kepler (see below) for celestial observations) he determined the positions of the sun, the moon, and the planets with a precision of less than one minute of arc, thus surpassing by far the exactitude of all measurements carried out before. He achieved in practice the theoretical limit for observations with the unaided eye. Carefully he recorded the motions of the celestial bodies over several years [Greiner 1989, Zey 1997].

Tycho Brahe gathered data about our planetary system—huge amounts of data, at least from a 16th century point of view. However, he could not discern the underlying structure. He could not combine his data into a consistent scheme, partly because be adhered to the geocentric system. He could tell exactly in what position Mars had been on a specific day in 1585, but he could not relate the positions on different days in such a way as to fit his highly accurate observational data. All his hypotheses were fruitless. He developed the so-called Tychonic planetary model, according to which the sun and the moon revolve around the earth and all other planets revolve around the sun, but this model, though popular in the 17th century, did not stand the test of time. Today we may say that Tycho Brahe had a "data mining" or "knowledge discovery" problem. He had the necessary data, but he could not extract the knowledge contained in it.

Johannes Kepler (1571–1630) was a German astronomer and mathematician and assistant to Tycho Brahe. He advocated the Copernican planetary model and during his whole life he endeavored to find the laws that govern the motions of the celestial bodies. He strived to find a mathematical description, which, in his time, was a virtually radical approach. His starting point were the catalogs of data Tycho Brahe had set up and which he continued in later years. After several unsuccessful trials and long and tedious calculations, Johannes Kepler finally managed to combine Tycho Brahe's data into three simple laws, which have been named after him. Having discovered in 1604 that the course of Mars is an ellipse, he published the first two laws in "Astronomia Nova" in 1609, the third ten years later in his principal work "Harmonica Mundi" [Greiner 1989, Zey 1997, Feynman *et al.* 1963].

1. Each planet moves around the sun in an ellipse, with the sun at one focus.
2. The radius vector from the sun to the planet sweeps out equal areas in equal intervals of time.
3. The squares of the periods of any two planets are proportional to the cubes of the semi-major axes of their respective orbits: $T \sim a^{\frac{3}{2}}$.

Tycho Brahe had collected a large amount of celestial data, Johannes Kepler found the laws by which they can be explained. He discovered the hidden knowledge and thus became one of the most famous "data miners" in history.

Today the works of Tycho Brahe are almost forgotten. His catalogs are merely of historical value. No textbook on astronomy contains extracts from his measurements. His observations and minute recordings are raw data and thus suffer from a decisive disadvantage: They do not provide us with any insight into the underlying mechanisms and therefore they do not allow us to make predictions. Kepler's laws, however, are treated in all textbooks on astronomy and physics, because they state the principles that govern the motions of planets as well as comets. They combine all of Brahe's measurements into three simple statements. In addition, they allow us to make predictions: If we know the position and the velocity of a planet at a given moment, we can compute, using Kepler's laws, its future course.

1.2 Knowledge Discovery and Data Mining

How did Johannes Kepler discover his laws? How did he manage to extract from Tycho Brahe's long tables and voluminous catalogs those simple laws that revolutionized astronomy? We know only a little about this. He must have tested a large number of hypotheses, most of them failing. He must have carried out long and complicated computations. Presumably, outstanding mathematical talent, tenacious work, and a considerable amount of good luck finally led to success. We may safely guess that he did not know a universal method to discover physical or astronomical laws.

Today we still do not know such a method. It is still much simpler to gather data, by which we are virtually swamped in today's "information society" (whatever that means), than to obtain knowledge. We even need not work diligently and perseveringly any more, as Tycho Brahe did, in order to collect data. Automatic measurement devices, scanners, digital cameras, and computers have taken this load from us. Modern database technology enables us to store an ever-increasing amount of data. It is indeed as John Naisbett remarked: We are drowning in information, but starving for knowledge.

If it took such a distinguished thinker like Johannes Kepler several years to evaluate the data gathered by Tycho Brahe, which today seem to be negligibly few and from which he even selected only the data on the course of Mars, how can we hope to cope with the huge amounts of data available today? "Manual" analysis has long ceased to be feasible. Simple aids like, for example, representations of data in charts and diagrams soon reach their limits. If we refuse to simply surrender to the flood of data, we are forced to look for intelligent computerized methods by which data analysis can be automated at least partially. These are the methods that are sought for in the research areas called *Knowledge Discovery in Databases (KDD)* and *Data Mining (DM)*. It is true, these methods are still very far from replacing people like Johannes Kepler, but it is not entirely implausible that he, if supported by these methods, would have reached his goal a little sooner.

Often the terms *Knowledge Discovery* and *Data Mining* are used interchangeably. However, we distinguish them here. By *Knowledge Discovery in Databases (KDD)* we mean a process consisting of several steps, which is usually characterized as follows [Fayyad *et al.* 1996]:

> Knowledge discovery in databases is the nontrivial process of identifying valid, novel, potentially useful, and ultimately understandable patterns in data.

One step of this process, though definitely one of the most important, is *Data Mining*. In this step modeling and discovery techniques are applied.

1.2.1 The KDD Process

In this section we structure the KDD process into two preliminary and five main steps or phases. However, the structure we discuss here is by no means binding. A unique scheme has not yet been agreed upon in the scientific community. A recent suggestion and detailed exposition of the KDD process, which is close to the scheme presented here and which can be expected to have considerable impact, since it is backed by several large companies like NCR and DaimlerChrysler, is the CRISP-DM model (CRoss Industry Standard Process for Data Mining) [Chapman *et al.* 1999].

Preliminary Steps

- estimation of potential benefit
- definition of goals, feasibility study

Main Steps

- check data availability, data selection, if necessary, data collection
- preprocessing (60-80% of total overhead)
  - unification and transformation of data formats
  - data cleaning
    (error correction, outlier detection, imputation of missing values)
  - reduction / focusing
    (sample drawing, feature selection, prototype generation)
- **Data Mining** (using a variety of methods)
- visualization
  (also in parallel to preprocessing, data mining, and interpretation)
- interpretation, evaluation, and test of results
- deployment and documentation

The preliminary steps mainly serve the purpose to decide whether the main steps should be carried out. Only if the potential benefit is high enough and the demands can be met by data mining methods, can it be expected that some profit results from the usually expensive main steps.

In the main steps the data to be analyzed for hidden knowledge are first collected (if necessary), appropriate subsets are selected, and they are transformed into a unique format that is suitable for applying data mining techniques. Then they are cleaned and reduced to improve the performance of the algorithms to be applied later. These preprocessing steps usually consume the greater part of the total costs. Depending on the data mining task that was identified in the goal definition step (see below for a list), data mining methods are applied (see below for a list), the results of which, in order to interpret and evaluate them, can be visualized. Since the desired goal is rarely achieved in the first go, usually several steps of the preprocessing phase (for example feature selection) and the application of data mining methods have to be reiterated in order to improve the result. If it has not been obvious before, it is clear now that KDD is not a completely automated, but an interactive process. A user has to evaluate the results, check them for plausibility, and test them against hold-out data. If necessary, he/she modifies the course of the process to make it meet his/her requirements.

1.2.2 Data Mining Tasks

In the course of time typical tasks have been identified, which data mining methods should be able to solve (although, of course, not every single method is required to be able to solve all of them—it is the combination of methods that makes them powerful). Among these are especially those named in the—surely incomplete—list below. We tried to characterize them not only by their name, but also by a typical question [Nakhaeizadeh 1998b].

- classification
  Is this customer credit-worthy?
- segmentation, clustering
  What groups of customers do I have?
- concept description
  Which properties characterize fault-prone vehicles?
- prediction, trend analysis
  What will the exchange rate of the dollar be tomorrow?
- dependence/association analysis
  Which products are frequently bought together?
- deviation analysis
  Are there seasonal or regional variations in turnover?

Classification and prediction are the most frequent tasks, since their solution can have a direct effect, for instance, on the turnover and the profit of a company. Dependence and association analysis come next, because they can be used, for example, to do shopping basket analysis, i.e. to discover which products are frequently bought together, and are therefore also of considerable commercial interest.

1.2.3 Data Mining Methods

Research in data mining is highly interdisciplinary. Methods to solve the tasks named in the preceding section have been developed in a large variety of research areas including, to name only the most important, statistics, artificial intelligence, machine learning, and soft computing. As a consequence there is an arsenal of methods that are based on a wide range of ideas. To give an overview, we list below some of the more prominent data mining methods. Each list entry refers to a few publications on the method and points out for which data mining tasks the method is especially suited. Of course, this list is far from being complete. The references are necessarily incomplete and may not always be the best ones possible, since we are clearly not experts for all of these methods and since, obviously, we cannot name everyone who has contributed to the one or the other.

- classical statistics (discriminant analysis, time series analysis etc.)
  [Larsen and Marx 1986, Everitt 1998]
  classification, prediction, trend analysis
- decision/classification and regression trees
  [Breiman *et al.* 1984, Quinlan 1986, Quinlan 1993]
  classification, prediction
- naive Bayes classifiers
  [Good 1965, Duda and Hart 1973, Langley *et al.* 1992]
  classification, prediction
- probabilistic networks (Bayesian networks/Markov networks)
  [Pearl 1988, Lauritzen and Spiegelhalter 1988, Heckerman *et al.* 1995]
  classification, dependence analysis
- artificial neural networks
  [Anderson 1995, Rojas 1993, Haykin 1994, Zell 1994]
  classification, prediction, clustering (Kohonen feature maps)
- neuro-fuzzy rule induction
  [Wang and Mendel 1992, Nauck and Kruse 1997, Nauck *et al.* 1997]
  classification, prediction
- k-nearest neighbor/case-based reasoning
  [Dasarathy 1990, Aha 1992, Kolodner 1993, Wettschereck 1994]
  classification, prediction

- inductive logic programming
  [Muggleton 1992, de Raedt and Bruynooghe 1993]
  classification, association analysis, concept description
- association rules
  [Agrawal *et al.* 1993, Agrawal *et al.* 1996, Srikant and Agrawal 1996]
  association analysis
- hierarchical and probabilistic cluster analysis
  [Bock 1974, Everitt 1981, Cheeseman *et al.* 1988, Mucha 1992]
  segmentation, clustering
- fuzzy cluster analysis
  [Bezdek and Pal 1992, Bezdek *et al.* 1999, Höppner *et al.* 1999]
  segmentation, clustering
- conceptual clustering
  [Michalski and Stepp 1983, Fisher 1987, Hanson and Bauer 1989]
  segmentation, concept description
- and many more

Although for each data mining task there are several reliable methods to solve it, there is, as already indicated above, no single method that can solve all tasks. Most methods are tailored to solve a specific task and each of them exhibits different strengths and weaknesses. In addition, usually several methods must be combined in order to achieve good results. Therefore commercial data mining products like, for instance, Clementine (Integral Solutions Ltd./SPSS, Basingstoke, United Kingdom), DataEngine (Management Intelligenter Technologien GmbH, Aachen, Germany), or Kepler (Dialogis GmbH, Sankt Augustin, Germany) offer several of the above methods under an easy to use graphical interface.[1] However, as far as we know there is currently no tool that contains all of the methods mentioned above. A survey and evaluation of data mining tools can be found in [Gentsch 1999] (unfortunately, this report is written in German, and thus may not be accessible for all readers). Extensive lists of currently available data mining tools can be found, for instance, at the Web sites

`http://www.statserv.com/datamsoft.html` and
`http://www.xore.com/prodtable.html`.

These sites provide checklists of the incorporated methods and links to the Web sites of the individual tools and the companies offering them.

[1] We selected these systems as examples, because the first author of this book is well acquainted with them. For Clementine he developed the association rule program underlying its "apriori node" and the DataEngine plug-in "DecisionXpert" is based on a decision tree induction program that he wrote [Borgelt 1998, Borgelt and Timm 2000]. The latter program has also been incorporated into Kepler. Of course, this does not imply that these systems are superior to any other on the market.

1.3 Graphical Models

This book deals with two data mining tasks, namely *dependence analysis* and *classification.* These tasks are, of course, closely related, since classification can be seen as a special case of dependence analysis: It concentrates on specific dependences, namely on how a distinguished attribute—the class attribute—depends on other, descriptive attributes. It tries to exploit these dependences to classify new cases. Within the set of methods that can be used to solve these tasks, we focus on techniques to induce *graphical models* or, as we will also call them, *inference networks* from data.

[Lauritzen 1996] traces the ideas of graphical models back to three origins, namely statistical mechanics [Gibbs 1902], genetics [Wright 1921], and the analysis of contingency tables [Bartlett 1935]. Originally, they were developed as means to build models of a domain of interest. The rationale underlying them is that, since high-dimensional domains tend to be unmanageable as a whole (and the more so if imprecision and uncertainty are involved), it is necessary to *decompose* the available information. In *graphical modeling* [Whittaker 1990, Kruse *et al.* 1991, Lauritzen 1996] such a decomposition is based on (conditional) dependence and independence relations between the attributes used to describe the domain under consideration. The structure of these relations is represented as a network or graph (hence the names graphical model and inference network), often called a *conditional independence graph.* In such a graph each node stands for an attribute and each edge for a direct dependence between two attributes.

However, this graph turns out to be not only a convenient way to represent the content of a model. It can also be used to facilitate reasoning in high-dimensional domains, since it reduces inferences to computations in lower-dimensional subspaces. Propagating evidence about the values of observed attributes to unobserved ones can be implemented by locally communicating node processors and therefore is very efficient. As a consequence, graphical models are often used in expert and decision support systems [Neapolitan 1990, Kruse *et al.* 1991, Cowell 1992, Castillo *et al.* 1997]. In such a context, i.e. if graphical models are used to draw inferences, we prefer to call them *inference networks* in order to emphasize their objective.

Using inference networks to facilitate reasoning in high-dimensional domains has originated in the probabilistic setting. *Bayesian networks* [Pearl 1986, Pearl 1988, Jensen 1996], which are based on directed conditional independence graphs, and *Markov networks* [Isham 1981, Lauritzen and Spiegelhalter 1988, Pearl 1988, Lauritzen 1996], which are based on undirected graphs, are the most prominent examples. Early efficient implementations of these types of networks include HUGIN [Andersen *et al.* 1989] and PATHFINDER [Heckerman 1991]. Among the best-known applications of probabilistic graphical models are the interpretation of electromyographic findings (the MUNIN program) [Andreassen *et al.* 1987], blood group determination of Danish Jersey cattle for parentage verification (the BOBLO network) [Rasmussen 1992], and troubleshooting non-functioning devices like printers and photocopiers [Heckerman *et al.* 1994].

However, graphical modeling has also been generalized to be usable with other uncertainty calculi than probability theory [Shafer and Shenoy 1988, Shenoy 1992b, Shenoy 1993], for instance in the so-called valuation-based networks [Shenoy 1992a], and has been implemented, for example, in PULCINELLA [Saffiotti and Umkehrer 1991]. Due to their connection to fuzzy systems, which in the past have successfully been applied to solve control problems, recently possibilistic networks gained some attention, too. They can be based on the context model interpretation of a degree of possibility, which focuses on imprecision [Gebhardt and Kruse 1993a, Gebhardt and Kruse 1993b], and have been implemented, for example, in POSSINFER [Gebhardt and Kruse 1996a, Kruse *et al.* 1994].

For some time the standard approach to construct a graphical model has been to let a human domain expert specify the dependences in the considered domain. This provided the network structure. Then the human domain expert had to estimate the necessary conditional or marginal distribution functions that represent the quantitative information about the domain. This approach, however, can be tedious and time consuming, especially if the domain under consideration is large. In some situations it may even be impossible to carry it out, because no or only vague expert knowledge is available about the (conditional) dependence and independence relations that hold in the considered domain. Therefore recent research has concentrated on learning graphical models from databases of sample cases (cf. [Herskovits and Cooper 1990, Cooper and Herskovits 1992, Buntine 1994, Heckerman *et al.* 1995, Jordan 1998] for learning probabilistic networks and [Gebhardt and Kruse 1995, Gebhardt and Kruse 1996b, Gebhardt and Kruse 1996c, Borgelt and Kruse 1997a, Borgelt and Kruse 1997b, Borgelt and Gebhardt 1997] for learning possibilistic networks). As a consequence, graphical models entered the realm of data mining methods.

Graphical models have several advantages when applied to knowledge discovery and data mining problems. In the first place, as already pointed out above, the network representation provides a comprehensible qualitative (network structure) and quantitative description (associated distribution functions) of the domain under consideration, so that the learning result can be checked for plausibility against the intuition of human experts. Secondly, learning algorithms for inference networks can easily be extended to incorporate the background knowledge of human experts. In the simplest case a human domain expert specifies the dependence structure of the domain to be modeled and automatic learning is used only to determine the distribution functions from a database of sample cases. More sophisticated approaches take a prior model of the domain and modify it (add or remove edges, change the distribution functions) w.r.t. the evidence provided by a database of sample cases [Heckerman *et al.* 1995]. Finally, although the learning task has been shown to be NP-complete in the general case [Chickering *et al.* 1994, Chickering 1995], there are several good heuristic approaches that have proven to be successful in practice and that lead to very efficient learning algorithms.

In addition to these practical advantages, graphical models provide a framework for some of the data mining methods named above: Naive Bayes classifiers are probabilistic networks with a special, star-like structure (cf. Chapter 6). Decision trees can be seen as a special type of probabilistic network in which there is only one child attribute and the emphasis is on learning the local structure of the network (cf. Chapter 8). Furthermore there are some interesting connections to fuzzy clustering [Borgelt *et al.* 2001] and neuro-fuzzy rule induction [Nürnberger *et al.* 1999] through naive Bayes classifiers, which may lead to powerful hybrid systems.

1.4 Outline of this Book

This book covers three types of graphical models: relational, probabilistic, and possibilistic networks, with relational networks mainly being drawn on to provide more comprehensible analogies. In the following we give a brief outline of the chapters.

In Chapter 2 we review very briefly relational and probabilistic reasoning and then concentrate on possibility theory, for which we provide a detailed semantical introduction based on the context model [Gebhardt and Kruse 1993a, Gebhardt and Kruse 1993b]. In this chapter we clarify and at some points modify the context model interpretation of a degree of possibility where we found its foundations to be weak or not spelled out clearly enough.

In Chapter 3 we study how relations and probability and possibility distributions, under certain conditions, can be decomposed. By starting from the simple relational networks, which are usually neglected entirely in introductions to graphical modeling, we try to make the theory of graphical models and reasoning in graphical models more easily accessible [Borgelt and Kruse 1998a]. In addition, by developing a peculiar formalization of relational networks a very strong formal similarity can be achieved to possibilistic networks. In this way possibilistic networks can be introduced as simple "fuzzyfications" of relational networks.

In Chapter 4 we explain the connection of decompositions of distributions to graphs that is brought about by the notion of *conditional independence*. In addition we briefly review two of the best-known propagation algorithms for inference networks. However, although we provide a derivation of the polytree propagation formulae that is based on the notion of evidence factors, this chapter does not contain a full exposition of evidence propagation. This topic has been covered extensively in other books, and thus we only focus on those components that we need for later chapters.

With Chapter 5 we turn to learning graphical models from data. We study a fundamental learning operation, namely how to estimate projections, i.e. marginal distributions, from a database of sample cases. Although trivial for the relational and the probabilistic case, this operation is a severe problem in the possibilistic case. Therefore we explain and formally justify an efficient method for computing maximum projections of database-induced possibility distributions [Borgelt and Kruse 1998c].

In Chapter 6 we derive a possibilistic classifier in direct analogy to a naive Bayes classifier [Borgelt and Gebhardt 1999].

In Chapter 7 we proceed to *qualitative* or *structural learning*, i.e. how to induce a network structure from a database of sample cases. Following an introduction to the principles of global structure learning, which is intended to provide an intuitive background (like the greater part of Chapter 3), we discuss some well-known and suggest some new evaluation measures for learning probabilistic as well as possibilistic networks [Borgelt and Kruse 1997a, Borgelt and Kruse 1998b, Borgelt and Kruse 2001]. Furthermore, we review some well-known and suggest some new search methods.

In Chapter 8 we extend qualitative network induction to learning local structure. We explain the connection to decision trees and decision graphs and suggest a modification of an approach by [Chickering *et al.* 1997] to local structure learning for Bayesian networks [Borgelt and Kruse 1998d].

In Chapter 9 we study the causal interpretation of learned Bayesian networks and in particular the so-called *inductive causation algorithm* [Pearl and Verma 1991a, Pearl and Verma 1991b]. Based on [Borgelt and Kruse 1999] we study carefully the assumptions underlying this approach and reach the conclusion that the strong claims made about this algorithm cannot be justified, although it provides useful heuristics.

In Chapter 10 we report about some successful applications of graphical models in the telecommunications and automotive industry.

Software and additional material that is related to the contents of this book can be found at the website

`http://fuzzy.cs.uni-magdeburg.de/books/gm/.`

Chapter 2

Imprecision and Uncertainty

Since this book is about graphical models and reasoning with them, we start by saying a few words about reasoning in general, with a focus on inferences under *imprecision* and *uncertainty* and the calculi to model these (cf. [Borgelt *et al.* 1998a]). The standard calculus to model imprecision is, of course, *relational algebra* and its special case (multidimensional) interval arithmetics. However, these calculi neglect that the available information may be uncertain. On the other hand, the standard calculi to model uncertainty for decision making purposes are *probability theory* and its extension utility theory. However, these calculi cannot deal very well with imprecise information—seen as set-valued information—in the absence of knowledge about the certainty of the possible alternatives. Therefore, in this chapter, we also provide an introduction to *possibility theory* in a specific interpretation that is based on the context model [Gebhardt and Kruse 1992, Gebhardt and Kruse 1993a, Gebhardt and Kruse 1993b]. In this interpretation possibility theory can handle imprecise as well as uncertain information.

2.1 Modeling Inferences

The essential feature of inference is that a certain type of knowledge—knowledge about truth, probability, (degree of) possibility etc.—is transferred from given propositions, events, states etc. to other propositions, events, states etc. For example, in a logical argument the knowledge about the truth of the premises is transferred to the conclusion; in probabilistic inference the knowledge about the probability of an event is used to calculate the probability of other, related events and is thus transferred to these.

For the transfer carried out in an inference three things are necessary:

knowledge to start from (for instance, the knowledge that a given proposition is true), knowledge that provides a path for the transfer (for example, an implication), and a mechanism to follow the path (for instance, the rule of *modus ponens* to establish the truth of the consequent of an implication of which the antecedent is known). Only if all three are given and fit together, an inference can be carried out. Of course, the transfer need not always be direct. In logic, for example, arguments can be chained by using the conclusion of one as the premise for another, and several such steps may be necessary to arrive at a desired conclusion.

From this description the main problems of modeling inferences are obvious. They consist in finding the paths along which knowledge can be transferred and in providing the proper mechanisms for following them. (In contrast to this, the knowledge to start from is usually readily available, for example from observations.) Indeed, it is well known that automatic theorem provers spend most of their time searching for a path from the given facts to the desired conclusion. The idea underlying *graphical models* or *inference networks* is to structure the paths along which knowledge can be transferred or *propagated* as a *network* or a *graph* in order to simplify and, of course, speed up the reasoning process. Such a representation can usually be achieved if the knowledge about the modeled domain can be *decomposed*, with the network or graph representing the decomposition.

Definitely symbolic logic (see, for example, [Reichenbach 1947, Carnap 1958, Salmon 1963]) is one of the most prominent calculi to represent knowledge and to draw inferences. Its standard method to decompose knowledge is to identify (universally or existentially quantified) propositions consisting of only few atomic propositions or predicates. These propositions can often be organized as a graph, which reflects possible chains of arguments that can be formed using these propositions and observed facts. However, classical symbolic logic is not always the best calculus to represent knowledge and to model inferences. If we confine ourselves to a specific reasoning task and if we have to deal with imprecision, it is often more convenient to use a different calculus. If we have to deal with uncertainty, it is necessary.

The specific reasoning task we confine ourselves to is to identify the true state ω_0 of a given section of the world within a set Ω of possible states. The set Ω of all possible states we call the *frame of discernment* or the *universe of discourse*. Throughout this book we assume that possible states $\omega \in \Omega$ of the domain under consideration can be described by stating the values of a finite set of *attributes*. Often we identify the description of a state ω by a tuple of attribute values with the state ω itself, since its description is usually the only way by which we can refer to a specific state ω.

The task to identify the true state ω_0 consists in combining *prior* or *generic knowledge* about the relations between the values of different attributes (derived from background expert knowledge or from databases of sample cases) and *evidence* about the current values of some of the attributes (obtained, for instance, from observations).[1] The goal is to find a description of the true state ω_0 that is as specific as possible, that is, a description which restricts the set of possible states as much as possible.

As an example consider medical diagnosis. Here the true state ω_0 is the current state of health of a given patient. All possible states can be described by attributes describing properties of patients (like sex or age) or symptoms (like fever or high blood pressure) or the presentness or absence of diseases. The generic knowledge reflects the medical competence of a physician, who knows about the relations between symptoms and diseases in the context of other properties of the patient. It may be gathered from medical textbooks or reports. The evidence is obtained from medical examination and answers given by the patient, which, for example, reveal that she is 42 years old and has a temperature of 39°C. The goal is to derive a full description of her state of health in order to determine which disease or diseases she suffers from.

2.2 Imprecision and Relational Algebra

Statements like "This ball is green or blue or turquoise" or "The velocity of the car was between 50 and 60 km/h" we call *imprecise*. What makes them imprecise is that they do not state one value for a property, but a *set of possible alternatives*. In contrast to this, a statement that names only one value for a property we call *precise*. An example of a precise statement is "The patient has a temperature of 39.3°C".[2]

Imprecision enters our considerations for two reasons. In the first place the generic knowledge about the dependences between attributes can be relational rather than functional, so that knowing exact values for the observed attributes does not allow us to infer exact values for the other attributes, but only sets of possible values. For example, in medical diagnosis a given body temperature is compatible with several physiological states. Secondly, the available information about the observed attributes can itself be imprecise. That is, it may not enable us to fix a specific value, but only a set of alternatives. For example, we may have a measurement device that can determine the value of an attribute only with a fixed error bound, so that all values within the interval determined by the error bound have to be considered possible. In such situations we can only infer that the current state ω_0 lies within a set of alternative states, but without further information we cannot single out the true state ω_0 from this set.

---

[1] Instead of "evidence" often the term "evidential knowledge" is used to complement the term "generic knowledge". However, this term is not in line with the distinction between data and knowledge we made in the first chapter, since it refers to individual properties of a single instance. Therefore, in this book, we use the term "evidence".

[2] Of course, there is also what may be called an implicit imprecision due to the fact that the temperature is stated with a finite precision, i.e. actually all values between 39.25°C and 39.35°C are possible. However, we neglect such subtleties here.

It is obvious that imprecision, interpreted as set-valued information, can easily be handled by symbolic logic: For finite sets of alternatives we can simply write a disjunction of predicates (for example "color(ball) = blue $\vee$ color(ball) = green $\vee$ color(ball) = turquoise" to represent the first example statement given above). For intervals, we may introduce a predicate to compare values (for example, the predicate $\leq$ in "50 km/h $\leq$ velocity(car) $\wedge$ velocity(car) $\leq$ 60 km/h" to represent the second example statement).

Trivially, since imprecise statements can be represented in symbolic logic, we can draw inferences using the mechanisms of symbolic logic. However, w.r.t. the specific reasoning task we consider here, it is often more convenient to use relational algebra for inferences with imprecise statements. The reason is that the operations of relational algebra can be seen as a special case of logical inference rules that draw several inferences in one step.

Consider a set of geometrical objects about which we know the rules

$$\begin{array}{lll} \forall x: & \text{color}(x) = \text{green} & \rightarrow \text{shape}(x) = \text{triangle}, \\ \forall x: & \text{color}(x) = \text{red} & \rightarrow \text{shape}(x) = \text{circle}, \\ \forall x: & \text{color}(x) = \text{blue} & \rightarrow \text{shape}(x) = \text{circle}, \\ \forall x: & \text{color}(x) = \text{yellow} & \rightarrow \text{shape}(x) = \text{square}. \end{array}$$

In addition, suppose that any object must be either green, red, blue, or yellow and that it must be either a triangle, a circle, or a square. That is, we know the domains of the attributes "color" and "shape". All these pieces of information together form the generic knowledge.

Suppose also that we know that the object o is red, blue, or yellow, i.e. color(o) = red $\vee$ color(o) = blue $\vee$ color(o) = yellow. This is the evidence. If we combine it with the generic knowledge above, we can infer that the object must be a circle or a square, i.e. shape(o) = circle $\vee$ shape(o) = square. However, it takes several steps to arrive at this conclusion.

In relational algebra (see, for example, [Ullman 1988]) the generic knowledge as well as the evidence is represented as relations, namely:

Generic Knowledge:

color	shape
green	triangle
red	circle
blue	circle
yellow	square

Evidence:

color
red
blue
yellow

Each tuple is seen as a conjunction, each term of which corresponds to an attribute (represented by a column) and asserts that the attribute has the value stated in the tuple. The tuples of a relation form a disjunction. For the evidence this is obvious. For the generic knowledge this becomes clear by realizing that from the available generic knowledge we can infer

$$\begin{array}{rlcl} \forall x: & (\text{color}(x) = \text{green} & \wedge & \text{shape}(x) = \text{triangle}) \\ \vee & (\text{color}(x) = \text{red} & \wedge & \text{shape}(x) = \text{circle}) \\ \vee & (\text{color}(x) = \text{blue} & \wedge & \text{shape}(x) = \text{circle}) \\ \vee & (\text{color}(x) = \text{yellow} & \wedge & \text{shape}(x) = \text{square}). \end{array}$$

That is, the generic knowledge reflects the possible combinations of attribute values. Note that the generic knowledge is universally quantified, whereas the evidence refers to a single instance, namely the observed object o.

The inference is drawn by projecting the natural join of the two relations to the column representing the shape of the object. The result is:

Inferred Result:

shape
circle
square

That is, we only need two operations, independent of the number of terms in the disjunction. The reason is that the logical inferences that need to be carried out are similar in structure and thus they can be combined. In Section 3.2 reasoning with relations is studied in more detail.

2.3 Uncertainty and Probability Theory

In the preceding section we implicitly assumed that all statements are *certain*, i.e. that all alternatives not named in the statements can be excluded. For example, we assumed that the ball in the first example is definitely not red and that the car in the second example was definitely faster than 40 km/h. If these alternatives cannot be excluded, then the statements are *uncertain*, because they are false if one of the alternatives not named in the statement describes the actual situation. Note that both precise and imprecise statements can be uncertain. What makes a statement certain or uncertain is whether all possible alternatives are listed in the statement or not.

The reason why we assumed up to now that all statements are certain is that the inference rules of classical symbolic logic (and, consequently, the operations of relational algebra) can be applied only, if the statements they are applied to are known to be definitely true: The prerequisite of all logical inferences is that the premises are true.

In applications, however, we rarely find ourselves in such a favorable position. To cite a well-known example, even the commonplace statement "If an animal is a bird, then it can fly" is not absolutely certain, because there are exceptions like penguins, ostriches etc. Nevertheless we would like to draw inferences with such statements, since they are "normally" or "often" correct.

Table 2.1 Generic knowledge about the relation of sex and color-blindness.

		sex		
		female	male	$\sum$
color-blind	yes	0.001	0.025	0.026
	no	0.499	0.475	0.974
	$\sum$	0.5	0.5	1

In addition, we frequently find ourselves in the position that we cannot exclude all but one alternative, but nevertheless have to decide on one. In such a situation, of course, we would like to decide on that precise statement that is "most likely" to be true. If, for example, the symptom *fever* is observed, then various disorders may be its cause and usually we cannot exclude all but one alternative. Nevertheless, in the absence of other information a physician may prefer a severe cold as a diagnosis, because it is a fairly common disorder.

To handle uncertain statements in (formal) inferences, we need a way to assess the certainty of a statement. This assessment may be purely comparative, resulting only in preferences between the alternative statements. More sophisticated approaches quantify these preferences and assign *degrees of certainty*, *degrees of confidence*, or *degrees of possibility* to the alternative statements, which are then treated in an adequate calculus.

The most prominent approach to quantify the certainty or the possibility of statements is, of course, *probability theory* (see, for example, [Feller 1968]). Probabilistic reasoning usually consists in *conditioning* a given probability distribution, which represents the generic knowledge. The conditions are supplied by observations made, i.e. by the evidence about the domain.

As an example consider Table 2.1, which shows a probability distribution about the relation between the sex of a human and whether he or she is color-blind. Suppose that we have a male patient. From Table 2.1 we can compute that the probability that he is color-blind as

$$\begin{aligned} &P(\text{color-blind}(x) = \text{yes} \mid \text{sex}(x) = \text{male}) \\ &= \frac{P(\text{color-blind}(x) = \text{yes} \wedge \text{sex}(x) = \text{male})}{P(\text{sex}(x) = \text{male})} = \frac{0.025}{0.5} = 0.05. \end{aligned}$$

Often the generic knowledge is not given as a joint distribution, but as marginal and conditional distributions. For example, we may know that the two sexes are equally likely and that the probabilities for a female and a male to be color-blind are 0.002 and 0.05, respectively. In this case the result of the inference considered above can be read directly from the generic knowledge.

If, however, we know that a person is color-blind, we have to compute the probability that the person is male using *Bayes' rule*:

$$\begin{aligned} &P(\text{sex}(x) = \text{male} \mid \text{color-blind}(x) = \text{yes}) \\ &= \frac{P(\text{color-blind}(x) = \text{yes} \mid \text{sex}(x) = \text{male}) \cdot P(\text{sex}(x) = \text{male})}{P(\text{color-blind}(x) = \text{yes})} \\ &= \frac{0.05 \cdot 0.5}{0.026} \approx 0.96, \end{aligned}$$

where $P(\text{color-blind} = \text{yes})$ is computed as

$$\begin{aligned} &P(\text{color-blind}(x) = \text{yes}) \\ &= P(\text{color-blind}(x) = \text{yes} \mid \text{sex}(x) = \text{female}) \cdot P(\text{sex}(x) = \text{female}) \\ &+ P(\text{color-blind}(x) = \text{yes} \mid \text{sex}(x) = \text{male}) \cdot P(\text{sex}(x) = \text{male}) \\ &= 0.002 \cdot 0.5 + 0.05 \cdot 5 = 0.026. \end{aligned}$$

In Section 3.3 reasoning with (multivariate) probability distributions is studied in more detail.

As a final remark let us point out that with a quantitative assessment of certainty, certainty and precision are usually complementary properties. A statement can often be made more certain by making it less precise and it can be made more precise by making it less certain.

2.4 Possibility Theory and the Context Model

Relational algebra and probability theory are well-known calculi, so we refrained from providing an introduction and confined ourselves to recalling what it means to draw inferences in these calculi. The case of *possibility theory*, however, is different. Although it has been aired for some time now, it is less well known than probability theory. In addition, there is still an intense discussion going on about its interpretation. Therefore this section provides an introduction to possibility theory that focuses on the interpretation of its key concept, namely a *degree of possibility*.

In colloquial language the notion (or, to be more precise, the modality, cf. modal logic) "possibility", like "truth", is two-valued: either an event, a circumstance etc. is possible or it is impossible. However, to define *degrees* of possibility, we need a quantitative notion. Thus our intuition, exemplified by how the word "possible" is used in colloquial language, does not help us much if we want to understand what may be meant by a degree of possibility. Unfortunately, this fact is often treated too lightly in publications on possibility theory. It is rarely easy to pin down the exact meaning that is given to a degree of possibility. To avoid such problems, we explain in detail a specific interpretation that is based on the *context model* [Gebhardt and Kruse

Figure 2.1 Five dice with different ranges of possible numbers.

1992, Gebhardt and Kruse 1993a, Gebhardt and Kruse 1993b]. In doing so, we distinguish carefully between a degree of possibility and the related notion of a probability, both of which can be seen as quantifications of possibility. Of course, there are also several other interpretations of degrees of possibility, like the epistemic interpretation of fuzzy sets [Zadeh 1978], the theory of epistemic states [Spohn 1990], and the theory of likelihoods [Dubois *et al.* 1993], but these are beyond the scope of this book.

2.4.1 Experiments with Dice

As a first example, consider five dice shakers containing different kinds of dice as indicated in Table 2.2. The dice, which are Platonic bodies, are shown in Figure 2.1. Shaker 1 contains a tetrahedron (a regular four-faced body) with its faces labeled with the numbers 1 through 4 (when throwing the die, the number on the face the tetrahedron lies on counts). Shaker 2 contains a hexahedron (a regular six-faced body, usually called a cube) with its faces labeled with the numbers 1 through 6. Shaker 3 contains an octahedron (a regular eight-faced body) the faces of which are labeled with the numbers 1 through 8. Shaker 4 contains an icosahedron (a regular twenty-faced body). On this die, opposite faces are labeled with the same number, so that the die shows the numbers 1 through 10. Finally, shaker 5 contains a dodecahedron (a regular twelve-faced body) with its faces labeled with the numbers 1 through 12.[3] In addition to the dice in the shakers there is another icosahedron on which groups of four faces are labeled with the same number, so that the die shows the numbers 1 through 5. Suppose the following random experiment is carried out: First the additional icosahedron is thrown. The number it shows indicates a shaker to be used. The number thrown with the die from this shaker is the result of the experiment.

Let us consider the *possibility* that a certain number is the result of this experiment. Obviously, before the shaker is fixed, any of the numbers 1 through 12 is possible. Although smaller numbers are more probable (see below), it is not impossible that the number 5 is thrown in the first step, which enables us

[3]Dice as these are not as unusual as one may think. They are commonly used in fantasy role games and can be bought at many major department stores.

Table 2.2 The five dice shown in Figure 2.1 in five shakers.

shaker 1	shaker 2	shaker 3	shaker 4	shaker 5
tetrahedron 1–4	hexahedron 1–6	octahedron 1–8	icosahedron 1–10	dodecahedron 1–12

Table 2.3 Degrees of possibility in the first dice example.

numbers	degree of possibility	normalized to sum 1
1–4	$\frac{1}{5}+\frac{1}{5}+\frac{1}{5}+\frac{1}{5}+\frac{1}{5}=1$	$\frac{1}{8}=0.125$
5–6	$\frac{1}{5}+\frac{1}{5}+\frac{1}{5}+\frac{1}{5}=\frac{4}{5}$	$\frac{4}{40}=0.1$
7–8	$\frac{1}{5}+\frac{1}{5}+\frac{1}{5}=\frac{3}{5}$	$\frac{3}{40}=0.075$
9–10	$\frac{1}{5}+\frac{1}{5}=\frac{2}{5}$	$\frac{2}{40}=0.05$
11–12	$\frac{1}{5}=\frac{1}{5}$	$\frac{1}{40}=0.025$

to use the dodecahedron in the second. However, if the additional icosahedron has already been thrown and thus the shaker is fixed, certain results may no longer be possible. For example, if the number 2 has been thrown, we have to use the hexahedron (the cube) and thus only the numbers 1 through 6 are possible. Because of this restriction of the set of possible outcomes by the result of the first step, it is reasonable to define as the *degree of possibility* of a number the *probability that it is* still *possible* after the additional icosahedron has been thrown.

Obviously, we have to distinguish five cases, namely those associated with the five possible results of throwing the additional icosahedron. The numbers 11 and 12 are only possible as the final result if we throw the number 5 in the first step. The probability of this event is $\frac{1}{5}$. Therefore the probability of their possibility, i.e. their degree of possibility, is $\frac{1}{5}$. The numbers 9 and 10 are possible if throwing the additional tetrahedron resulted in one of the numbers 4 or 5. It follows that their degree of possibility is $\frac{2}{5}$. Analogously we can determine the degrees of possibility of the numbers 7 and 8 to be $\frac{3}{5}$, those of the numbers 5 and 6 to be $\frac{4}{5}$, and those of the numbers 1 through 4 to be 1, since the latter are possible independent of the result of throwing the additional icosahedron. These degrees of possibility are listed in the center column of Table 2.3.

The function that assigns a degree of possibility to each elementary event of a given sample space (in this case to the twelve possible outcomes of the described experiment) is often called a *possibility distribution* and the degree of possibility it assigns to an elementary event E is written $\pi(E)$. However, if this definition of a possibility distribution is checked against the axiomatic approach to possibility theory [Dubois and Prade 1988], which is directly analogous to the axiomatic approach to probability theory,[4] it turns out that it leads to several conceptual and formal problems. The main reasons are that in the axiomatic approach a possibility distribution is defined for a random variable, but we only have a sample space yet, and that there are, of course, random variables for which the possibility distribution is not an assignment of degrees of possibility to the elementary events of the underlying sample space. Therefore we deviate from the terminology mentioned above and call the function that assigns a degree of possibility to each elementary event of a sample space the *basic* or *elementary possibility assignment.* Analogously, we speak of a *basic* or *elementary probability assignment.* This deviation in terminology goes less far, though, than one might think at first sight, since a basic possibility or probability assignment is, obviously, identical to a specific possibility or probability distribution, namely the one of the random variable that has the sample space as its range of values. Therefore we keep the notation $\pi(E)$ for the degree of possibility that is assigned to an elementary event E by a basic possibility assignment. In analogy to this, we use the notation $p(E)$ for the probability that is assigned to an elementary event by a basic probability assignment (note the lowercase p).

The function that assigns a degree of possibility to all (general) events, i.e. to all subsets of the sample space, is called a *possibility measure.* This term, fortunately, is compatible with the axiomatic approach to possibility theory and thus no change of terminology is necessary here. A possibility measure is usually denoted by a Π, i.e. by an uppercase π. This is directly analogous to a probability measure, which is usually denoted by a P.

In the following we demonstrate, using the simple dice experiment, the difference between a degree of possibility and a probability in two steps. In the first step we compute the probabilities of the numbers for the dice experiment and compare them to the degrees of possibility. Here the most striking difference is the way in which the degree of possibility of (general) events—i.e. sets of elementary events—is computed. In the second step we modify the dice experiment in such a way that the basic probability assignment changes significantly, whereas the basic possibility assignment stays the same. This shows that the two concepts are not very strongly related to each other.

The probabilities of the outcomes of the dice experiment are easily computed using the product rule of probability $P(A \cap B) = P(A \mid B)P(B)$ where

[4] This axiomatic approach is developed for binary possibility measures in Section 3.2.5 and can be carried over directly to general possibility measures.

Table 2.4 Probabilities in the first dice example.

numbers	probability
1–4	$\frac{1}{5}\cdot\left(\frac{1}{4}+\frac{1}{6}+\frac{1}{8}+\frac{1}{10}+\frac{1}{12}\right)=\frac{29}{200}=0.145$
5–6	$\frac{1}{5}\cdot\left(\frac{1}{6}+\frac{1}{8}+\frac{1}{10}+\frac{1}{12}\right)=\frac{19}{200}=0.095$
7–8	$\frac{1}{5}\cdot\left(\frac{1}{8}+\frac{1}{10}+\frac{1}{12}\right)=\frac{37}{600}=0.061\overline{6}$
9–10	$\frac{1}{5}\cdot\left(\frac{1}{10}+\frac{1}{12}\right)=\frac{11}{300}=0.03\overline{6}$
11–12	$\frac{1}{5}\cdot\left(\frac{1}{12}\right)=\frac{1}{60}=0.01\overline{6}$

A and B are events. Let O_i, $i = 1, \ldots, 12$, be the events that the final outcome of the dice experiment is the number i and let S_j, $j = 1, \ldots, 5$, be the event that the shaker j was selected in the first step. Then

$$P(O_i) = \sum_{j=1}^{5} P(O_i \wedge S_j) = \sum_{j=1}^{5} P(O_i \mid S_j)P(S_j).$$

Since we determine the shaker by throwing the additional icosahedron, it is $P(S_j) = \frac{1}{5}$, independent of the number j of the shaker. Hence

$$P(O_i \mid S_j) = \begin{cases} \dfrac{1}{2j+2}, & \text{if } 1 \leq i \leq 2j+2, \\ 0, & \text{otherwise.} \end{cases}$$

The reason is that shaker j contains a die labeled with the numbers 1 through $2j+2$. The resulting probabilities are listed in Table 2.4. The difference from the degrees of possibility of Table 2.3 is evident.

However, one may conjecture that the difference results from the fact that a basic probability assignment is normalized to sum 1 (since $4 \cdot \frac{29}{200} + 2 \cdot \frac{19}{200} + 2 \cdot \frac{37}{600} + 2 \cdot \frac{11}{600} + 2 \cdot \frac{1}{60} = 1$) and that the difference vanishes if the basic possibility assignment is normalized by dividing each degree of possibility by $s = 4 \cdot 1 + 2 \cdot \frac{4}{5} + 2 \cdot \frac{3}{5} + 2 \cdot \frac{2}{5} + 2 \cdot \frac{1}{5} = 8$. The right column of Table 2.3 shows, though, that this is not the case, although the differences are not very large. In addition, such a normalization is not meaningful—at least from the point of view adopted above. The normalized numbers can no longer be interpreted as the probabilities of the possibility of events.

The difference between the two concepts becomes even more noticeable if we consider the probability and the degree of possibility of (general) events, i.e. of sets of elementary events. For instance, we may consider the event "The final outcome of the experiment is a 5 *or* a 6". Probabilities are additive in

Table 2.5 Probabilities in the second dice example.

number	probability
1	$\frac{1}{5} \cdot \left(\frac{1}{16} + \frac{1}{36} + \frac{1}{64} + \frac{1}{100} + \frac{1}{144}\right) = \frac{1769}{72000} \approx 0.0246$
2	$\frac{3}{5} \cdot \left(\frac{1}{16} + \frac{1}{36} + \frac{1}{64} + \frac{1}{100} + \frac{1}{144}\right) = \frac{1769}{24000} \approx 0.0737$
3	$\frac{5}{5} \cdot \left(\frac{1}{16} + \frac{1}{36} + \frac{1}{64} + \frac{1}{100} + \frac{1}{144}\right) = \frac{1769}{14400} \approx 0.1228$
4	$\frac{7}{5} \cdot \left(\frac{1}{16} + \frac{1}{36} + \frac{1}{64} + \frac{1}{100} + \frac{1}{144}\right) = \frac{12383}{72000} \approx 0.1720$
5	$\frac{9}{5} \cdot \left(\frac{1}{36} + \frac{1}{64} + \frac{1}{100} + \frac{1}{144}\right) = \frac{869}{8000} \approx 0.1086$
6	$\frac{11}{5} \cdot \left(\frac{1}{36} + \frac{1}{64} + \frac{1}{100} + \frac{1}{144}\right) = \frac{9559}{72000} \approx 0.1328$
7	$\frac{13}{5} \cdot \left(\frac{1}{64} + \frac{1}{100} + \frac{1}{144}\right) = \frac{6097}{72000} \approx 0.0847$
8	$\frac{15}{5} \cdot \left(\frac{1}{64} + \frac{1}{100} + \frac{1}{144}\right) = \frac{469}{4800} \approx 0.0977$
9	$\frac{17}{5} \cdot \left(\frac{1}{100} + \frac{1}{144}\right) = \frac{1037}{18000} \approx 0.0576$
10	$\frac{19}{5} \cdot \left(\frac{1}{100} + \frac{1}{144}\right) = \frac{1159}{18000} \approx 0.0644$
11	$\frac{21}{5} \cdot \left(\frac{1}{144}\right) = \frac{7}{240} \approx 0.0292$
12	$\frac{23}{5} \cdot \left(\frac{1}{144}\right) = \frac{23}{720} \approx 0.0319$

this case (cf. Kolmogorov's axioms), i.e. the probability of the above event is $P(O_5 \cup O_6) = P(O_5) + P(O_6) = \frac{38}{200} = 0.19$. A degree of possibility, on the other hand, behaves in an entirely different way. According to the interpretation we laid down above, one has to ask: What is the probability that after throwing the additional icosahedron it is still possible to get a 5 *or* a 6 as the final outcome? Obviously, a 5 or a 6 are still possible, if throwing the additional icosahedron resulted in a 2, 3, 4, or 5. Therefore the degree of possibility is $\Pi(O_5 \cup O_6) = \frac{4}{5}$ and thus the same as the degree of possibility of each of the two elementary events.

It is easy to verify that in the dice example the degree of possibility of a set of elementary events always is the maximum of the degrees of possibility of the elementary events contained in it. However, the reason for this lies in the specific structure of this experiment. In general, this need not be the case. When discussing measures of possibility in more detail below, we show what conditions have to hold.

In the following second step we slightly modify the experiment to demonstrate the relative independence of a basic probability and a basic possibility assignment. Suppose that instead of only one die, there are now two dice in each of the shakers, but two dice of the same kind. It is laid down that the higher number thrown counts. Of course, with this arrangement we get a different basic probability assignment, because we now have

$$P(O_i \mid S_j) = \begin{cases} \frac{2i-1}{(2j+2)^2}, & \text{if } 1 \leq i \leq 2j+2, \\ 0, & \text{otherwise.} \end{cases}$$

To see this, notice that there are $2i - 1$ pairs (r, s) with $1 \leq r, s \leq 2j+2$ and $\max\{r, s\} = i$, and that in all there are $(2j+2)^2$ pairs, all of which are equally likely. The resulting basic probability assignment is shown in Table 2.5. The basic possibility assignment, on the other hand, remains unchanged (cf. Table 2.3). This is not surprising, because the two dice in each shaker are of the same kind and thus the same range of numbers is possible as in the original experiment. From this example it should be clear that the basic possibility assignment entirely disregards any information about the shakers that goes beyond the range of possible numbers.

If we compare the probabilities and the degrees of possibility in Tables 2.3, 2.4, and 2.5, we can see the following interesting fact: Whereas in the first experiment the rankings of the outcomes are the same for the basic probability and the basic possibility assignment, they differ significantly for the second experiment. Although the number with the highest probability (the number 4) is still among those having the highest degree of possibility (numbers 1, 2, 3, and 4), the number with the lowest probability (the number 1) is—surprisingly enough—also among them. It follows that from a high degree of possibility one cannot infer a high probability.

It is intuitively clear, though, that the degree of possibility of an event, in the interpretation adopted here, can never be less than its probability. The reason is that computing a degree of possibility can also be seen as neglecting the conditional probability of an event given the context. Hence a degree of possibility of an event can be seen as an upper bound for the probability of this event [Dubois and Prade 1992], which is derived by distinguishing a certain set of cases. In other words, we have at least the converse of the statement found to be invalid above, namely that from a low degree of possibility we can infer a low probability. However, beyond this weak statement no generally valid conclusions can be drawn.

2.4.2 The Context Model

It is obvious that the degree of possibility assigned to an event depends on the set of cases or *contexts* that are distinguished. These contexts are responsible for the name of the already mentioned *context model* [Gebhardt and Kruse 1992, Gebhardt and Kruse 1993a, Gebhardt and Kruse 1993b]. In this model the degree of possibility of an event is the probability of the set of those contexts in which it is possible—in accordance to the interpretation used above: It is the probability of the possibility of an event.

Table 2.6 Degrees of possibility derived from grouped dice.

numbers	degree of possibility
1–6	$\frac{1}{5} + \frac{3}{5} + \frac{1}{5} = 1$
7–10	$\frac{3}{5} + \frac{1}{5} = \frac{4}{5}$
11–12	$\frac{1}{5} = \frac{1}{5}$

In the above example we chose the shakers as contexts. However, we may choose differently, relying on some other aspect of the experiment. For example, three of the Platonic bodies—tetrahedron, octahedron, and icosahedron—have triangular faces, so we may decide to group them, whereas each of the other two forms a context by itself. Thus we have three contexts, with the group of tetrahedron, octahedron, and icosahedron having probability $\frac{3}{5}$ and each of the other two contexts having probability $\frac{1}{5}$. The resulting basic possibility assignment is shown in Table 2.6. This choice of contexts also shows that the contexts need not be equally likely. As a third alternative we could use the initial situation as the only context and thus assign a degree of possibility of 1 to all numbers 1 through 12.

It follows that it is very important to specify which contexts are used to compute the degrees of possibility. Different sets of contexts lead, in general, to different basic possibility assignments. Of course, if the choice of the contexts is so important, the question arises how the contexts should be chosen. From the examples just discussed, it is plausible that we should make the contexts as fine-grained as possible to preserve as much information as possible. If the contexts are coarse, as with the grouped dice, fewer distinctions are possible between the elementary events and thus information is lost. This can be seen clearly from Tables 2.3 and 2.6 where the former allows us to distinguish between a larger number of different situations.

From these considerations it becomes clear that we actually cheated a bit (for didactical reasons) by choosing the shakers as contexts. With the available information, it is possible to define a much more fine-grained set of contexts. Indeed, since we have full information, each possible course of the experiment can be made its own context. That is, we can have one context for the selection of shaker 1 and the throw of a 1 with the die from this shaker, a second context for the selection of shaker 1 and the throw of a 2, and so on, then a context for the selection of shaker 2 and the throw of a 1 etc. We can choose these contexts, because with the available information they can easily be distinguished and assigned a probability (cf. the formulae used to compute Table 2.4). It is obvious that with this set of contexts the resulting basic possibility assignment coincides with the basic probability assignment, because there is only one possible outcome per context.

These considerations illustrate in more detail the fact that the degree of possibility of an event can also be seen as an upper bound for the probability of this event, derived from a distinction of cases. Obviously, the bound is tighter if the sets of possible values per context are smaller. In the limit, for one possible value per context, it reaches the underlying basic probability assignment. They also show that degrees of possibility essentially model *negative information*: Our knowledge about the underlying unknown probability gets more precise the more values can be *excluded* per context, whereas the possible values do not convey any information (indeed: we already know from the domain definition that they are possible). Therefore, we must endeavor to exclude as many values as possible in the contexts to make the bound on the probability as tight as possible.

As just argued, the basic possibility assignment coincides with the basic probability assignment if we use a set of contexts that is sufficiently fine-grained, so that there is only one possible value per context. If we use a coarser set instead, so that several values are possible per context, the resulting basic possibility assignment gets less specific (is only a loose bound on the probability) and the basic probability assignment is clearly to be preferred. This is most obvious if we compare Tables 2.3 and 2.5, where the ranking of the degrees of possibility actually misleads us. So why bother about degrees of possibility in the first place? Would we not be better off by sticking to probability theory? A superficial evaluation of the above considerations suggests that the answer must be a definite "yes".

However, we have to admit that we could compute the probabilities in the dice example only, because we had full information about the experimental setup. In applications, we rarely find ourselves in such a favorable position. Therefore let us consider an experiment in which we do not have full information. Suppose that we still have five shakers, one of which is selected by throwing an icosahedron. Let us assume that we know about the *kind* of the dice that are contained in each shaker, i.e. tetrahedrons in shaker 1, hexahedrons in shaker 2 and so on. However, let it be unknown *how many* dice there are in each shaker and what the *rule* is, by which the final outcome is determined, i.e. whether the maximum number counts, or the minimum, or whether the outcome is computed as an average rounded to the nearest integer number, or whether it is determined in some other, more complicated, way. Let it only be known that the rule is such that the outcome is in the range of numbers present on the faces of the dice.

With this state of information, we can no longer compute a basic probability assignment on the set of possible outcomes, because we do not have an essential piece of information needed in these computations, namely the conditional probabilities of the outcomes given the shaker. However, since we know the possible outcomes, we *can* compute a basic possibility assignment if we choose the shakers as contexts (cf. Table 2.3 on page 23). Note that in this case choosing the shakers as contexts is the best we can do. We cannot choose a finer-grained set of contexts, because we lack the necessary information. Of course, the basic possibility assignment is less specific than the best one for the original experiment, but this is not surprising, since we have much less information about the experimental setup.

2.4.3 The Insufficient Reason Principle

We said above that we cannot compute a basic probability assignment with the state of information we assumed in the modified dice example. A more precise formulation is, of course, that we cannot compute a basic probability assignment *without adding information about the setup*. It is clear that we can always *define* conditional probabilities for the outcomes given the shaker and thus place ourselves in a position in which we have all that is needed to compute a basic probability assignment. The problem with this approach is, obviously, that the conditional probabilities we lay down appear out of the blue. In contrast to this, basic possibility assignments can be computed without inventing information (but at the price of being less specific).

It has to be admitted, though, that for the probabilistic setting there is a well-known principle, namely the so-called *insufficient reason principle*, which prescribes a specific way of fixing the conditional probabilities within the contexts and for which a very strong case can be made. The insufficient reason principle states that if you can specify neither probabilities (quantitative information) nor preferences (comparative information) for a given set of (mutually exclusive) events, then you should assign equal probabilities to the events in the set, because you have insufficient reasons to assign to one event a higher probability than to another. Hence the conditional probabilities of the events possible in a context should be the same.

A standard argument in favor of the insufficient reason principle is the *permutation invariance argument*: In the absence of any information about the (relative) probability of a given set of (mutually exclusive) events, permuting the event labels should not change anything. However, the only assignment of probabilities that remains unchanged under such a permutation is the uniform assignment (all events are assigned the same probability).

Note that the structure of the permutation invariance argument is the same as the structure of the argument by which we usually convince ourselves that the probability of an ace when throwing a (normal, i.e. cube-shaped) die is $\frac{1}{6}$. We argue that the die is symmetric and thus permuting the numbers should not change the probabilities of the numbers. This structural equivalence is the basis for the persuasiveness of the insufficient reason principle. However, it should be noted that the two situations are fundamentally different. In the case of the die we consider the physical parameters that influence the probability of the outcomes and find them to be invariant under a permutation of the face labels. In the case of the insufficient reason principle we consider, as we may say, our ignorance about the obtaining situation and find that it stays the same if we permute the event labels. That is, in the former case, we conclude from the *presence* of (physical) information that the probabilities must be the same, in the latter we conclude from the *absence* of information, that equal probabilities are the best choice.[5]

Of course, this does not invalidate the insufficient reason principle, but it warns us against an unreflected application. Note also that the permutation invariance argument assumes that probabilities reflect our knowledge about a situation and are no "physical" parameters, i.e. it presupposes a subjective or personalistic interpretation of probability. Hence it may be unacceptable to an empirical or frequentist interpretation of probability.[6] Although an approach based on this principle is the toughest competitor of a possibilistic approach, we do not consider the insufficient reason principle much further. The reason is that whether the methods discussed in this book are useful or not does not depend on the answer given to the question whether an insufficient reason principle approach or a possibilistic approach is to be preferred. Although we study possibilistic networks, we argue that the decomposition and propagation operations used in these networks can be justified for a purely probabilistic approach as well. However, we take up the insufficient reason principle again in Section 2.4.12, in which we point out open problems in connection with a possibilistic approach.

2.4.4 Overlapping Contexts

The examples discussed up to now may have conveyed the idea that the contexts must be (physically) disjoint, because the selection of a shaker excludes all events associated with another shaker. (Note that we should not be deceived by the fact that the same outcome can be produced with two different shakers. In this case only the *outcome* is the same, but the *courses of the experiment* leading to it are different.) However, this is not necessary. In order to illustrate what we mean by non-disjoint or *overlapping* contexts, we consider yet another modification of the dice experiment.

Suppose that the shakers are marked with colors. On each shaker there are two colored spots: one on the front and one on the back. There are five colors: red, green, yellow, blue, and white, each of which has been used twice. The full assignment of colors to the shakers is shown in Table 2.7. Obviously, in connection with a red or a green spot all numbers in the range 1 through

[5]Note, however, that in the case of the die the theoretically introduced probability is, strictly speaking, only a hypothesis that has to be verified empirically, since there is no *a priori* knowledge about reality [Reichenbach 1944].

[6]For a brief comparison of the three major interpretations of probability—logical, empirical (or frequentist), and subjective (or personalistic)—see, for instance, [Savage 1954, von Weizsäcker 1992].

Table 2.7 The shakers marked with colors.

shaker	1	2	3	4	5
front	red	white	blue	yellow	green
back	green	yellow	white	blue	red

Table 2.8 The probabilities of all possible sets of colors.

event	P	event	P	event	P
red	$\frac{2}{5}$	red or green	$\frac{2}{5}$	red or green or yellow	$\frac{4}{5}$
green	$\frac{2}{5}$	red or yellow	$\frac{4}{5}$	red or green or blue	$\frac{4}{5}$
yellow	$\frac{2}{5}$	red or blue	$\frac{4}{5}$	red or green or white	$\frac{4}{5}$
blue	$\frac{2}{5}$	red or white	$\frac{4}{5}$	red or yellow or blue	1
white	$\frac{2}{5}$	green or yellow	$\frac{4}{5}$	red or yellow or white	1
		green or blue	$\frac{4}{5}$	red or blue or white	1
		green or white	$\frac{4}{5}$	green or yellow or blue	1
		yellow or blue	$\frac{3}{5}$	green or yellow or white	1
any four	1	yellow or white	$\frac{3}{5}$	green or blue or white	1
all five	1	blue or white	$\frac{3}{5}$	yellow or blue or white	$\frac{3}{5}$

Table 2.9 Degrees of possibility in the example with color-marked shakers.

numbers	degree of possibility
1–8	1
9–10	$\frac{4}{5}$
11–12	$\frac{2}{5}$

Table 2.10 Another possible color assignment.

shaker	1	2	3	4	5
front	blue	red	white	yellow	green
back	green	white	yellow	blue	red

12 are possible, whereas in connection with a yellow or a blue spot only the numbers 1 through 10 are possible, and in connection with a white spot only the numbers 1 through 8 are possible. Let us assume that we do not know about the shakers and how they are selected, but that we do know the probabilities that can be derived from the selection process for sets of colors. These probabilities are shown in Table 2.8.

In this situation we should choose the colors as contexts, because they provide the finest-grained distinction of cases that is available to us. Apart from this the only difference is that we can no longer compute the degree of possibility of an outcome as the sum of the probabilities of the contexts in which it is possible. Instead, we have to look up the probability of this set of contexts in the table of probabilities. The resulting basic possibility assignment is shown in Table 2.9. Note that this basic possibility assignment is less specific than the one shown in Table 2.6 (it is less restrictive for the numbers 7 and 8 and the numbers 11 and 12), which reflects that we have less specific information about the experimental setup.

Although in the previous example we could handle overlapping contexts, it points out a problem. If the contexts overlap and we do *not* know the full probability measure on the set of contexts, but only, say, the probability of single contexts, we can no longer compute the degree of possibility of an outcome. Suppose, for instance, that w.r.t. the connection of colors and outcomes we have the same information as in the previous example, but of the probability measure on the set of colors we only know that each color has a probability of $\frac{2}{5}$. Unfortunately, there is an assignment of colors to the shakers that is consistent with this information, but in which the contexts overlap in a different way, so that the probability measure on the set of colors differs. This assignment is shown in Table 2.10, the corresponding probabilities of sets of colors are shown in Table 2.11. The resulting basic possibility assignment—which, of course, we could compute only if we knew the probabilities of the sets of colors shown in Table 2.11, which we assume not to know—is shown in Table 2.12.

A simple way to cope with the problem that we do not know the probabilities of sets of colors is the following: If we know the probabilities of the contexts, we can compute upper bounds for the probabilities of sets of con-

Table 2.11 The probabilities of all possible sets of colors as they can be derived from the second possible color assignment.

event	P	event	P	event	P
red	$\frac{2}{5}$	red or green	$\frac{3}{5}$	red or green or yellow	1
green	$\frac{2}{5}$	red or yellow	$\frac{4}{5}$	red or green or blue	$\frac{4}{5}$
yellow	$\frac{2}{5}$	red or blue	$\frac{4}{5}$	red or green or white	$\frac{4}{5}$
blue	$\frac{2}{5}$	red or white	$\frac{3}{5}$	red or yellow or blue	1
white	$\frac{2}{5}$	green or yellow	$\frac{4}{5}$	red or yellow or white	$\frac{4}{5}$
		green or blue	$\frac{3}{5}$	red or blue or white	1
		green or white	$\frac{4}{5}$	green or yellow or blue	$\frac{4}{5}$
		yellow or blue	$\frac{3}{5}$	green or yellow or white	1
any four	1	yellow or white	$\frac{3}{5}$	green or blue or white	1
all five	1	blue or white	$\frac{4}{5}$	yellow or blue or white	$\frac{4}{5}$

Table 2.12 Degrees of possibility in the second example with color-marked shakers.

numbers	degree of possibility
1–10	1
11–12	$\frac{3}{5}$

Table 2.13 Degrees of possibility computed from upper bounds on context probabilities.

numbers	degree of possibility
1–10	1
11–12	$\frac{4}{5}$

text using $P(A \cup B) \leq P(A) + P(B)$. That is, we may use the sum of the probabilities of the contexts in a set (bounded by 1, of course) instead of the (unknown) probability of the set. However, in this case we have to reinterpret a degree of possibility as an *upper bound on the probability of the possibility* of an event. This is acceptable, since a degree of possibility can be seen as an upper bound for the probability of an event anyway (see above) and using an upper bound for the probability of a context only loosens the bound. In the example at hand such a computation of the degrees of possibilities leads to the basic possibility assignment shown in Table 2.13. Of course, due to the simplicity of the example and the scarcity of the information left, which is not sufficient to derive strong restrictions, this basic possibility assignment is highly unspecific and thus almost useless. In more complex situation the basic possibility assignment resulting from such an approach may still be useful, though, because more information may be left to base a decision on.

In the following we do not consider the approach using upper bounds any further, but assume that the contexts are disjoint or that the probability measure on the set of contexts is known. For the domains of application we are concerned with in this book, this is usually a reasonable assumption, so that we can spare ourselves the formal and semantical complications that would result from an upper bounds approach. Note, by the way, that the situation just described cannot be handled by the insufficient reason principle alone, since the missing information about the probabilities of sets of contexts cannot be filled in using it.

2.4.5 Mathematical Formalization

The context model interpretation of a degree of possibility we adopted in the above examples can be formalized by the notion of a *random set* [Nguyen 1978, Hestir *et al.* 1991]. A random set is simply a set-valued random variable: In analogy to a standard, usually real-valued random variable, which maps elementary events to numbers, a random set maps elementary events to subsets of a given reference set. Applied to the dice experiments this means: The sample space consists of five elementary events, namely the five shakers. Each of them is mapped by a random set to the set of numbers that can be thrown with the die or dice contained in it. Formally, a random set can be defined as follows [Nguyen 1978, Nguyen 1984, Kruse *et al.* 1994]:

Definition 2.4.1 *Let $(C, 2^C, P)$ be a finite probability space and let Ω be a nonempty set. A set-valued mapping $\Gamma : C \to 2^\Omega$ is called a* **random set**.[7] *The sets $\Gamma(c)$, $c \in C$, are called the* **focal sets** *of Γ.*

[7] For reasons unknown to us a random set is often defined as the pair (P, Γ) whereas a standard random variable is identified with the mapping. However, a random set is merely a set-valued random variable, so there seems to be no need for a distinction. Therefore we identify a random set with the set-valued mapping.

The set C, i.e. the sample space of the finite probability space $(C, 2^C, P)$, is intended to model the contexts. The focal set $\Gamma(c)$ contains the values that are possible in context c. It is often useful to require the focal set $\Gamma(c)$ to be nonempty for all contexts c.

A context c may be defined, as illustrated above, by physical or observational frame conditions. It may also be, for example, an observer or a measurement device. The probability P can state the probability of the occurrence (how often shaker i is used) or the probability of the selection of a context (how often a certain color is chosen to accompany the outcome), or a combination of both. An interpretation that is less bound to probability theory may also choose to see in the probability measure P a kind of quantification of the relative importance or reliability of observers or other information sources [Gebhardt and Kruse 1998].

Note that this definition implicitly assumes that the contexts are disjoint, since they are made the elementary events of a probability space. At first sight, this seems to prevent us from using overlapping contexts, like those of the color-marked shakers example. However, it should be noted that overlapping contexts can always be handled by introducing artificial contexts, combinations of which then form the actual contexts. These artificial contexts are only a mathematical device to get the probability arithmetics right. For example, in the color-marked shakers example, we may introduce five contexts c_1 to c_5, each having a probability of $\frac{1}{5}$, and then define that the contexts "red" and "green" both correspond to the set $\{c_1, c_5\}$, "yellow" corresponds to $\{c_2, c_4\}$, "blue" corresponds to $\{c_3, c_4\}$, and "white" corresponds to $\{c_2, c_3\}$. Actually, with these assignments, the contexts c_1 to c_5 can be interpreted as the shakers 1 to 5. However, it is clear that they can also be constructed without knowing about the shakers as a mathematical tool. In the same way, appropriate focal sets can be found.

A basic possibility assignment is formally derived from a random set by computing the *contour function* [Shafer 1976] or the *falling shadow* [Wang 1983a] of a random set on the set Ω. That is, to each element $\omega \in \Omega$ the probability of the set of those contexts is assigned that are mapped by Γ to a set containing ω [Kruse *et al.* 1994].

Definition 2.4.2 *Let $\Gamma : C \to 2^\Omega$ be a random set. The* **basic possibility assignment** *induced by Γ is the mapping*

$$\begin{aligned} \pi : \Omega &\to [0, 1], \\ \omega &\mapsto P(\{c \in C \mid \omega \in \Gamma(c)\}). \end{aligned}$$

With this definition the informal definition given above is made formally precise: The degree of possibility of an event is the probability of the possibility of the event, i.e. the probability of the contexts in which it is possible.

In the following we will consider mainly basic possibility assignments on finite sets Ω to avoid some technical complications.

2.4.6 Normalization and Consistency

Often it is required that a basic possibility assignment is normalized, where normalization is defined as follows:

Definition 2.4.3 *A basic possibility assignment is called* **normalized** *iff*

$$\exists \omega \in \Omega : \pi(\omega) = 1.$$

The reason for this requirement can be seen from the corresponding requirement for the underlying random set [Kruse *et al.* 1994]:

Definition 2.4.4 *A random set* $\Gamma : C \to 2^{\Omega}$ *is called* **consistent** *iff*

$$\bigcap_{c \in C} \Gamma(c) \neq \emptyset.$$

That is, a random set is consistent if there is at least one element of Ω that is contained in all focal sets. Obviously, the basic possibility assignment induced by a random set is normalized only if the random set is consistent. With an inconsistent random set no element of Ω can have a degree of possibility of 1, because each element is excluded from at least one context.

What is the intention underlying these requirements? Let us consider a simple example: Two observers, which form the contexts, are asked to estimate the value of some measure, say, a length. Observer A replies that the value lies in the interval $[a, b]$, whereas observer B replies that it lies in the interval $[c, d]$. If $a < b < c < d$, then there is no value that is considered to be possible by both observers, i.e. the two observers contradict each other. We can infer directly from the structure of the available information—without knowing the true value of the measure—that (at least) one observer must be wrong. Hence requiring consistency is meant to ensure the compatibility of the information available in the different observation contexts.

However, the question arises whether this is always reasonable. In the first place, it should be noted that requiring consistency is plausible only if the contexts are observers or measurement devices or other sources of information. All of these contexts should also refer to the *same underlying physical situation*, in order to make it reasonable to assume that there is one fixed true value for the considered measure, which, at least, can be expected to be considered as possible in all contexts. Otherwise it would not be clear why it is a contradiction if no element of Ω is contained in all focal sets. For example, if the contexts are determined by physical frame conditions, we may face a situation in which under certain circumstances (of which we do not know whether they obtain) one set of values is possible, whereas under other circumstances (which we also have to take into account) a different set of values is possible. In this case there is no contradiction if the two sets are disjoint, but rather an either/or situation.

Table 2.14 Degrees of possibility of an inconsistent random set.

numbers	degree of possibility
1–4	$\frac{1}{5}+\frac{1}{5}+\frac{1}{5}+\frac{1}{5} = \frac{4}{5}$
5–6	$\frac{1}{5}+\frac{1}{5}+\frac{1}{5} = \frac{3}{5}$
7–8	$\frac{1}{5}+\frac{1}{5} = \frac{2}{5}$
9–10	$\frac{1}{5}+\frac{1}{5} = \frac{2}{5}$
11–12	$\frac{1}{5} = \frac{1}{5}$

To make this more precise, let us consider yet another version of the dice experiment. Suppose that we have the usual five shakers with an unknown number of tetrahedrons in the first, hexahedrons in the second, and so on, but with the dodecahedrons in the fifth shaker replaced by tetrahedrons. These tetrahedrons differ, though, from those in shaker 1 as their faces are labeled with the numbers 9 through 12. From this information we can compute the basic possibility assignment shown in Table 2.14.

Since there is no outcome that is possible independent of the shaker that gets selected, the corresponding random set is inconsistent. However, it is not clear where there is a contradiction in this case. The "inconsistency" only reflects that for each outcome there is at least one context in which it is impossible. But this is not surprising, since the contexts are defined by physical frame conditions. The events that are considered in each context are entirely unrelated. Under these conditions it would actually be more surprising if there were an outcome that is possible in all cases.

However, even if we restrict the requirement of a normalized basic possibility assignment and a consistent random set to observations and measurements, they are semantically dubious. Although the intention to avoid contradictions—if the contexts are observers that refer to the same physical situation, we actually have a contradiction—is understandable, we should check what is achieved by these requirements. In symbolic logic contradictions are avoided, because contradictions point out errors in the theory. On the other hand, if there are no contradictions, the theory is correct—at least formally. Can we say the same about a consistent random set?

To answer this question, let us reconsider the simple example of the two observers, who gave the intervals $[a, b]$ and $[c, d]$ as sets of possible values for some measure. Let $a < c < b < d$. Since both observers consider it to be possible that the value of the measure lies in the interval $[c, b]$, the corresponding random set is consistent. However, it is clear that this does not exclude errors. If the actual value lies in the interval $[a, c)$, then observer 1 was right and observer 2 erred. If the actual value lies in the interval $(b, d]$, it is the other way round. In addition, both observers could be wrong. The fact that errors are still possible is not important, though. A consistent theory can also be proven wrong, namely if it turns out that it does not fit the experimental facts (obviously, formal correctness does not imply factual correctness). Likewise, a consistently estimating set of observers may be proven (partly) wrong by reality.

The important point to notice here is that by using basic possibility assignments we always *expect* such errors. Why do we assign a degree of possibility greater than zero to the intervals $[a, c)$ and $(b, d]$, although one observer must have been wrong if the actual value lies within these intervals? Obviously, because we are not sure that neither made a mistake. If we were sure of their reliability, we could intersect the two intervals and thus restrict our considerations to the interval $[c, b]$. Actually, by the probabilities we assign to the observers we model their reliability, i.e. we state the probability with which we expect their estimates to be correct. However, if we always expect errors to be found, it is not clear why we try to exclude those cases in which we know from the structure of the available information that there must be a (hidden) error. The ways in which we have to deal with this case and with those cases in which there is no structural indication of an error—at least not right from the start—should not be so different. Thus, in this respect, the normalization and consistency requirements turn out to be highly dubious. Therefore, and since we use contexts mainly to distinguish between physical frame conditions, we disregard these requirements and work with non-normalized basic possibility assignments (cf. also Chapter 5, in which we use the context model to define database-induced possibility distributions).

2.4.7 Possibility Measures

Up to now we have considered mainly degrees of possibility for elementary events, which are combined in a basic possibility assignment. However, in order to draw inferences with degrees of possibility a ***possibility measure*** is needed, which assigns degrees of possibility to (general) events, i.e. sets of elementary events. From the definition of a degree of possibility we have adopted, namely that it is the probability of the possibility of an event, it is clear what values we have to assign. In direct analogy to the definition of a basic possibility assignment (Definition 2.4.2 on page 36) we can make the following definition:

Definition 2.4.5 *Let $\Gamma : C \to 2^\Omega$ be a random set. The* **possibility measure** *induced by Γ is the mapping*

$$\begin{aligned} \Pi : 2^\Omega &\to [0, 1], \\ E &\mapsto P(\{c \in C \mid E \cap \Gamma(c) \neq \emptyset\}). \end{aligned}$$

That is, a possibility measure is simply the extension of a basic possibility assignment to the powerset of the set Ω and, conversely, a basic possibility assignment is a possibility measure restricted to single element sets.

In probability theory the most interesting thing about a basic probability assignment and the accompanying probability measure is that the measure can be constructed from the assignment. That is, we need not store all probabilities that are assigned to (general) events, but it suffices to store the probabilities of the elementary events. From these the probabilities of any event E can be recovered by a simple application of Kolmogorov's axioms, which prescribe to add the probabilities of all elementary events contained in the event E. Therefore the question arises whether similar circumstances obtain for possibility measures and basic possibility assignments.

Unfortunately, w.r.t. the above definition, this is not the case. Although in standard possibility theory it is defined that [Zadeh 1978]

$$\Pi(E) = \max_{\omega \in E} \Pi(\{\omega\}) = \max_{\omega \in E} \pi(\omega),$$

where E is a (general) event, and, indeed, this equation is valid in the first dice experiment we discussed on page 22, it is easy to see that it need not be true for general random sets: Reconsider the dice experiment discussed on page 37, in which the dodecahedrons of the fifth shaker are replaced by tetrahedrons labeled with the numbers 9 to 12 (by which we illustrated inconsistent random sets). Let us compute the degree of possibility of the event $E = \{O_7, O_{11}\}$, where O_i is the elementary event that the outcome of the experiment is the number i. To compute the degree of possibility of this event, we have to ask: What is the probability of the set of contexts in which at least one of the outcomes 7 and 11 is possible? Obviously, there are three contexts, in which at least one of the numbers is possible, namely the third and the fourth shaker (here the 7 is possible) and the fifth shaker (here the 11 is possible). Since the three contexts are disjoint and have a probability of $\frac{1}{5}$ each, the answer is $\Pi(E) = \frac{3}{5}$. However, $\max\{\pi(O_7), \pi(O_{11})\} = \max\{\frac{2}{5}, \frac{1}{5}\} = \frac{2}{5}$ (cf. Table 2.14 on page 38).

Note that the failure of the above equation is not due to the fact that the considered random set is inconsistent. Requiring a random set to be consistent is not sufficient to make the above equation hold in general. To prove this, let us relabel the faces of the tetrahedrons of the fifth shaker, so that they show the numbers 1, 10, 11, and 12. In this case the corresponding random set is consistent, since the outcome 1 is now possible in all contexts. However, the degree of possibility of the event E is still $\frac{3}{5}$ and the maximum of the degrees of possibility of the elementary events contained in E is still $\frac{2}{5}$.

As can easily be verified, a condition that is necessary and sufficient to make $\pi(E) = \max_{\omega \in E} \pi(\omega)$ hold for all events E is that the focal sets of the random set are *consonant*, which is defined as follows [Kruse *et al.* 1994]:

Definition 2.4.6 *Let* $\Gamma : C \to 2^\Omega$ *be a random set with* $C = \{c_1, \ldots, c_n\}$. *The focal sets* $\Gamma(c_i)$, $1 \leq i \leq n$, *are called* **consonant** *iff there exists a sequence* $c_{i_1}, c_{i_2}, \ldots, c_{i_n}$, $1 \leq i_1, \ldots, i_n \leq n$, $\forall 1 \leq j < k \leq n : i_j \neq i_k$, *so that*

$$\Gamma(c_{i_1}) \subseteq \Gamma(c_{i_2}) \subseteq \ldots \subseteq \Gamma(c_{i_n}).$$

Intuitively, it must be possible to arrange the focal sets so that they form a "(stair) pyramid" or a "(stair) cone" of "possibility mass" on Ω (with the focal sets corresponding to horizontal "slices", the thickness of which represents their probability). With this picture in mind it is easy to see that requiring consonant focal sets is sufficient for $\forall E \subseteq \Omega : \Pi(E) = \max_{\omega \in E} \pi(\omega)$. In addition, it is immediately clear that a random set with consonant nonempty focal sets must be consistent, because all elements of the first focal set in the inclusion sequence are possible in all contexts. (The opposite, however, is not true, as shown above.) On the other hand, with non-consonant focal sets only "pyramids" or "cones" with "holes" can be formed, and using the elementary event underlying such a "hole" and an appropriately selected other elementary event it is always possible to construct a counterexample. (Compare Table 2.14 on page 38 and consider how the counterexample discussed above is constructed from it.)

For possibility measures induced by general random sets, we only have

$$\forall E \subseteq \Omega : \quad \max_{\omega \in E} \pi(\omega) \;\leq\; \Pi(E) \;\leq\; \min\Big\{1, \sum_{\omega \in E} \pi(\omega)\Big\}.$$

On the one hand, $\Pi(E)$ is equal to the left hand side if there is an elementary event $\omega \in E$, such that no elementary event in E is possible in a context in which ω is impossible. Formally,

$$\exists \omega \in E : \forall \rho \in E : \forall c \in C : \quad \rho \in \Gamma(c) \;\Rightarrow\; \omega \in \Gamma(c).$$

On the other hand, $\Pi(E)$ is equal to the right hand side if no two elementary events contained in E are possible in the same context, or formally,

$$\forall \omega, \rho \in E : \forall c \in C : \quad (\omega \in \Gamma(c) \wedge \rho \in \Gamma(c)) \;\Rightarrow\; \omega = \rho.$$

If we are given a basic possibility assignment, but not the underlying random set, we only know the inequality stated above. However, this inequality only restricts $\pi(E)$ to a range of values. Therefore it is not possible in general to compute the degree of possibility of a (general) event from a basic possibility assignment. The reason is that computing a contour function (cf. the paragraph preceding Definition 2.4.2 on page 36) loses information. (Euphemistically, we may also say that a basic possibility assignment is an *information-compressed* representation of a random set.) This is illustrated in Table 2.15. It shows two random sets over $\Omega = \{1, 2, 3, 4, 5\}$, both of which lead to the same basic possibility assignment. However, with the left random

Table 2.15 Two random sets that induce the same basic possibility assignment. The numbers marked with a • are possible in the contexts.

	1	2	3	4	5
$c_1 : \frac{1}{4}$		•	•	•	•
$c_2 : \frac{1}{4}$	•	•	•		•
$c_3 : \frac{1}{2}$			•	•	
π	$\frac{1}{4}$	$\frac{1}{2}$	1	$\frac{3}{4}$	$\frac{1}{2}$

	1	2	3	4	5
$c_1 : \frac{1}{4}$	•	•	•		
$c_2 : \frac{1}{4}$		•	•	•	
$c_3 : \frac{1}{2}$			•	•	•
π	$\frac{1}{4}$	$\frac{1}{2}$	1	$\frac{3}{4}$	$\frac{1}{2}$

set, $\Pi(\{1,5\}) = \frac{1}{2}$ (maximum of the degrees of possibility of the elementary events), but with the right random set, $\Pi(\{1,5\}) = \frac{3}{4}$ (sum of the degrees of possibility of the elementary events).

The considerations of this section leave us in an unfortunate position. As it seems, we have to choose between two alternatives, both of which have serious drawbacks. (Note that approaches like storing the *possibility measure* instead of the *basic possibility assignment*—which is the approach underlying the so-called Dempster–Shafer theory [Dempster 1967, Dempster 1968, Shafer 1976]—or trying to represent the contexts explicitly are clearly out of the question, because for problems in the real world both approaches require too much storage space.) In the first place, we could try to stick to the standard way of completing a possibility measure from a basic possibility assignment, namely by taking the maximum over the elementary events. However, this seems to force us to accept the requirement that the focal sets of the random set must be consonant. In our opinion this requirement is entirely unacceptable. We cannot think of an application in which this requirement is actually met. (Symmetric confidence levels on statistically estimated parameters may be the only exception, but to handle these, statistical methods should be preferred.) Nevertheless, this is the approach underlying the so-called mass assignment theory [Baldwin *et al.* 1995], which we discuss briefly in the next section, and which is fairly popular.

In passing we mention that a very simple argument put forth in favor of the maximum operation in [Gebhardt 1997], namely that we should choose it because it is the most pessimistic choice possible, is hardly acceptable. This argument overlooks that a possibility measure in the interpretation of the context model is an upper bound on the underlying probability measure and that degrees of possibility model negative information. That is, the more we know about the experimental setup, the tighter the bound on the probability will be (see above). But by choosing the maximum operation, we make the bound tighter than the information available from the basic possibility assignment permits us to. It is even possible that by choosing the maximum we go below the probability and thus make the bound lower than it can be made. Therefore choosing the maximum operation is obviously not the most *pessimistic*, but, contrariwise, the most *optimistic* choice.

On the other hand, we could accept the weak upper bound given by the right hand side of the above inequality, i.e. $\Pi(E) \leq \min\left\{1, \sum_{\omega \in E} \pi(\omega)\right\}$ (which actually is the most *pessimistic* choice), thus seeking refuge in an approach already mentioned in a different context above, namely to redefine a degree of possibility as an *upper bound* on the probability of the possibility of an event. Although this bound is usually greater than necessary, we can keep up the interpretation that a degree of possibility is an upper bound for the probability of an event. However, for this bound to be useful, there must be very few contexts with more than one possible value, to keep the sum below the cutoff value 1. Clearly, if the cutoff value 1 is reached for too many sets, the measure is useless, since it conveys too little information. This will become clearer in the next chapter where we consider multidimensional possibility distributions. In our opinion this drawback disqualifies this approach, because it practically eliminates the capability of possibility theory to handle situations with imprecise, i.e. set-valued information.

Nevertheless, there is a (surprisingly simple) way out of this dilemma, which we discuss in the next but one section. It involves a reinterpretation of a degree of possibility for general, non-elementary events while keeping the adopted interpretation for elementary events.

2.4.8 Mass Assignment Theory

Although we already indicated above that we reject the assumption of consonant focal sets, we have to admit that a basic possibility assignment induced by a random set with consonant nonempty focal sets has an important advantage, namely that from it we can recover all relevant properties of the inducing random set by computing a so-called *mass assignment.*

Definition 2.4.7 *The* **mass assignment** *induced by a basic possibility assignment π is the mapping*

$$\begin{aligned} m: 2^{\Omega} &\rightarrow [0,1], \\ E &\mapsto \min_{\omega \in E} \pi(\omega) - \max_{\omega \in \Omega - E} \pi(\omega). \end{aligned}$$

This mapping is easily understood if one recalls the intuitive picture described above, namely that for a random set having consonant nonempty focal sets we can imagine the possibility mass as a "(stair) pyramid" or a "(stair) cone" on Ω. Then the formula in the above definition simply measures the "height" of the step that corresponds to the event E.

A mass assignment, restricted to those sets $E \subseteq \Omega$ for which $m(E) > 0$, can be seen as a representation of some kind of standardized random set, which results if for a given random set we merge all contexts with identical focal sets. That is, a mass assignment can be computed from a random set $\Gamma : C \to 2^\Omega$ having consonant nonempty focal sets as

$$\begin{aligned} m : 2^\Omega &\to [0,1], \\ E &\mapsto P(\{c \in C \mid \Gamma(c) = E\}). \end{aligned}$$

It is evident that no information is lost if contexts with identical focal sets are merged: w.r.t. the values that can be excluded, they are equivalent and thus do not provide any information to distinguish between the values in their focal sets. The masses of a mass assignment are the probabilities of the merged contexts. Mass assignments are often used to compute a so-called *least prejudiced basic probability assignment* by applying the insufficient reason principle (cf. Section 2.4.3) to all sets $E \subseteq \Omega$ with $|E| > 1$ and $m(E) > 0$ (see, for example, [Baldwin *et al.* 1995]).

The mass assignment theory is based, as discussed above, on the assumption that the focal sets of the random set underlying a given basic possibility assignment are consonant. In [Baldwin *et al.* 1995] the following argument is put forward in favor of this assumption: Consider a set of observers (corresponding to the contexts we use), each of which states some set of values (for a symbolic or a discrete attribute) or an interval (for a continuous attribute). It is plausible that there are some observers who boldly state a small interval or a small set of values and some who are more cautious and thus state a larger interval or a larger set of values. If there is one true value, it is plausible that the different estimates given by the observers can be arranged into an inclusion sequence.

However, in our opinion, this *voting model* (as we may say that each observer "votes" for an interval or a set of values and the degree of possibility measures the number of votes that fall to a value) is not convincing. It only establishes that usually there will be smaller and larger focal sets resulting from observers who boldly or cautiously estimate a given measure. It cannot justify the assumption that the focal sets are consonant. The reason is that it is not clear why two observers must not state intervals $[a, b]$ and $[c, d]$ with, for instance, $a < c$ and $b < d$. Obviously, in such a situation the two observers only express differing expectations: One states that the true value of the measure is, in his opinion, likely to be smaller, whereas the other assumes that it is likely to be larger. Whether they estimate boldly or cautiously does not influence this. That there is one true value can lead at most to the requirement that $b > c$, although with degrees of possibility even this weak requirement is hard to accept (cf. Section 2.4.6).

To establish the consonance of the sets voted for, we need strong assumptions about *how* the observers arrive at their estimates. We could, for example, assume that all observers use the same estimation method, which depends only on the available information and some kind of "cautiousness parameter". One such method (actually the only plausible one known to us) is the statistical estimation of a confidence interval with the additional requirement that it must be symmetric for numeric attributes and it must be greedy for symbolic attributes (i.e. that the most probable values are selected first), where the "cautiousness parameter" is the accepted error bound. However, confidence intervals should be handled with the proper statistical methods. On the other hand, even if one accepts this approach, it is not clear why one should make such strong assumptions about the behavior of the observers, since in applications these almost never hold.

As a consequence, we conclude that the voting model does not suffice to establish the consonance of a random set. The additional assumptions needed are hard to accept, though. In addition, the voting model cannot be applied to situations where the contexts are formed by physical frame conditions, because—as shown above—in such situations it is not even reasonable to require a basic possibility assignment to be normalized, let alone the underlying random set to have consonant focal sets.

2.4.9 Degrees of Possibility for Decision Making

Since we rejected the approaches that suggest themselves immediately, we face the task to provide an alternative. There actually is one and it is surprisingly simple [Borgelt 1995]. The rationale underlying it is that in most applications calculi to handle imprecision and uncertainty are employed to support *decision making*. That is, it is often the goal to decide on *one* course of action and to decide in such a way as to optimize the expected benefit. The best course of action, obviously, depends on the obtaining situation, but usually there is only imperfect (i.e. incomplete or uncertain) knowledge, so that the obtaining situation cannot be identified with certainty.

In a probabilistic setting it is plausible that we should decide on that action that corresponds to the most probable situation—at least, if each situation requires a different course of action and if all costs are equal—because this decision strategy guarantees that in the long run we make the smallest number of mistakes. In the possibilistic setting the commonly used decision rule is directly analogous, namely to decide on that course of action that corresponds to the situation with the highest degree of possibility. It may be argued that this situation can be "least excluded", since for it the probability of those contexts in which it can be excluded is smallest. However, this possibilistic decision rule is open to criticism, especially, if the decision is compared to one derived from the basic probability assignment computed by applying the insufficient reason principle (cf. Section 2.4.3). In Section 2.4.12, in which we discuss open problems, we briefly consider possible points of criticism. For the other parts of this book, however, we accept this decision rule, although we share most doubts about its being reasonable.

If we take the goal to make a decision into account right from the start, it modifies our view of the modeling and reasoning process and thus leads to different demands on a measure assigned to *sets* of elementary events. The reason is that with this goal in mind we no longer care about, say, the probability of a set of elementary events, because in the end we have to decide on *one* (at least under certain circumstances, for example, if no two events require the same course of action). We only care about the *probability of the most probable elementary event* contained in the set. As a consequence, if we want to rank two (general) events, we rank them according to the best decision we can make by selecting an elementary event contained in them. Thus it is reasonable to assign to a (general) event the maximum of the measures assigned to the elementary events contained in it, since it directly reflects the best decision possible (or at least the best decision w.r.t. the uncertainty measure used), if we are constrained to select from this event.

As a consequence, we immediately get the formula to compute the degrees of possibility of a (general) event E, namely

$$\Pi(E) = \max_{\omega \in E} \Pi(\{\omega\}) = \max_{\omega \in E} \pi(\omega).$$

Thus we have the following redefinition that replaces Definition 2.4.5:

Definition 2.4.8 *Let $\Gamma : C \to 2^\Omega$ be a random set. The* **possibility measure** *induced by Γ is the mapping*

$$\begin{aligned} \pi : 2^\Omega &\to [0, 1], \\ E &\mapsto \max_{\omega \in E} P(\{c \in C \mid \omega \in \Gamma(c)\}). \end{aligned}$$

Note that the interpretation we adopt here is not restricted to possibility theory. It is perfectly reasonable for probability theory, too, since it is justified by the goal of the reasoning process, namely to identify the true state ω_0 of a section of the world, and not by the underlying calculus. This is very important, because it decouples the methods examined in connection with possibility theory in the following chapters from whether one considers possibility theory to be a reasonable uncertainty calculus or not and thus makes them noteworthy even for those who reject possibility theory.

A problem that remains is why anyone should bother about the degrees of possibility of (general) events defined in this way, i.e. as the maximum over the degrees of possibility of the contained elementary events. Actually, for one-dimensional problems, they are quite useless, since we can work with the basic possibility assignment and need not consider any sets. However, if we have multidimensional possibility (or probability) distributions, which we need to decompose in order to handle them, they turn out to be useful—at least on certain sets. This is considered in more detail in the next chapter, in which we discuss decompositions of relations and of multivariate probability and possibility distributions.

2.4.10 Conditional Degrees of Possibility

Possibilistic reasoning is directly analogous to probabilistic reasoning. It consists in *conditioning* a given (multivariate) possibility distribution on a set Ω of possible states (or events), which represents the generic knowledge about the considered domain. The conditions are supplied by observations made, i.e. by the evidence about the domain. A conditional degree of possibility is defined as follows:

Definition 2.4.9 *Let Π be a possibility measure on Ω and $E_1, E_2 \subseteq \Omega$. Then*

$$\Pi(E_1 \mid E_2) = \Pi(E_1 \cap E_2)$$

is called the **conditional degree of possibility** *of E_1 given E_2.*

The reason for this definition will become clear in the next chapter, where possibilistic reasoning is studied in more detail in Section 3.4.

This definition of a conditional degree of possibility is not the only one that has been suggested. Others include the approach by [Hisdal 1978] that is based on the equation

$$\Pi(E_1 \cap E_2) = \min\{\Pi(E_1 \mid E_2), \Pi(E_2)\}.$$

A definition of a conditional degree of possibility is derived from this equation by choosing its greatest solution:

$$\Pi(E_1 \mid E_2) = \begin{cases} 1, & \text{if } \Pi(E_1 \cap E_2) = \Pi(E_2), \\ \Pi(E_1 \cap E_2), & \text{otherwise.} \end{cases}$$

Obviously the difference to the above definition consists only in the first case, which ensures the normalization condition. Since we rejected the normalization condition in Section 2.4.6 and thus need not make sure that it is satisfied, we do not consider this definition any further.

Another approach simply defines a conditional degree of possibility in direct analogy to a conditional probability as

$$\Pi(E_1 \mid E_2) = \frac{\Pi(E_1 \cap E_2)}{\Pi(E_2)},$$

provided that $\Pi(E_2) > 0$. This definition leads, obviously, to an entirely different theory, since it involves a renormalization of the degrees of possibility (due to the division by $\Pi(E_2)$), whereas Definition 2.4.9 leaves the degrees of possibility unchanged by the conditioning operation.

The renormalization, however, poses some problems. As it seems, it can be justified on the basis of the context model only if the focal sets of the random set underlying a possibility distribution are required to be consonant (cf. Definition 2.4.6 on page 41). In this case the probability-oriented definition is perfectly sound semantically, because it takes care of the reduction of the number of contexts with nonempty focal sets that is brought about by the conditioning on E_2. However, if one rejects the requirement for consonant focal sets—as we did in Section 2.4.8 on page 43—then it seems to be very difficult to justify it, if possible at all. Therefore we do not go into the details of this approach, but focus on the approach underlying Definition 2.4.9.

2.4.11 Imprecision and Uncertainty

From the description of the context model we gave in the preceding sections it should be clear that possibility theory, if it is based on this model, can handle imprecise as well as uncertain information: The focal set of each context represents an imprecise, i.e. set-valued statement about what values are possible in this context. The probability measure on the set of contexts represents the uncertainty about which context is the one to chose when one wants to describe the obtaining situation.

The reason for this division—imprecision within contexts, uncertainty about the obtaining context—is that "pure" imprecision is obviously disadvantageous to decision making: If we do not have any preferences between the possible alternatives, we do not have any indication which decision is the best. However, we often face situations in which we cannot avoid "pure" imprecision. With the context model we try to make the best of such an unfavorable situation by "encapsulating" the imprecision and making the set of contexts as fine-grained as the available information allows us to.

From this point of view it is not surprising that both relational algebra and probability theory can be seen as special cases of possibility theory: If there is only one context, no uncertainty information is represented and we have a purely relational model. On the other hand, as already indicated above, if there is only one possible value per context, we have a precise model and the basic possibility assignment coincides with the basic probability assignment. This also explains what is meant by saying that relational algebra can handle imprecise, but certain information, whereas probability theory can handle uncertain, but precise information: Since there must be only one context for relational algebra, the information may be imprecise, but must be certain, and since there must be exactly one possible value per context for probability theory, the information may be uncertain, but must be precise.

2.4.12 Open Problems

Possibility theory (in the interpretation considered here) is intended to provide means to deal with imprecision, seen as set-valued data, under uncertainty. However, as already discussed above, possibility theory is not the only approach to handle imprecision in such cases. If we accept the context model as a starting point, its toughest competitor is the *insufficient reason principle*

Focal Sets:
$\{(a_1, b_1)\}$
$\{(a_1, b_1)\}$
$\{(a_3, b_1), (a_3, b_2), (a_3, b_3)\}$
$\{(a_1, b_3), (a_2, b_3), (a_3, b_3)\}$
$\{(a_2, b_2), (a_2, b_3), (a_3, b_2), (a_3, b_3)\}$

π	a_1	a_2	a_3
b_3	$\frac{12}{60}$	$\frac{24}{60}$	$\frac{36}{60}$
b_2	–	$\frac{12}{60}$	$\frac{24}{60}$
b_1	$\frac{24}{60}$	–	$\frac{12}{60}$

P	a_1	a_2	a_3
b_3	$\frac{4}{60}$	$\frac{7}{60}$	$\frac{11}{60}$
b_2	–	$\frac{3}{60}$	$\frac{7}{60}$
b_1	$\frac{24}{60}$	–	$\frac{4}{60}$

Figure 2.2 Possibility versus probability computed using the insufficient reason principle. In this example they lead to different decisions.

(cf. Section 2.4.3), which is often seen as superior to a possibilistic approach. To see why, consider the simple set of (two-dimensional) focal sets shown in Figure 3.17, which are defined over two attributes A and B, with respective domains $\text{dom}(A) = \{a_1, a_2, a_3\}$ and $\text{dom}(B) = \{b_1, b_2, b_3\}$.

If we compute the degrees of possibility according to the context model (assuming each context has the same weight), we get the possibility distribution shown in the middle of Figure 2.2. For example, a degree of possibility of $\frac{2}{5} = \frac{24}{60}$ is assigned to the point (a_3, b_2), because this point is covered by two of the five focal sets (the third and the fifth). Alternatively, we may use the insufficient reason principle to distribute the probability mass of a context uniformly to the points in $\text{dom}(A) \times \text{dom}(B)$ covered by its focal set. In this way a probability mass of $\frac{1}{4}$ is assigned to each of the four points covered by the fifth focal set. Thus we arrive at the probability distribution on the right of Figure 2.2. For example, to the point (a_3, b_2) the probability $\frac{1}{3} \cdot \frac{1}{5} + \frac{1}{4} \cdot \frac{1}{5} = \frac{7}{60}$ is assigned, because it is covered by the third focal set, which in all covers three points, and the fourth focal set, which in all covers four points. For simplicity, we call this probability distribution the IRP probability distribution (for "insufficient reason principle").

Suppose now that the focal sets and the distributions computed from them represent your knowledge about some domain and that you have to decide on a point of the domain—without being able to gather further information about the obtaining state. You could base your decision on the possibility distribution, which suggests the point (a_3, b_3), or on the IRP probability distribution, which suggests the point (a_1, b_1). We believe that in this example most people would agree that the point (a_1, b_1) is the better choice, because there is evidence of two contexts, in which the point (a_1, b_1) *definitely* is the correct description of the prevailing state. In contrast to this, the point (a_3, b_3) is covered by the focal sets of three contexts, but in all of them it is only *possible* among others. Although it is possible that the point (a_3, b_3) is the correct one for all three contexts, we usually consider this to be unlikely and therefore decide against it. As a consequence in this situation the insufficient reason principle approach seems to be superior to the possibilistic one.

On the other hand, consider the set of focal sets shown in Figure 2.2 with the first focal set replaced by $\{(a_1, b_2)\}$. This only changes the values in the two lower left squares in both the possibility distribution and IRP probability distribution tables from $\frac{24}{60}$ or 0, respectively, to $\frac{12}{60}$. Still the possibility distribution suggests to decide on the point (a_3, b_3), whereas the IRP probability distribution still expresses a preference (though reduced) for the point (a_1, b_1). However, with this modification it is less clear that the decision based on the IRP probability distribution is actually better than the one based on the IRP probability distribution. The reason seems to be that the upper bound estimate for the probability of the point (a_3, b_3) (which is identical to its degree of possibility) is so much higher than the IRP probability estimate. Maybe intuitively we take this into account by (too optimistically?) "correcting" the relative values of the probability estimates, so that the point (a_3, b_3) is preferred to the point (a_1, b_1), for which the upper bound estimate coincides with the IRP probability estimate.

It is clear that one can easily make the situation more extreme by considering more contexts with focal sets overlapping on a specific point of the joint domain. To make the IRP probability distribution prefer a point for which there was only one example, we only have to make the overlapping focal sets sufficiently large (this can easily be achieved by adding values to the domains of the attributes A and B, or, more naturally, by considering additional attributes). We guess that there is a point at which one rejects the decision suggested by the IRP probability distribution.

We are not sure, though, whether this is an argument in favor of possibility theory (in the interpretation considered here). In the first place, there are several situations, in which the decision based on the IRP probability estimates is clearly better (see above). Secondly, even if it could be shown that on average people would prefer the decision resulting from possibility theory in cases as the one discussed, this would not prove that this decision is reasonable. Several examples are known in which the involved probabilities are precisely defined and hence the insufficient reason principle need not be called upon, and nevertheless people on average decide in such a way as does not maximize their utility. Hence, this may only point out a(nother) deficiency in commonsense reasoning.

Another open problem that is connected to the above considerations results from the fact that possibility distributions (in the interpretation considered here) essentially model negative information. The reason is that in each context it is not important which values are possible (this we know from the domain definitions of the attributes), but which values are impossible, i.e. can be excluded. As a consequence, a possibility distribution is some kind of upper bound for the probability (cf. Sections 2.4.1 and 2.4.2). It would be worthwhile to consider whether this negative information can be complemented by positive information, which could take the form of a *necessity distribution*. Of course, this idea is not new. Research on possibility theory has already come up with a definition of a necessity measure. However, it is usually defined as $N(E) = 1 - \Pi(\overline{E})$ and this definition depends heavily on the underlying possibility distribution being normalized—a prerequisite we rejected in Section 2.4.6.

In contrast to this, we prefer to rely on the context model, which directly suggests an idea for a necessity distribution: Assign to each elementary event the sum of the weights of all contexts in which it is the only possible event, i.e. in which it necessary. The natural extension operation to sets of elementary events would be the minimum, the natural interpretation a lower bound for the probability. However, if the focal sets are sufficiently imprecise—it suffices if there is no single element set among them—then the necessity distribution as defined above will simply be zero everywhere and thus entirely useless. What is missing is a consistent way to take into account the amount of imprecision that is present in the contexts. Obviously, this amount of imprecision can easily be measured as the number of events that are possible in a given context. This points directly to the insufficient reason principle, which distributes the probability mass of a context equally on the events possible in it, thus providing a direct account of the "extent" of the imprecision.

However, relying exclusively on the insufficient reason principle suffers from the drawback that any information about the possible variance in the probability of an elementary event is lost. To mitigate this drawback it may be worthwhile to study a hybrid model that employs three types of distributions: possibility, necessity, and IRP probability distributions. In this model the IRP probability distributions would model the "best" expectation, while the possibility and necessity distributions provide information on the probability bounds, which in certain situations may change the decision made (compare the example we provided above). Maybe these considerations point out some paths for future research, especially since one can draw on already existing work on upper and lower probabilities (for example [Walley 1991]).

Chapter 3

Decomposition

In this and the next chapter we introduce the basic ideas underlying inference networks. Since at least probabilistic inference networks, especially Bayesian networks and Markov networks, have been well known for some time now, such an introduction may appear to be superfluous or at least should be kept brief. However, there are several approaches to the theory of inference networks and not all of them are equally well suited as a basis for the later chapters of this book. In addition, we feel that in some existing introductions the intuitive background is somewhat neglected.[1]

By this we do not mean that these introductions do not provide illustrative examples—of course they do. But in our opinion these examples fail to create a well-founded intuition of the formal mechanisms underlying decompositions and reasoning with decompositions. We believe that this failure is mainly due to two reasons: In the first place, the exposition often starts immediately with the probabilistic case, in which the numbers can disguise the simplicity of the underlying ideas, although relational networks provide means to explain the basic ideas without this disguise. Secondly, introductions to probabilistic networks often do not distinguish clearly between causal and stochastic dependence, deriving their examples from a causal model of some domain. This is understandable, since causal real-world structures are much easier to comprehend than abstract formal structures. In addition, if probabilistic networks are constructed "manually", one often starts from a causal model. However, such an approach bears the danger that assumptions about causality, which have nothing to do with the idea of decomposition and reasoning, unjustifiably enter our thinking about the matter. Therefore we tried to provide an introduction that does not refer to causality in any way, but is, as we hope, nevertheless easy to understand.

[1]We use the term "intuitive background" in the same way as it is used in the rightly praised book on probability theory by [Feller 1968], who carefully distinguishes three aspects of a theory: the formal logical content, the intuitive background, and the applications.

3.1 Decomposition and Reasoning

Stated as concisely as possible, the basic ideas underlying inference networks are these: Under certain conditions a distribution δ (for example a probability distribution) on a multidimensional domain, which encodes *prior* or *generic knowledge* about this domain, can be decomposed into a set $\{\delta_1, \ldots, \delta_s\}$ of (overlapping) distributions on lower-dimensional subspaces. If such a decomposition is possible, it is sufficient to know the distributions on the subspaces to draw all inferences in the domain under consideration that can be drawn using the original distribution δ. Since such a decomposition is represented as a network and since it is used to draw inferences, we call it an *inference network*. Another popular name is *graphical model*, indicating that it is based on a graph in the sense of graph theory.

Although this description of the ideas underlying inference networks mentions all essential ingredients, it is—necessarily—too condensed to be easily comprehensible, so let us explain first in a little more detail the main notions used in it. Later we will provide some illustrative examples.

By *multidimensional domain* we mean that each state of the section of the world to be modeled can be described by stating the values of a set of attributes (cf. page 16). For example, if we want to describe cars, we may choose to state the manufacturer, the model, the color, whether the car has certain special equipment items or not etc. Each such attribute—or, more precisely, the set of its possible values—forms a dimension of the domain. Of course, to form a dimension, the possible values have to be *exhaustive* and *mutually exclusive*. That is, for instance, there must be for each car one and only one manufacturer, one and only one model name etc. With these restrictions each state of the world section to be modeled (in the example: each car) corresponds to a single point of the multidimensional domain.

Of course, there may be several cars that correspond to the same point—simply because they have the same values for all attributes (same manufacturer, color etc.). On the other hand, there may be points to which no existing car corresponds—for example, because some special equipment items are not available for a certain model. Such information is represented by a distribution on the multidimensional domain. A *distribution* δ assigns to each point of the domain a number in the interval $[0, 1]$, which indicates the possibility or measures the (prior) probability that the modeled section of the world is in a state corresponding to that point. These numbers are usually estimated by human domain experts or computed from a statistical analysis of available data. In the car example they may simply indicate the relative number of cars of a certain type that have been sold.

By *decomposition* we mean that the distribution δ on the domain as a whole can be reconstructed from the distributions $\{\delta_1, \ldots, \delta_s\}$ on subspaces. Such a decomposition has several advantages, the most important being that it can usually be stored much more efficiently and with less redundancy than the original distribution. These advantages are the main motive for studying decompositions of relations in database theory [Maier 1983, Date 1986, Ullman 1988]. Not surprisingly, database theory is closely connected to the theory of inference networks. The only difference is that the theory of inference networks focuses on reasoning, while database theory focuses on storing, maintaining, and retrieving data.

However, being able to store a distribution more efficiently would not be of much use for reasoning tasks, were it not for the possibility to draw inferences in the underlying domain using only the distributions $\{\delta_1, \ldots, \delta_s\}$ on the subspaces *without having to reconstruct the original distribution* δ. The basic idea is to pass information from subspace distribution to subspace distribution until all have been updated. This process is usually called *evidence propagation.* How it works is probably explained best by a simple example, which we present in the relational setting first. Later we study the probabilistic and finally the possibilistic case. There are, of course, even more types of inference networks, for example, Dempster–Shafer networks. These are, however, beyond the scope of this book.

3.2 Relational Decomposition

In relational decomposition and relational networks [Dechter 1990, Kruse and Schwecke 1990, Dechter and Pearl 1992, Kruse *et al.* 1994] one distinguishes only between possible and impossible states of the world. In other words, one confines oneself to distributions that assign to each point of the underlying domain either a 1 (if it is possible) or a 0 (if it is impossible). This is made formally precise below, after the simple example we are going to discuss has provided the intuitive background.

3.2.1 A Simple Example

Consider three attributes, A, B, and C, with respective domains $\text{dom}(A) = \{a_1, a_2, a_3, a_4\}$, $\text{dom}(B) = \{b_1, b_2, b_3\}$, and $\text{dom}(C) = \{c_1, c_2, c_3\}$. Thus the underlying joint domain of this example is the Cartesian product $\text{dom}(A) \times \text{dom}(B) \times \text{dom}(C)$ or, abbreviated, the three-dimensional space $\{A, B, C\}$, or, even more abbreviated, ABC. Table 3.1 states prior knowledge about the possible combinations of attribute values in the form of a relation R_{ABC}: only the value combinations contained in R_{ABC} are possible. (This relation is to be interpreted under the closed-world assumption, i.e. all value combinations not contained in R_{ABC} are impossible.) An interpretation of this simple relation is shown on the left in Figure 3.1. In this interpretation each attribute corresponds to a property of a geometrical object: attribute A is the color/hatching, attribute B is the shape, and attribute C is the size. The table on the right in Figure 3.1 restates the relation R_{ABC} in this interpretation.

Table 3.1 The relation R_{ABC} states prior knowledge about the possible combinations of attribute values. Value combinations not contained in the above table are considered to be impossible.

A	a_1	a_1	a_2	a_2	a_2	a_2	a_3	a_4	a_4	a_4
B	b_1	b_1	b_1	b_1	b_3	b_3	b_2	b_2	b_3	b_3
C	c_1	c_2	c_1	c_2	c_2	c_3	c_2	c_2	c_2	c_3

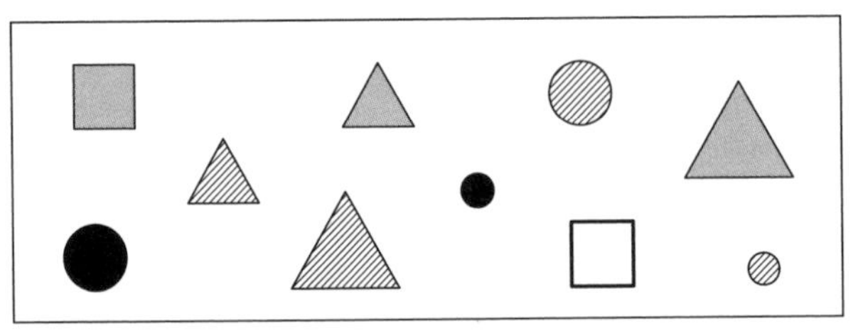

Figure 3.1 A set of geometrical objects that is an interpretation of the relation R_{ABC}. Attribute A is the color/hatching of an object, attribute B is its shape, and attribute C is its size. The table on the right restates the relation R_{ABC} in this interpretation.

color	shape	size
■	○	small
■	○	medium
▨	○	small
▨	○	medium
▨	△	medium
▨	△	large
□	□	medium
▩	□	medium
▩	△	medium
▩	△	large

Suppose that an object of the set shown in Figure 3.1 is selected at random and that one of its three properties is observed, but not the other two. For instance, suppose that this object is drawn from a box, but the box is at some distance or may be observed only through a pane of frosted glass, so that the color of the object can be identified, while its size and shape are too blurred. In this situation, what can we infer about the other two properties and how can we do so? How can we combine the *evidence* obtained from our observation with the *prior* or *generic knowledge* that there are only ten possible combinations of attribute values?

Such tasks often occur in applications—reconsider, for example, medical diagnosis as it was described on page 17. The same holds, obviously, for any other diagnosis problem, for example for a mechanic who faces the task to repair a broken engine. Of course, these tasks are much more complex, because there are much more properties that have to be taken into account. In the geometrical objects example, we could discard all objects not incompatible with the observation and scan the rest for possible shapes and sizes. However, it is obvious that such an approach is no longer feasible if the number of rele-

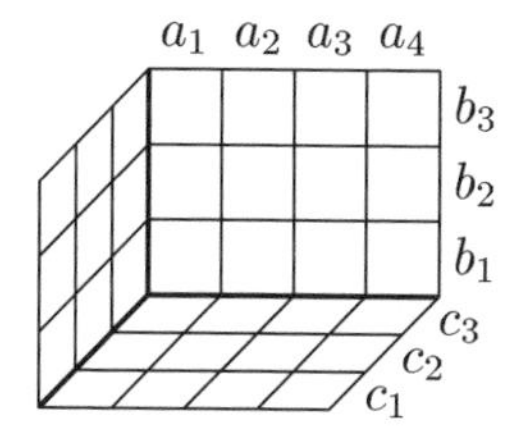

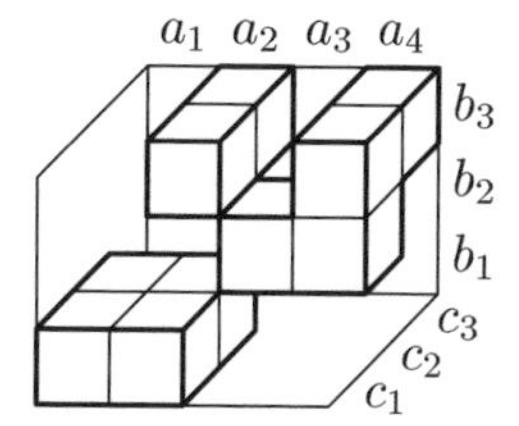

Figure 3.2 The reasoning space and a graphical representation of the relation R_{ABC} in this space. Each cube represents one tuple of the relation.

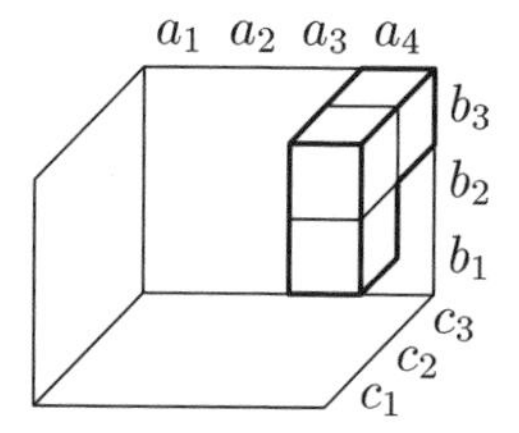

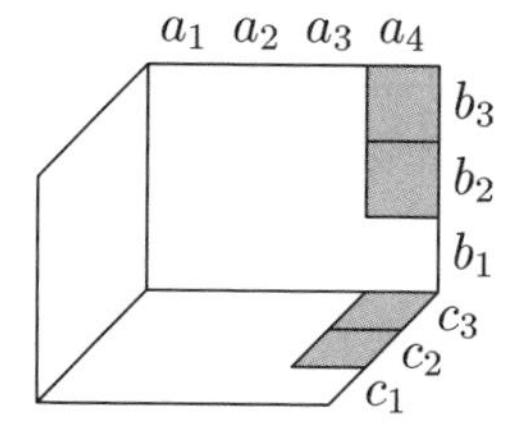

Figure 3.3 Reasoning in the domain as a whole.

vant properties is large. In this case, more sophisticated methods are needed. One such method are graphical models, with which it is tried to decompose the generic knowledge by exploiting conditional independence relations. Although this method aims at making reasoning in high-dimensional domains feasible, its basic idea can be explained with only three attributes. Nevertheless, it should be kept in mind that decomposition techniques, which may appear to be superfluous in the geometrical objects example, are essential for applications in the real world.

3.2.2 Reasoning in the Simple Example

In order to understand what is meant by reasoning in the tasks indicated above, let us take a closer look at the space in which it is carried out. The three-dimensional reasoning space underlying the geometrical objects example is shown on the left in Figure 3.2. Each attribute—or, more precisely, the set of its possible values—forms a dimension of this space. Each combination of attribute values corresponds to a small cube in this space. That only ten combinations of values are actually possible is the *prior* or *generic knowledge*. It is represented graphically by marking those cubes of the reasoning space which correspond to existing objects. This is demonstrated on the right in Figure 3.2: each cube indicates a possible value combination.

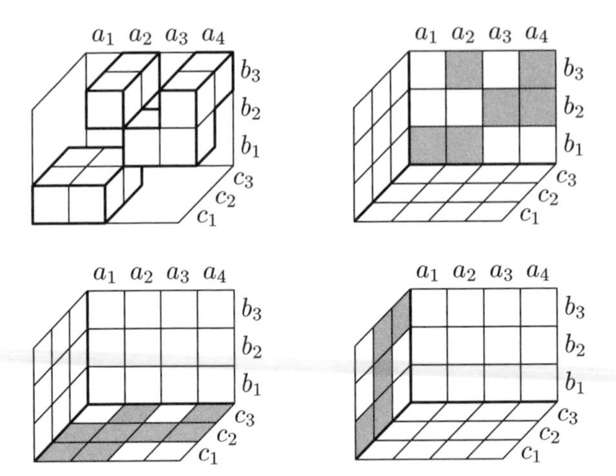

Figure 3.4 Graphical representation of the relation R_{ABC} and of all three possible projections to two-dimensional subspaces.

Suppose we observe that the object drawn has the value a_4 for attribute A, i.e. that its color is grey. With the visualization shown in Figure 3.2 it is very simple to draw inferences about the values of the other two attributes: Simply cut out the "slice" that corresponds to $A = a_4$. This is demonstrated on the left in Figure 3.3. Cutting out this "slice" can be seen either as *intersecting* the generic knowledge and the evidence, the latter of which corresponds to all possible cubes in the "slice" corresponding to $A = a_4$, or as *conditioning* the generic knowledge on the observation $A = a_4$ by restricting it to the "slice" corresponding to the observation that $A = a_4$. The values for the attributes B and C that are compatible with the evidence $A = a_4$ can be read from the result by *projecting* it to the domains of these attributes. This is demonstrated on the right in Figure 3.3. We thus conclude that the object drawn cannot be a circle (b_1), but must be a triangle (b_2) or a square (b_3), and that it cannot be small (c_1), but must be medium sized (c_2) or large (c_3).

This method of reasoning is, of course, trivial and can always be used—at least theoretically. However, the relation R_{ABC} has an interesting property, which allows us to derive the same result in an entirely different fashion: It can be decomposed into two smaller relations, from which it can be reconstructed. This is demonstrated in Figures 3.4 and 3.5. In Figure 3.4 the relation R_{ABC} is shown together with all possible *projections* to two-dimensional subspaces. These projections are the "shadows" thrown by the cubes if light sources are imagined (in sufficient distance) in front of, above, and to the right of the graphical representation of the relation R_{ABC}. The relation R_{ABC} can be decomposed into the two projections to the subspaces $\{A, B\}$ and $\{B, C\}$, both shown in the right half of Figure 3.4.

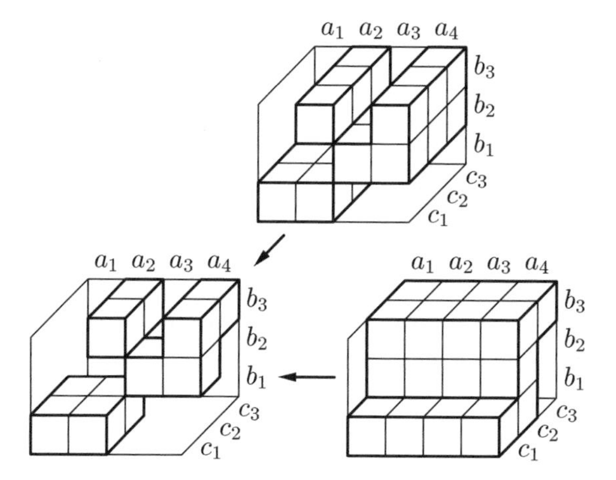

Figure 3.5 Cylindrical extensions of two projections of the relation R_{ABC} shown in Figure 3.4 and their intersection. Obviously the result is the original relation (compare the top left of Figure 3.4).

That these two projections are sufficient to reconstruct the relation R_{ABC} is demonstrated in Figure 3.5. In the top half and on the right the *cylindrical extensions* of the two projections R_{AB} and R_{BC} to the subspaces $\{A, B\}$ and $\{B, C\}$, respectively, are shown. The cylindrical extension of a projection is obtained by simply adding to the tuples in the projection all possible values of the missing dimension. That is, to the tuples in the relation R_{AB} all possible values of the attribute C are added and to the tuples in the relation R_{BC} all possible values of the attribute A are added. (That this operation is called "cylindrical extension" is due to the common practice to sketch sets as circles or disks: Adding a dimension to a disk yields a cylinder.) In a second step the two cylindrical extensions are intersected, by which the the original relation R_{ABC} is reconstructed, as shown on the bottom left of Figure 3.5.

Since relational networks are closely related to database theory, it is not surprising that the decomposition property just studied is well known: If a relation can be decomposed into projections to subspaces (i.e. to subsets of attributes), it is called *join-decomposable* [Maier 1983, Date 1986, Ullman 1988], because the intersection of the cylindrical extensions is identical to a *natural join* of the projections. In database theory join-decomposition is studied mainly in order to avoid redundancy (which can lead to update anomalies) and to exploit the resulting storage savings. Note that even in this very simple example some savings result: To store the three-dimensional relation, we need 36 bits—one for each combination of attribute values in the reasoning space, indicating whether the combination is possible or not. To store the two projections, however, we need only $9 + 12 = 21$ bits.

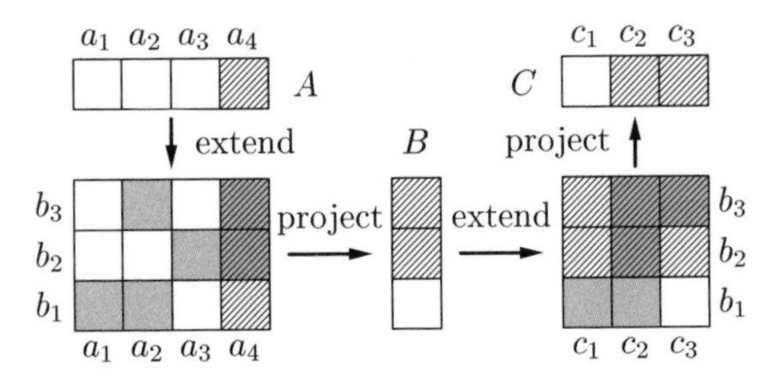

Figure 3.6 Propagation of the evidence that attribute A has value a_4 in the three-dimensional relation shown in Figure 3.4 using the projections to the subspaces $\{A, B\}$ and $\{B, C\}$.

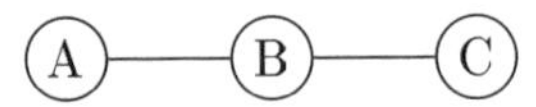

Figure 3.7 The relational propagation scheme shown in Figure 3.6 justifies a network representation of the reasoning space. The edges indicate the projections needed for the reconstruction and the evidence propagation.

W.r.t. inference networks these storage savings are important, too. However, they would be worth nothing if to draw inferences the three-dimensional relation R_{ABC} had to be reconstructed first. Fortunately, this is not necessary. We can draw the same inferences as in the whole reasoning space using only the projections, each one in turn. This is demonstrated in Figure 3.6, which illustrates relational *evidence propagation*:

Suppose again that from an observation we know that the attribute A has value a_4, i.e. that the object drawn at random is grey. This is indicated in Figure 3.6 by the hatched square for the value a_4 in the top line. In a first step we extend this information cylindrically to the subspace $\{A, B\}$ (indicated by the hatched column) and intersect it with the projection of the relation R_{ABC} to this subspace (grey squares). The resulting intersection (the squares that are grey and hatched) is then projected to the subspace that consists only of attribute B. In this way we can infer that the object drawn cannot be a circle (b_1), but must be a triangle (b_2) or a square (b_3).

In a second step, the knowledge obtained about the possible values of attribute B is extended cylindrically to the subspace $\{B, C\}$ (rows of hatched squares) and intersected with the projection of the relation R_{ABC} to this subspace (grey squares). In analogy to the above, the resulting intersection (the squares that are grey and hatched) is then projected to the subspace that consists only of the attribute C. In this way we can infer that the object drawn cannot be small (c_1), but must be medium sized (c_2) or large (c_3).

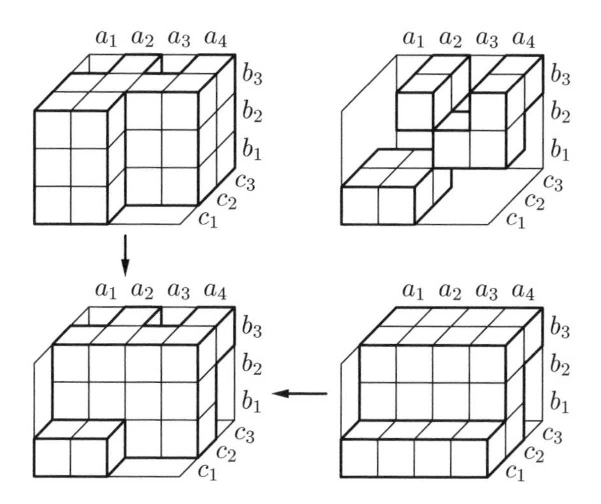

Figure 3.8 Using other projections.

Obviously these results are identical to those we obtained from projecting the "slice" of the relation R_{ABC} that corresponds to $A = a_4$ to the attributes B and C. It is easily verified for this example that this propagation scheme always (i.e. for all possible observations) leads to the same result as an inference based on the original three-dimensional relation. Therefore this propagation scheme justifies the network representation of the reasoning space shown in Figure 3.7 in which there is a node for each attribute and an edge for each projection used. This connection of decomposition and network representation is studied in more detail in Chapter 4.

3.2.3 Decomposability of Relations

Having demonstrated the usefulness of relational decomposition, the next question is whether any selection of a set of projections of a given relation provides a decomposition. Unfortunately this is not the case as is demonstrated in Figure 3.8. Whereas in Figure 3.5 we used the projections to the subspaces $\{A, B\}$ and $\{B, C\}$, in this figure we replaced the projection to the subspace $\{A, B\}$ by the projection to the subspace $\{A, C\}$. As in Figure 3.5 the cylindrical extensions of these two projections are determined and intersected, which yields the relation R'_{ABC} shown in the bottom left of Figure 3.8. Obviously, this relation differs considerably from the original relation R_{ABC}, which is repeated in the top right of Figure 3.8: Whereas R_{ABC} contains only 10 tuples, R'_{ABC} contains 16 and therefore the two subspaces $\{A, C\}$ and $\{B, C\}$ are an especially bad choice. Note that from this example it is also immediately clear that the intersection of the cylindrical extension of

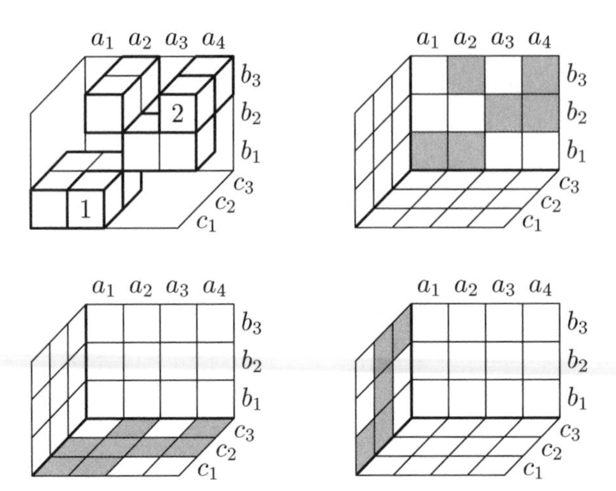

Figure 3.9 Is decomposition always possible?

two projections can never have fewer tuples than the original relation. For a decomposition, the number of tuples must be equal.

Another question is whether, although the projections must be chosen with care, it is at least always possible to find a set of projections that is a decomposition. Unfortunately this question has also to be answered in the negative. To understand this, consider Figure 3.9. In the top left of this figure, the relation R_{ABC} is shown with two cubes (two tuples) marked with numbers.

Consider first that the cube marked with a 1 were missing, which corresponds to the tuple (a_2, b_1, c_1), i.e. to a small hatched circle. It is immediately clear that without this tuple the relation is no longer decomposable into the projections to the subspaces $\{A, B\}$ and $\{B, C\}$. The reason is that removing the tuple (a_2, b_1, c_1) does not change these projections, because the cubes corresponding to the tuples (a_2, b_1, c_2) and (a_1, b_1, c_1) still throw "shadows" onto the subspaces. Therefore the intersection of the cylindrical extensions of these projections still contains the tuple (a_2, b_1, c_1). However, if *all three* projections to two-dimensional subspaces are used, the modified relation can be reconstructed. This is due to the fact that removing the tuple (a_2, b_1, c_1) changes the projection to the subspace $\{A, C\}$. If we first intersect the cylindrical extensions of the projections to the subspaces $\{A, B\}$ and then intersect the result with the cylindrical extension of the projection to the subspace $\{A, C\}$, the tuple (a_2, b_1, c_1) is cut away in the second intersection.

However, although it is intuitively compelling that any three-dimensional relation can be reconstructed if all three projections to two-dimensional subspaces are used (as we have experienced several times when we explained relational networks to students), this assumption is false. Suppose that the cube marked with a 2 in Figure 3.9 were missing, which corresponds to the tuple (a_4, b_3, c_2), i.e. to a grey medium sized triangle. In this case all projections to two-dimensional subspaces are unchanged, because in all three possible directions there is still another cube which throws the "shadow". Therefore the intersection of the cylindrical extensions of the projections still contains the tuple (a_4, b_3, c_2), although it is missing from the relation. It follows that, without this tuple, the relation is not decomposable.

Unfortunately, decomposable relations are fairly rare. (The geometrical objects example is, of course, especially *constructed* to be decomposable.) However, in applications a certain loss of information is often acceptable if it is accompanied by a reduction in complexity. In this case precision is traded for time and space. Thus one may choose a set of projections, although the intersection of their cylindrical extensions contains more tuples than the original relation, simply because the projections are smaller and can be processed more rapidly. Note that, if a certain loss of information is acceptable, the number of additional tuples in the intersection of the cylindrical extensions of a set of projections provides a direct measure of the quality of a set of projections, which can be used to rank different sets of projections [Dechter 1990] (see also Chapter 7).

3.2.4 Tuple-Based Formalization

Up to now our explanation of relational networks has been very informal. Our rationale was to convey a clear intuition first, without which we believe it is very hard, if not impossible, to cope with mathematical formalism. In the following we turn to making mathematically precise the notions informally introduced above. We do so in two steps. The first step is more oriented at the classical notions used in connection with relations. In a second step we modify this description and use a notion of possibility to describe relations, which can be defined in close analogy to the notion of probability. The reason for the second step is that it simplifies the notation, strengthens the parallelism to probabilistic networks, and provides an ideal starting point for introducing possibilistic networks in Section 3.4.

We start by defining the basic notions, i.e. the notions of a *tuple* and a *relation*. Although these notions are trivial, we provide definitions here, because they differ somewhat from those most commonly used.

Definition 3.2.1 *Let* $U = \{A_1, \ldots, A_n\}$ *be a (finite) set of attributes with respective domains* $\text{dom}(A_i)$, $i = 1, \ldots, n$. *An* **instantiation** *of the attributes in* U *or a* **tuple** *over* U *is a mapping*

$$t_U : U \to \bigcup_{A \in U} \text{dom}(A)$$

satisfying $\forall A \in U : t_U(A) \in \text{dom}(A)$. *The set of all tuples over* U *is denoted* T_U. *A* **relation** R_U *over* U *is a set of tuples over* U, *i.e.* $R_U \subseteq T_U$.

If the set of attributes is clear from the context, we drop the index U. To indicate that U is the domain of definition of a tuple t, i.e. that t is a tuple over U, we sometimes also write $\text{dom}(t) = U$. We write tuples similar to the usual vector notation. For example, a tuple t over $\{A, B, C\}$ which maps attribute A to a_1, attribute B to b_2, and attribute C to c_2 is written as $t = (A \mapsto a_1, B \mapsto b_2, C \mapsto c_2)$. If an implicit order is fixed, the attributes may be omitted, i.e. the tuple may then be written $t = (a_1, b_2, c_2)$.

At first sight the above definition of a tuple may seem a little strange. It is more common to define a tuple as an element of the Cartesian product of the domains of the underlying attributes. We refrain from using the standard definition, since it causes problems if projections of tuples have to be defined: A projection, in general, changes the position of the attributes in the Cartesian product, since some attributes are removed. Usually this is taken care of by index mapping functions, which can get confusing if two or more projections have to be carried out in sequence or if two projections obtained in different ways have to be compared. This problem instantly disappears if tuples are defined as functions as in the above definition. Then a projection to a subset of attributes is simply a restriction of a function—a well-known concept in mathematics. No index transformations are necessary and projections can easily be compared by comparing their domains of definition and the values they map the attributes to. In addition, the above definition can easily be extended to imprecise tuples, which we need in Chapter 5.

Definition 3.2.2 *If t_X is a tuple over a set X of attributes and $Y \subseteq X$, then $t_X|_Y$ denotes the* **restriction** *or* **projection** *of the tuple t_X to Y. That is, the mapping $t_X|_Y$ assigns values only to the attributes in Y. Hence $\text{dom}(t_X|_Y) = Y$, i.e. $t_X|_Y$ is a tuple over Y.*

Definition 3.2.3 *Let R_X be a relation over a set X of attributes and $Y \subseteq X$. The* **projection** *$\text{proj}_Y^X(R_X)$ of the relation R_X from X to Y is defined as*

$$\text{proj}_Y^X(R_X) \stackrel{\text{def}}{=} \{t_Y \in T_Y \mid \exists t_X \in R_X : t_X \equiv t_X|_Y\}.$$

If R_X is a relation over X and $Z \supseteq X$, then the **cylindrical extension** *$\text{cext}_Y^X(R_X)$ of the relation R_X from X to Z is defined as*

$$\text{cext}_X^Z(R_X) \stackrel{\text{def}}{=} \{t_Z \in T_Z \mid \exists t_X \in R_X : t_Z|_X \equiv t_X\}.$$

With this definition, we can write the two projections used in the decomposition of the example relation R_{ABC} as (cf. Figure 3.4)

$$R_{AB} = \text{proj}_{AB}^{ABC}(R_{ABC}) \qquad \text{and} \qquad R_{BC} = \text{proj}_{BC}^{ABC}(R_{ABC}).$$

That these two projections are a decomposition of the relation R_{ABC} can be written as (cf. Figure 3.5)

$$R_{ABC} = \text{cext}_{AB}^{ABC}(R_{AB}) \cap \text{cext}_{BC}^{ABC}(R_{BC})$$

if we use the cylindrical extension operator and as

$$R_{ABC} = R_{AB} \bowtie R_{BC}$$

if we use the natural join operator $\bowtie$, which is well known from relational algebra. Generalizing, we can define relational decomposition as follows:

Definition 3.2.4 *Let U be a set of attributes and R_U a relation over U. Furthermore, let $\mathcal{M} = \{M_1, \ldots, M_m\} \subseteq 2^U$ be a (finite) set of nonempty (but not necessarily disjoint) subsets of U satisfying*

$$\bigcup_{M \in \mathcal{M}} M = U.$$

R_U is called **decomposable** *w.r.t. $\mathcal{M}$, iff*

$$R_U = \bigcap_{M \in \mathcal{M}} \operatorname{cext}_M^U \left(\operatorname{proj}_M^U(R_U)\right).$$

If R_U is decomposable w.r.t. $\mathcal{M}$, the set of projections

$$\mathcal{R}_{\mathcal{M}} = \left\{\operatorname{proj}_{M_1}^U(R_U), \ldots, \operatorname{proj}_{M_m}^U(R_U)\right\}$$

is called the **decomposition** *of R_U w.r.t. $\mathcal{M}$.*

Applying this definition to the example relation R_{ABC}, we can say that R_{ABC} is decomposable w.r.t. $\{\{A, B\}, \{B, C\}\}$ and that $\{R_{AB}, R_{BC}\}$ is the corresponding decomposition.

It is obvious that it is very simple to find decompositions in this sense: Any set $\mathcal{M}$ that contains the set U of all attributes leads to a decomposition of a given relation R_U over U. However, it is also obvious that such a decomposition would be useless, because an element of the decomposition is the relation itself. Therefore restrictions have to be introduced in order to characterize "good" decompositions. It is clear that the savings result mainly from the fact that the subspaces the relation is projected to are "small". In addition, there should not be any "unnecessary" projections.

Definition 3.2.5 *Let $\mathcal{R}_{\mathcal{M}}$ be a decomposition of a relation R_U over a set U of attributes w.r.t. a set $\mathcal{M} \subseteq 2^U$. $\mathcal{R}_{\mathcal{M}}$ is called* **trivial** *iff $U \in \mathcal{M}$ (and thus $R_U \in \mathcal{R}_{\mathcal{M}}$). $\mathcal{R}_{\mathcal{M}}$ is called* **irredundant** *iff no set of attributes in $\mathcal{M}$ is contained in another set of attributes in $\mathcal{M}$, i.e. iff*

$$\forall M_1 \in \mathcal{M} : \neg\exists M_2 \in \mathcal{M} - \{M_1\} : M_1 \subseteq M_2.$$

Otherwise, $\mathcal{R}_{\mathcal{M}}$ is called **redundant**.

Let $\mathcal{R}_{\mathcal{N}}$ be another decomposition of the relation R_U w.r.t. a set $\mathcal{N} \subseteq 2^U$. $\mathcal{R}_{\mathcal{M}}$ is called **at least as fine as** *$\mathcal{R}_{\mathcal{N}}$, written $\mathcal{R}_{\mathcal{M}} \preceq \mathcal{R}_{\mathcal{N}}$, iff*

$$\forall M \in \mathcal{M} : \exists N \in \mathcal{N} : M \subseteq N.$$

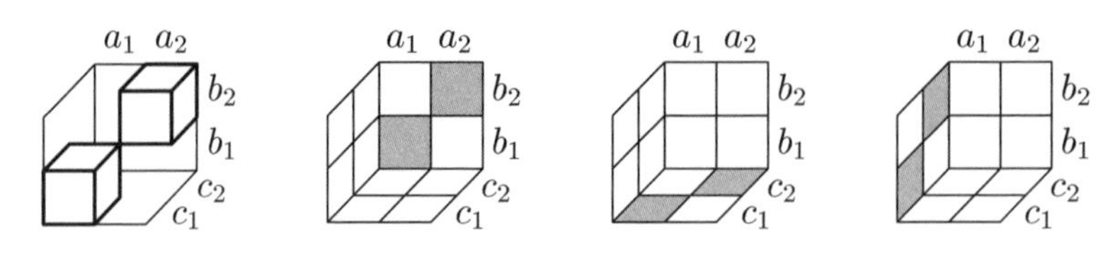

Figure 3.10 Minimal decompositions need not be unique.

$\mathcal{R}_\mathcal{M}$ is called **finer** *than $\mathcal{R}_\mathcal{N}$, written $\mathcal{R}_\mathcal{M} \prec \mathcal{R}_\mathcal{N}$, iff*

$$(\mathcal{R}_\mathcal{M} \preceq \mathcal{R}_\mathcal{N}) \quad \wedge \quad \neg(\mathcal{R}_\mathcal{N} \preceq \mathcal{R}_\mathcal{M}).$$

A decomposition $\mathcal{R}_\mathcal{M}$ is called **minimal** *iff it is irredundant and there is no irredundant decomposition that is finer than $\mathcal{R}_\mathcal{M}$.*

Clearly, we do not want redundant decompositions. If a set $M_1 \in \mathcal{M} \subseteq 2^U$ is a subset of another set $M_2 \in \mathcal{M}$, then

$$\text{cext}_{M_1}^U(\text{proj}_{M_1}^U(R_U)) \cap \text{cext}_{M_2}^U(\text{proj}_{M_2}^U(R_U)) = \text{cext}_{M_2}^U(\text{proj}_{M_2}^U(R_U)),$$

so we can remove the projection to the set M_1 without destroying the decomposition property. The notions of a decomposition being finer than another and of a decomposition being minimal serve the purpose to make as small as possible the sets of attributes defining the decomposition. If there are two irredundant decompositions $\mathcal{R}_\mathcal{M}$ and $\mathcal{R}_\mathcal{N}$ of a relation R_U with

$$M \in \mathcal{M}, N \in \mathcal{N} \quad \text{and} \quad \mathcal{M} - \{M\} = \mathcal{N} - \{N\} \quad \text{and} \quad M \subset N,$$

then obviously $\mathcal{R}_\mathcal{M} \prec \mathcal{R}_\mathcal{N}$. Hence in minimal decompositions the sets of attributes underlying the projections are as small as possible.

It would be convenient if there always were a unique minimal decomposition, because then we could always find a single best decomposition. However, in general there can be several minimal decompositions. This is demonstrated in Figure 3.10, which shows a very simple relation R_{ABC} over three binary attributes A, B, and C. As can easily be seen from the projections also shown in Figure 3.10, R_{ABC} can be decomposed into $\{R_{AB}, R_{BC}\}$, into $\{R_{AB}, R_{AC}\}$, or into $\{R_{AC}, R_{BC}\}$, all of which are minimal.

3.2.5 Possibility-Based Formalization

In the following we turn to a characterization that uses the notion of a binary possibility measure R to represent relations. Such a measure can be defined as a function satisfying certain axioms—as a probability measure P is defined as a function satisfying Kolmogorov's axioms [Kolmogorov 1933]. This characterization, as already indicated above, strengthens the connection between relational and probabilistic networks and provides an excellent starting point for the transition to (general) possibilistic networks.

Definition 3.2.6 *Let Ω be a (finite) sample space.*[2] *A* **binary possibility measure** R *on* Ω *is a function* $R : 2^\Omega \to \{0, 1\}$ *satisfying*

1. $R(\emptyset) = 0$ *and*
2. $\forall E_1, E_2 \subseteq \Omega : R(E_1 \cup E_2) = \max\{R(E_1), R(E_2)\}$.

The intuitive interpretation of a binary possibility measure is obvious: If an event E can occur (if it is possible), then $R(E) = 1$, otherwise (if E cannot occur/is impossible), then $R(E) = 0$. With this intuition the axioms are evident: The empty event is impossible and if at least one of two events is possible, then their union is possible. The term "binary" indicates that the measure can assume only the values 0 and 1—in contrast to a general possibility measure (defined semantically in Chapter 2 and to be defined axiomatically below), which can assume all values in the interval $[0, 1]$. Note, by the way, that the (general) possibility measure defined in Definition 2.4.8 on page 46 satisfies these axioms if there is only one context and Ω is finite.

It is useful to note that from the above definition it follows $\forall E_1, E_2 \subseteq \Omega :$

$$\begin{array}{rrl}
(a) & & R(E_1) = R(E_1 \cup (E_1 \cap E_2)) = \max\{R(E_1), R(E_1 \cap E_2)\} \\
& \Rightarrow & R(E_1) \geq R(E_1 \cap E_2) \\
(b) & & R(E_2) = R(E_2 \cup (E_1 \cap E_2)) = \max\{R(E_2), R(E_1 \cap E_2)\} \\
& \Rightarrow & R(E_2) \geq R(E_1 \cap E_2) \\
(a) + (b) & \Rightarrow & R(E_1 \cap E_2) \leq \min\{R(E_1), R(E_2)\}
\end{array}$$

In general $R(E_1 \cap E_2) = \min\{R(E_1), R(E_2)\}$ does not hold, because the elements that give $R(E_1)$ and $R(E_2)$ the value 1 need not be in $E_1 \cap E_2$.

In Definition 3.2.1 on page 63 a relation was defined over a set of attributes, so we had attributes right from the start. With the above definition of a binary possibility measure, however, attributes have to be added as a secondary concept. As in probability theory they are defined as random variables, i.e. as functions mapping from the sample space to some domain. We use attributes in the usual way to describe events. For example, if A is an attribute, then the statement $A = a$ is a abbreviation for the event $\{\omega \in \Omega \mid A(\omega) = a\}$ and thus one may write $R(A = a)$.[3]

The difference between the two approaches is worth noting: In the tuple-based approach, the attributes are represented by (mathematical) objects that are mapped to values by tuples, which represent objects or cases or events etc. The possibility-based approach models it the other way round: objects,

[2] For reasons of simplicity this definition is restricted to finite sample spaces. It is clear that it can easily be extended to general sample spaces by replacing 2^Ω by a suitable σ-algebra, but we do not consider this extension here.

[3] Although this should be well known, we repeat it here, because it is easily forgotten. Indeed, this was an issue in a discussion about the term "random variable" on the Uncertainty in Artificial Intelligence (UAI) mailing list in 1998.

cases, or events are represented by (mathematical) objects that are mapped to values by (random) variables, which represent attributes. It is, however, obvious that both approaches are equivalent.

Based on a binary possibility measure, relations can be introduced as probability distributions are introduced based on a probability measure. For a single attribute A a probability distribution is defined as a function

$$\begin{aligned} p : \mathrm{dom}(A) &\rightarrow [0,1], \\ p(a) &\mapsto P(\{\omega \in \Omega \mid A(\omega) = a\}). \end{aligned}$$

This definition is extended to sets of attributes by considering vectors, i.e. elements of the Cartesian product of the domains of the attributes. However, using elements of a Cartesian product introduces problems if projections have to be considered, as we already pointed out on page 64. The main difficulty is that the standard definition associates an attribute and its value only through the position in the argument list of a distribution function and thus, when computing projections, index transformations are needed that keep track of the change of positions. In Section 3.2.4 these problems made us refrain from using the standard definition of a tuple. Thus it is not surprising that we deviate from the standard definition of a (probability) distribution, too.

The idea underlying our definition is as follows: The binary possibility measure R assigns a possibility to all elements of 2^Ω, but because of the axioms a binary possibility measure has to satisfy, not all of these values need to be stored. Certain subsets are sufficient to recover the whole measure. In particular, the subset of 2^Ω that consists of all one element sets is sufficient. Suppose we have an attribute A the domain of which is Ω. Then we can recover the whole measure from the *distribution* over A (in the sense defined above). However, this distribution is merely a restriction of the measure to a specific set of events (cf. the notions of a basic probability assignment and a basic possibility assignment in Section 2.4). Now, what if we defined all distributions simply as restrictions of a measure (a probability measure or a binary possibility measure) to certain sets of events? It turns out that this is a very convenient definition, which avoids all problems that a definition based on Cartesian products would introduce.

Definition 3.2.7 *Let $X = \{A_1, \ldots, A_n\}$ be a set of attributes defined on a (finite) sample space Ω with respective domains* $\mathrm{dom}(A_i)$, $i = 1, \ldots, n$. *A* **relation** r_X *over X is the restriction of a binary possibility measure R on Ω to the set of all events that can be defined by stating values for all attributes in X. That is, $r_X = R|_{\mathcal{E}_X}$, where*

$$\mathcal{E}_X = \Big\{ E \in 2^\Omega \;\Big|\; \exists a_1 \in \mathrm{dom}(A_1) : \ldots \exists a_n \in \mathrm{dom}(A_n) : E \,\widehat{=} \bigwedge_{A_j \in X} A_j = a_j \Big\}$$

$$= \Big\{E \in 2^\Omega \;\Big|\; \exists a_1 \in \mathrm{dom}(A_1) : \ldots \exists a_n \in \mathrm{dom}(A_n) :$$
$$E = \Big\{\omega \in \Omega \;\Big|\; \bigwedge_{A_j \in X} A_j(\omega) = a_j\Big\}\Big\}.$$

We use the term "relation" instead of "binary possibility distribution", because the restrictions of a binary possibility measure defined above correspond directly to the relations defined in Definition 3.2.1 on page 63. The only difference is that with Definition 3.2.1 a tuple is marked as possible by making it a member of a set, whereas with Definition 3.2.7 it is marked as possible by assigning to it the value 1. Alternatively, we may say that Definition 3.2.7 defines relations via their *indicator function*, i.e. the function that assumes the value 1 for all members of a set and the value 0 for all non-members.

Note that—deviating from Definition 3.2.1—relations are now denoted by a lowercase r in analogy to probability distributions which are usually denoted by a lowercase p. Note also that the events referred to by a relation are characterized by a conjunction of conditions that explicitly name the attributes. Since the terms of a conjunction can be reordered without changing its meaning, projections are no longer a problem: In a projection we only have fewer conditions in the conjunctions characterizing the events. We need not be concerned with the position of attributes or the associations of attributes and their values as we had to in the standard definition.

With the above definition of a relation we can redefine the notions of decomposability and decomposition (cf. Definition 3.2.4 on page 65) based on binary possibility measures:

Definition 3.2.8 *Let $U = \{A_1, \ldots, A_n\}$ be a set of attributes and r_U a relation over U. Furthermore, let $\mathcal{M} = \{M_1, \ldots, M_m\} \subseteq 2^U$ be a set of nonempty (but not necessarily disjoint) subsets of U satisfying*

$$\bigcup_{M \in \mathcal{M}} M = U.$$

r_U is called **decomposable** *w.r.t. $\mathcal{M}$, iff*

$$\forall a_1 \in \mathrm{dom}(A_1) : \ldots \forall a_n \in \mathrm{dom}(A_n) :$$
$$r_U\Big(\bigwedge_{A_i \in U} A_i = a_i\Big) = \min_{M \in \mathcal{M}} \Big\{r_M\Big(\bigwedge_{A_i \in M} A_i = a_i\Big)\Big\}.$$

If r_U is decomposable w.r.t. $\mathcal{M}$, the set of relations

$$\mathcal{R}_\mathcal{M} = \{r_{M_1}, \ldots, r_{M_m}\} = \{r_M \mid M \in \mathcal{M}\}$$

is called the **decomposition** *of r_U.*

The definitions of the properties of relational decompositions (trivial, redundant, finer, minimal etc.—cf. Definition 3.2.5 on page 65) carry over directly from the tuple-based formalization.

3.2.6 Conditional Possibility and Independence

The most important advantage of a binary possibility measure R over the tuple-based formalization of relations is that we can define a *conditional possibility* in analogy to a conditional probability.

Definition 3.2.9 *Let Ω be a (finite) sample space, R a binary possibility measure on Ω, and $E_1, E_2 \subseteq \Omega$ events. Then*

$$R(E_1 \mid E_2) = R(E_1 \cap E_2)$$

is called the **conditional possibility** *of E_1 given E_2.*

Note that the above definition does not require $R(E_2) > 0$. Since $R(E_2) = 0$ does not lead to an undefined mathematical operation, we can make the definition more general, which is very convenient.

The notion of a conditional possibility is needed for the definition of *conditional relational independence*, which is an important tool to characterize decompositions. In order to define conditional relational independence, it is most useful to realize first that (unconditional) relational independence is most naturally characterized as follows:

Definition 3.2.10 *Let Ω be a (finite) sample space, R a binary possibility measure on Ω, and $E_1, E_2 \subseteq \Omega$ events. E_1 and E_2 are called* **relationally independent** *iff*

$$R(E_1 \cap E_2) = \min\{R(E_1), R(E_2)\}.$$

That is, if either event can occur, then it must be possible that they occur together. In other words, neither event excludes the other, which would indicate a dependence of the events. (Compare also the generally true inequality $R(E_1 \cap E_2) \leq \min\{R(E_1), R(E_2)\}$ derived above.) Note that relational independence differs from probabilistic independence only by the fact that it uses the minimum instead of the product.

The above definition is easily extended to attributes:

Definition 3.2.11 *Let Ω be a (finite) sample space, R a possibility measure on Ω, and A and B attributes with respective domains* $\mathrm{dom}(A)$ *and* $\mathrm{dom}(B)$*. A and B are called* **relationally independent**, *written $A \perp\!\!\!\perp_R B$, iff*

$$\forall a \in \mathrm{dom}(A) : \forall b \in \mathrm{dom}(B) :$$
$$R(A = a, B = b) = \min\{R(A = a), R(B = b)\}.$$

Intuitively, A and B are independent if their possible values are freely combinable. That is, if A can have value a and B can have value b, then the combination of both, i.e. the tuple (a, b), must also be possible. Note that relational independence is obviously symmetric, i.e. from $A \perp\!\!\!\perp_R B$ it follows $B \perp\!\!\!\perp_R A$. Note also that the definition is easily adapted to sets of attributes.

With the notion of a conditional possibility we can now extend the notion of relational independence to conditional relational independence:

Definition 3.2.12 *Let Ω be a (finite) sample space, R a binary possibility measure on Ω, and A, B, and C attributes with respective domains* $\text{dom}(A)$, $\text{dom}(B)$, *and* $\text{dom}(C)$. *A and C are called* **conditionally relationally independent** *given B, written $A \perp\!\!\!\perp_R C \mid B$, iff*

$$\begin{aligned}&\forall a \in \text{dom}(A) : \forall b \in \text{dom}(B) : \forall c \in \text{dom}(C) :\\ &\qquad R(A = a, C = c \mid B = b) = \min\{R(A = a \mid B = b), R(C = c \mid B = b)\}.\end{aligned}$$

The intuitive interpretation is the same as above, namely that given the value of attribute C, the values that are possible for the attributes A and B under this condition are freely combinable. Obviously, conditional relational independence is symmetric, i.e. from $A \perp\!\!\!\perp_R B \mid C$ it follows $B \perp\!\!\!\perp_R A \mid C$.

The connection of conditional relational independence to decomposition can be seen directly if we replace the conditional possibilities in the above equation by their definition:

$$\begin{aligned}&\forall a \in \text{dom}(A) : \forall b \in \text{dom}(B) : \forall c \in \text{dom}(C) :\\ &\qquad R(A = a, B = b, C = c) = \min\{R(A = a, B = b), R(B = b, C = c)\}.\end{aligned}$$

We thus arrive at the decomposition formula for the geometrical objects example discussed above. In other words, the relation R_{ABC} of the geometrical objects example is decomposable into the relations R_{AB} and R_{BC}, because in it the attributes A and C are conditionally relationally independent given the attribute B. This can easily be checked in Figure 3.2 on page 57: In each horizontal "slice" (corresponding to a value of the attribute B) the values of the attributes A and C possible in that "slice" are freely combinable. Conditional independence and its connection to network representations of decompositions is studied in more detail in Chapter 4.

Another advantage of the definition of a conditional possibility is that with it we are finally in a position to justify formally the evidence propagation scheme used in Figure 3.6 on page 60 for the geometrical objects example, because this scheme basically computes conditional possibilities given the observations. It does so in two steps, one for each unobserved attribute.

In the first step, we have to compute the conditional possibilities for the values of attribute B given the observation that attribute A has the value $A = a_{\text{obs}}$. That is, we have to compute $\forall b \in \text{dom}(B)$:

$$\begin{aligned}&R(B = b \mid A = a_{\text{obs}})\\ &\quad = R\Big(\bigvee_{a \in \text{dom}(A)} A = a, B = b, \bigvee_{c \in \text{dom}(C)} C = c \,\Big|\, A = a_{\text{obs}}\Big)\end{aligned}$$

$$
\begin{aligned}
&\overset{(1)}{=} \max_{a\in\mathrm{dom}(A)} \{ \max_{c\in\mathrm{dom}(C)} \{R(A = a, B = b, C = c \mid A = a_{\mathrm{obs}})\}\} \\
&\overset{(2)}{=} \max_{a\in\mathrm{dom}(A)} \{ \max_{c\in\mathrm{dom}(C)} \{\min\{R(A = a, B = b, C = c), \\
&\qquad\qquad R(A = a \mid A = a_{\mathrm{obs}})\}\}\} \\
&\overset{(3)}{=} \max_{a\in\mathrm{dom}(A)} \{ \max_{c\in\mathrm{dom}(C)} \{\min\{R(A = a, B = b), R(B = b, C = c), \\
&\qquad\qquad R(A = a \mid A = a_{\mathrm{obs}})\}\}\} \\
&= \max_{a\in\mathrm{dom}(A)} \{\min\{R(A = a, B = b), R(A = a \mid A = a_{\mathrm{obs}}), \\
&\qquad\qquad \underbrace{\max_{c\in\mathrm{dom}(C)} \{R(B = b, C = c)\}}_{=R(B=b)\geq R(A=a,B=b)}\}\} \\
&= \max_{a\in\mathrm{dom}(A)} \{\min\{R(A = a, B = b), R(A = a \mid A = a_{\mathrm{obs}})\}\}.
\end{aligned}
$$

Here (1) holds because of the second axiom a binary possibility measure has to satisfy. (3) holds because of the fact that the relation R_{ABC} can be decomposed w.r.t. the set $\mathcal{M} = \{\{A, B\}, \{B, C\}\}$ according to the decomposition formula stated above. (2) holds, since in the first place

$$
\begin{aligned}
&R(A = a, B = b, C = c \mid A = a_{obs}) \\
&= R(A = a, B = b, C = c, A = a_{obs}) \\
&= \begin{cases} R(A = a, B = b, C = c), & \text{if } a = a_{\mathrm{obs}}, \\ 0, & \text{otherwise,} \end{cases}
\end{aligned}
$$

and secondly

$$
\begin{aligned}
R(A = a \mid A = a_{\mathrm{obs}}) &= R(A = a, A = a_{\mathrm{obs}}) \\
&= \begin{cases} R(A = a), & \text{if } a = a_{\mathrm{obs}}, \\ 0, & \text{otherwise,} \end{cases}
\end{aligned}
$$

and therefore, since trivially $R(A = a) \geq R(A = a, B = b, C = c)$,

$$
\begin{aligned}
&R(A = a, B = b, C = c \mid A = a_{obs}) \\
&= \min\{R(A = a, B = b, C = c), R(A = a \mid A = a_{\mathrm{obs}})\}.
\end{aligned}
$$

It is obvious that the left part of Figure 3.6 is a graphical representation, for each possible value of attribute B, of the above formula for computing the conditional possibility $R(B = b \mid A = a_{\mathrm{obs}})$.

In the second step, we have to compute the conditional possibilities for the values of attribute C given the observation that attribute A has the value a_{obs}. That is, we have to compute $\forall c \in \text{dom}(C)$:

$$
\begin{aligned}
&R(C = c \mid A = a_{\text{obs}}) \\
&= \quad R\Big(\bigvee_{a\in\text{dom}(A)} A = a, \bigvee_{b\in\text{dom}(B)} B = b, C = c \,\Big|\, A = a_{\text{obs}} \Big) \\
&\overset{(1)}{=} \quad \max_{a\in\text{dom}(A)} \{ \max_{b\in\text{dom}(B)} \{ R(A = a, B = b, C = c \mid A = a_{\text{obs}}) \} \} \\
&\overset{(2)}{=} \quad \max_{a\in\text{dom}(A)} \{ \max_{b\in\text{dom}(B)} \{ \min\{ R(A = a, B = b, C = c), \\
&\qquad\qquad R(A = a \mid A = a_{\text{obs}}) \} \} \} \\
&\overset{(3)}{=} \quad \max_{a\in\text{dom}(A)} \{ \max_{b\in\text{dom}(B)} \{ \min\{ R(A = a, B = b), R(B = b, C = c), \\
&\qquad\qquad R(A = a \mid A = a_{\text{obs}}) \} \} \} \\
&= \quad \max_{b\in\text{dom}(B)} \{ \min\{ R(B = b, C = c), \\
&\qquad\qquad \underbrace{\max_{a\in\text{dom}(A)} \{ \min\{ R(A = a, B = b), R(A = a \mid A = a_{\text{obs}}) \} \} \}}_{=R(B=b|A=a_{\text{obs}})} \\
&= \quad \max_{b\in\text{dom}(B)} \{ \min\{ R(B = b, C = c), R(B = b \mid A = a_{\text{obs}}) \} \}.
\end{aligned}
$$

Here (1), (2), and (3) hold for the same reasons as above. It is obvious that the right part of Figure 3.6 is a graphical representation, for each possible value of attribute C, of the above formula for computing the conditional possibility $R(C = c \mid A = a_{\text{obs}})$.

In the same fashion as above we can also compute the influence of observations of more than one attribute. Suppose, for example, that the values of the attributes A and C have both been observed and found to be a_{obs} and c_{obs}, respectively. To compute the resulting conditional possibilities for the values of attribute B given these observations, we have to compute $\forall b \in \text{dom}(B)$:

$$
\begin{aligned}
&R(B = b \mid A = a_{\text{obs}}, C = c_{\text{obs}}) \\
&= \quad R\Big(\bigvee_{a\in\text{dom}(A)} A = a, B = b, \bigvee_{c\in\text{dom}(C)} C = c \,\Big|\, A = a_{\text{obs}}, C = c_{\text{obs}} \Big) \\
&\overset{(1)}{=} \quad \max_{a\in\text{dom}(A)} \{ \max_{c\in\text{dom}(C)} \{ R(A = a, B = b, C = c \mid A = a_{\text{obs}}, C = c_{\text{obs}}) \} \} \\
&\overset{(2)}{=} \quad \max_{a\in\text{dom}(A)} \{ \max_{c\in\text{dom}(C)} \{ \min\{ R(A = a, B = b, C = c), \\
&\qquad\qquad R(A = a \mid A = a_{\text{obs}}), R(C = c \mid C = c_{\text{obs}}) \} \} \}
\end{aligned}
$$

$$\begin{aligned}
&\overset{(3)}{=} \max_{a \in \mathrm{dom}(A)} \{ \max_{c \in \mathrm{dom}(C)} \{\min\{R(A = a, B = b), \quad R(B = b, C = c), \\
&\qquad\qquad R(A = a \mid A = a_{\mathrm{obs}}), R(C = c \mid C = c_{\mathrm{obs}})\}\}\} \\
&= \min\{ \max_{a \in \mathrm{dom}(A)} \{\min\{R(A = a, B = b), R(A = a \mid A = a_{\mathrm{obs}})\}, \\
&\qquad \max_{c \in \mathrm{dom}(C)} \{\min\{R(B = b, C = c), R(C = c \mid C = c_{\mathrm{obs}})\}\}.
\end{aligned}$$

Here (1), (2), and (3) hold for similar/the same reasons as above. Again the evidence propagation process can easily be depicted in the style of Figure 3.6.

Note that from the basic principle applied to derive the above formulae, namely exploiting the decomposition property and shifting the maximum operators so that terms independent of their index variable are moved out of their range, generalizes easily to more than three attributes. How the terms can be reorganized, however, depends on the decomposition formula. Clearly, a decomposition into small terms, i.e. possibility distributions on subspaces scaffolded by few attributes, is desirable, because this facilitates the reorganization and leads to simple propagation operations.

Seen from the point of view of the formulae derived above, the network representation of the decomposition indicated in Figure 3.7 on page 60 can also be interpreted as the result of some kind of pre-execution of the first steps in the above derivations. The network structure pre-executes some of the computations that have to be carried out to compute conditional possibilities by exploiting the decomposition formula and shifting the aggregation (maximum) operators. This interpretation is discussed in more detail in Chapter 4, in particular in Section 4.2.1.

3.3 Probabilistic Decomposition

The method of decomposing a relation can easily be transferred to probability distributions. Only the definitions of projection, cylindrical extension, and intersection have to be modified. Projection now consists in calculating the marginal distribution on a subspace. Extension and intersection are combined and consist in multiplying the prior distribution with the quotient of posterior and prior marginal probability.

3.3.1 A Simple Example

The idea of probabilistic decomposition is best explained by a simple example. Figure 3.11 shows a probability distribution on the joint domain of the three attributes A, B, and C together with its marginal distributions (sums over rows/columns). It is closely related to the example of the preceding section, since in this distribution those value combinations that were contained in the relation R_{ABC} (were possible) have a high probability, while those that were

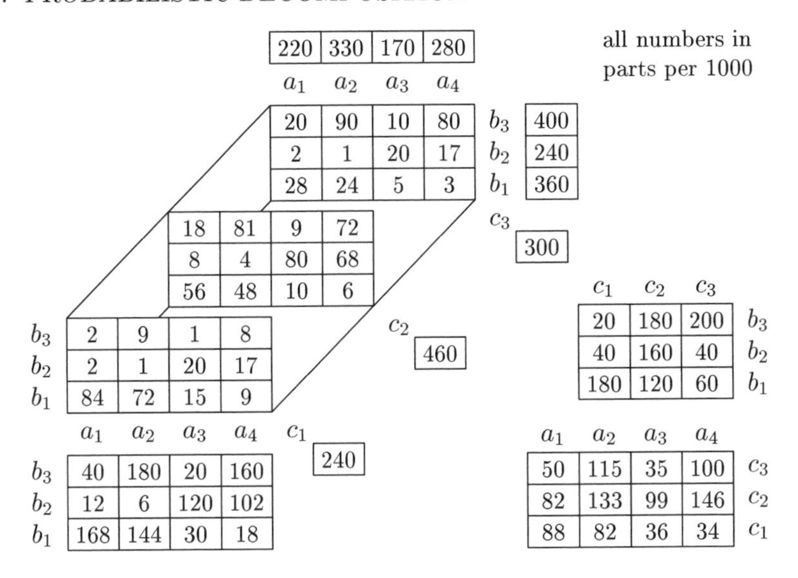

Figure 3.11 A three-dimensional probability distribution with its marginal distributions (sums over rows/columns). It can be decomposed into the marginal distributions on the subspaces $\{A, B\}$ and $\{B, C\}$.

missing (were impossible) have a low probability. The probabilities could, for example, state the relative frequencies of the objects in the box one of them is drawn from.

As the relation R_{ABC} can be decomposed into the relations R_{AB} and R_{BC}, the probability distribution in Figure 3.11 can be decomposed into the two marginal distributions on the subspaces $\{A, B\}$ and $\{B, C\}$. This is possible, because it can be reconstructed using the formula

$$\forall a \in \mathrm{dom}(A) : \forall b \in \mathrm{dom}(B) : \forall c \in \mathrm{dom}(C) :$$
$$P(A = a, B = b, C = c) = \frac{P(A = a, B = b) \cdot P(B = b, C = c)}{P(B = b)}.$$

This formula is the direct analog of the decomposition formula

$$\forall a \in \mathrm{dom}(A) : \forall b \in \mathrm{dom}(B) : \forall c \in \mathrm{dom}(C) :$$
$$R(A = a, B = b, C = c) = \min\{R(A = a, B = b), R(B = b, C = c)\}$$

for the relational case (cf. Figure 3.5 on page 59 and the formula on page 71). Note that in the probabilistic formula the minimum is replaced by the product, that there is an additional factor $\frac{1}{P(B=b)}$, and that (because of this factor) $P(B = b)$ must be positive.

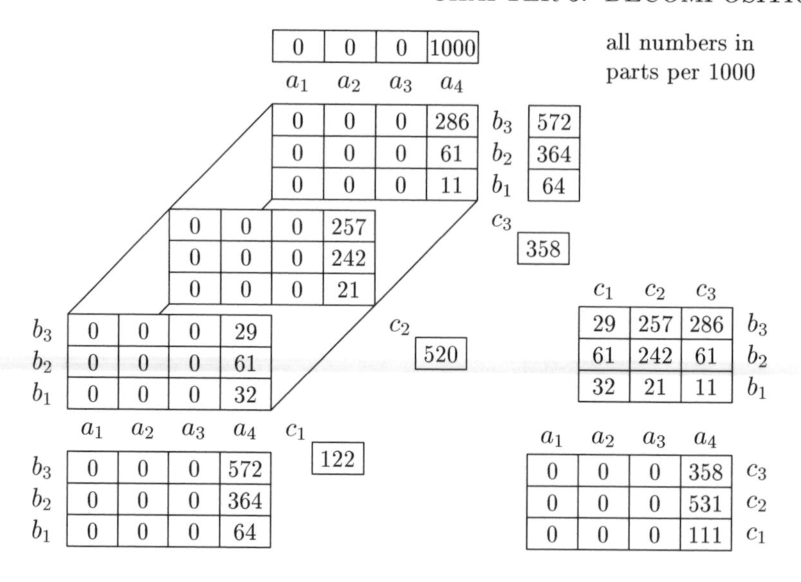

Figure 3.12 Reasoning in the domain as a whole.

3.3.2 Reasoning in the Simple Example

Let us assume—as in the relational example—that we know that attribute A has value a_4. Obviously the corresponding probability distributions for B and C can be determined from the three-dimensional distribution by restricting it to the "slice" that corresponds to $A = a_4$, i.e. by *conditioning* it on $A = a_4$, and computing the marginal distributions of that "slice". This is demonstrated in Figure 3.12. Note that all numbers in the "slices" corresponding to other values of attribute A are set to zero, because these are known now to be impossible. Note also that the probabilities in the "slice" corresponding to $A = a_4$ have been renormalized by multiplying them by $\frac{1}{P(A=a_4)} = \frac{1000}{280}$ in order to make them sum up to 1 (as required for a probability distribution).

However—as in the relational example—distributions on two-dimensional subspaces are also sufficient to draw this inference; see Figure 3.13. The information $A = a_4$ is extended to the subspace $\{A, B\}$ by multiplying the joint probabilities by the quotient of posterior and prior probability of $A = a_i$, $i = 1, 2, 3, 4$. Then the marginal distribution on B is determined by summing over the rows. In the same way the information of the new probability distribution on B is propagated to C: The joint distribution on $\{B, C\}$ is multiplied with the quotient of prior and posterior probability of $B = b_j$, $j = 1, 2, 3$, and the marginal distribution on C is computed by summing over the columns. It is easy to check that the results obtained in this way are the same as those of corresponding computations on the three-dimensional domain.

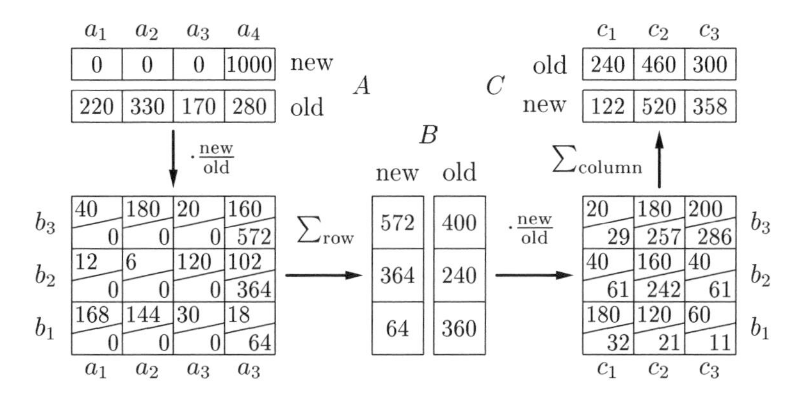

Figure 3.13 Propagation of the evidence that attribute A has value a_4 in the three-dimensional probability distribution shown in Figure 3.11 using the marginal probability distributions on the subspaces $\{A, B\}$ and $\{B, C\}$.

3.3.3 Factorization of Probability Distributions

Generalizing from the simple example discussed above, probabilistic decomposition can be defined in close analogy to the relational case. This leads to the following definition [Castillo *et al.* 1997]:

Definition 3.3.1 *Let $U = \{A_1, \ldots, A_n\}$ be a set of attributes and p_U a probability distribution over U. Furthermore, let $\mathcal{M} = \{M_1, \ldots, M_m\} \subseteq 2^U$ be a set of nonempty (but not necessarily disjoint) subsets of U satisfying*

$$\bigcup_{M \in \mathcal{M}} M = U.$$

p_U is called **decomposable** *or* **factorizable** *w.r.t. $\mathcal{M}$ iff it can be written as a product of m nonnegative functions $\phi_M : \mathcal{E}_M \to \mathbb{R}_0^+$, $M \in \mathcal{M}$, i.e. iff*

$$\forall a_1 \in \text{dom}(A_1) : \ldots \forall a_n \in \text{dom}(A_n) :$$
$$p_U\Big(\bigwedge_{A_i \in U} A_i = a_i\Big) = \prod_{M \in \mathcal{M}} \phi_M\Big(\bigwedge_{A_i \in M} A_i = a_i\Big).$$

If p_U is decomposable w.r.t. $\mathcal{M}$ the set of functions

$$\Phi_{\mathcal{M}} = \{\phi_{M_1}, \ldots, \phi_{M_m}\} = \{\phi_M \mid M \in \mathcal{M}\}$$

is called the **decomposition** *or the* **factorization** *of p_U. The functions in $\Phi_{\mathcal{M}}$ are called the* **factor potentials** *of p_U.*

In the simple example discussed above, in which the three-dimensional probability distribution on the joint domain $\{A, B, C\}$ can be decomposed into the marginal distributions on the subspaces $\{A, B\}$ and $\{B, C\}$, we may choose, for instance, two functions ϕ_{AB} and ϕ_{BC} in such a way that

$$\forall a \in \mathrm{dom}(A) : \forall b \in \mathrm{dom}(B) : \forall c \in \mathrm{dom}(C) :$$
$$\phi_{AB}(A = a, B = b) = P(A = a, B = b) \quad \text{and}$$
$$\phi_{BC}(B = b, C = c) = \begin{cases} P(C = c \mid B = b), & \text{if } P(B = b) \neq 0, \\ 0, & \text{otherwise.} \end{cases}$$

Note that using factor potentials instead of marginal probability distributions (which would be directly analogous to the relational case) is necessary, since we have to take care of the factor $\frac{1}{P(B=b)}$, which has to be incorporated into at least one factor potential of the decomposition.

The definitions of the properties of decompositions (trivial, redundant, finer, minimal etc.—cf. Definition 3.2.5 on page 65) carry over directly from the relational case if we replace the set $\mathcal{R}_\mathcal{M}$ of relations by the set $\Phi_\mathcal{M}$ of factor potentials. However, an important difference to the relational case is that for *strictly positive* probability distributions the minimal decomposition is unique w.r.t. the sets of attributes the factor potentials are defined on (cf. the notion of a minimal independence map in, for example, [Pearl 1988, Castillo *et al.* 1997]). The exact definition of the factor potentials may still differ, though, as can already be seen from the decomposition formula of the geometrical objects example: The factor $\frac{1}{P(B=b)}$ may be included in one factor potential or may be distributed to both, for example as $\frac{1}{\sqrt{P(B=b)}}$.

3.3.4 Conditional Probability and Independence

As stated above, the three-dimensional probability distribution shown in Figure 3.11 on page 75 can be reconstructed from the marginal distributions on the subspaces $\{A, B\}$ and $\{B, C\}$ using the formula

$$\forall a \in \mathrm{dom}(A) : \forall b \in \mathrm{dom}(B) : \forall c \in \mathrm{dom}(C) :$$
$$P(A = a, B = b, C = c) = \frac{P(A = a, B = b)\, P(B = b, C = c)}{P(B = b)}.$$

Drawing on the notion of a conditional probability, this formula can be derived from the (generally true) formula

$$\forall a \in \mathrm{dom}(A) : \forall b \in \mathrm{dom}(B) : \forall c \in \mathrm{dom}(C) :$$
$$P(A = a, B = b, C = c) = P(A = a \mid B = b, C = c)\, P(B = b, C = c)$$

by noting that in the probability distribution of the example A is *conditionally independent* of C given B, written $A \perp\!\!\!\perp_P C \mid B$. That is,

$$\forall a \in \mathrm{dom}(A) : \forall b \in \mathrm{dom}(B) : \forall c \in \mathrm{dom}(C) :$$
$$P(A = a \mid B = b, C = c) = P(A = a \mid B = b) = \frac{P(A = a, B = b)}{P(B = b)},$$

i.e. if the value of attribute B is known, the probabilities of the values of attribute A do not depend on the value of attribute C. Note that conditional independence is symmetric, i.e. if $A \perp\!\!\!\perp_P C \mid B$, then

$$\forall a \in \mathrm{dom}(A) : \forall b \in \mathrm{dom}(B) : \forall c \in \mathrm{dom}(C) :$$
$$P(C = c \mid B = b, A = a) = P(C = c \mid B = b) = \frac{P(C = c, B = b)}{P(B = b)},$$

also holds. In other words, $A \perp\!\!\!\perp_P C \mid B$ entails $C \perp\!\!\!\perp_P A \mid B$. This becomes most obvious if we state conditional probabilistic independence in its most common form, which is directly analogous to the standard definition of (unconditional) probabilistic independence, namely as

$$\forall a \in \mathrm{dom}(A) : \forall b \in \mathrm{dom}(B) : \forall c \in \mathrm{dom}(C) :$$
$$P(A = a, C = c \mid B = b) = P(A = a \mid B = b)\, P(C = c \mid B = b).$$

The notion of conditional probabilistic independence is often used to derive a factorization formula for a multivariate probability distribution that is more explicit about the factors than Definition 3.3.1. The idea is to start from the (generally true) *chain rule of probability*

$$\forall a_1 \in \mathrm{dom}(A_1) : \ldots \forall a_n \in \mathrm{dom}(A_n) :$$
$$P\Big(\bigwedge\nolimits_{i=1}^{n} A_i = a_i\Big) = \prod_{i=1}^{n} P\Big(A_i = a_i \;\Big|\; \bigwedge\nolimits_{j=1}^{i-1} A_j = a_j\Big)$$

and to simplify the factors on the right by exploiting conditional independences. As can be seen from the three attribute example above, conditional independences allow us to cancel some of the attributes appearing in the conditions of the conditional probabilities. In this way the factors refer to fewer conditional probability distributions and thus may be stored more efficiently. Since this type of factorization is based on the chain rule of probability, it is often called *chain rule factorization* (cf. [Castillo *et al.* 1997]).

The notion of a conditional probability also provides a justification of the reasoning scheme outlined in Section 3.3.2, which can be developed in direct analogy to the relational case (recall from Chapter 2 that probabilistic reasoning consists in computing conditional probabilities). In the first step we have to compute the conditional probabilities of the values of attribute B given the observation that attribute A has the value a_{obs}. That is, we have to compute $\forall b \in \mathrm{dom}(B)$:

$$P(B = b \mid A = a_{\mathrm{obs}})$$
$$= P\Big(\bigvee_{a \in \mathrm{dom}(A)} A = a, B = b, \bigvee_{c \in \mathrm{dom}(C)} C = c \;\Big|\; A = a_{\mathrm{obs}}\Big)$$

$$\stackrel{(1)}{=} \sum_{a\in\mathrm{dom}(A)} \sum_{c\in\mathrm{dom}(C)} P(A=a, B=b, C=c \mid A=a_{\mathrm{obs}})$$

$$\stackrel{(2)}{=} \sum_{a\in\mathrm{dom}(A)} \sum_{c\in\mathrm{dom}(C)} P(A=a, B=b, C=c) \cdot \frac{P(A=a \mid A=a_{\mathrm{obs}})}{P(A=a_i)}$$

$$\stackrel{(3)}{=} \sum_{a\in\mathrm{dom}(A)} \sum_{c\in\mathrm{dom}(C)} \frac{P(A=a, B=b)P(B=b, C=c)}{P(B=b)} \cdot \frac{P(A=a \mid A=a_{\mathrm{obs}})}{P(A=a)}$$

$$= \sum_{a\in\mathrm{dom}(A)} P(A=a, B=b) \cdot \frac{P(A=a \mid A=a_{\mathrm{obs}})}{P(A=a)} \cdot \underbrace{\sum_{c\in\mathrm{dom}(C)} P(C=c \mid B=b)}_{=1}$$

$$= \sum_{a\in\mathrm{dom}(A)} P(A=a, B=b) \cdot \frac{P(A=a \mid A=a_{\mathrm{obs}})}{P(A=a)}.$$

Here (1) holds because of Kolmogorov's axioms and (3) holds because of the conditional probabilistic independence of A and C given B, which allows us to decompose the joint probability distribution P_{ABC} according to the formula stated above. (2) holds, since in the first place

$$\begin{aligned} & P(A=a, B=b, C=c \mid A=a_{obs}) \\ &= \frac{P(A=a, B=b, C=c, A=a_{obs})}{P(A=a_{\mathrm{obs}})} \\ &= \begin{cases} \dfrac{P(A=a, B=b, C=c)}{P(A=a_{\mathrm{obs}})}, & \text{if } a=a_{\mathrm{obs}}, \\ 0, & \text{otherwise,} \end{cases} \end{aligned}$$

and secondly

$$P(A=a, A=a_{\mathrm{obs}}) = \begin{cases} P(A=a), & \text{if } a=a_{\mathrm{obs}}, \\ 0, & \text{otherwise,} \end{cases}$$

and therefore

$$\begin{aligned} & P(A=a, B=b, C=c \mid A=a_{obs}) \\ &= P(A=a, B=b, C=c) \cdot \frac{P(A=a \mid A=a_{\mathrm{obs}})}{P(A=a)}. \end{aligned}$$

It is obvious that the left part of Figure 3.13 on page 77 is only a graphical representation, for each possible value of attribute B, of the above formula.

Note that this propagation formula is directly analogous to the formula for the relational case (cf. page 71 in Section 3.2.6). The only difference (apart from the factor $\frac{1}{P(A=a)}$) is that the probabilistic formula uses the sum instead of the maximum and the product instead of the minimum.

In the second step of the propagation, we have to determine the conditional probabilities of the values of attribute C given the observation that attribute A has the value a_{obs}. That is, we have to compute $\forall c \in \text{dom}(C)$:

$$
\begin{aligned}
& P(C = c \mid A = a_{\text{obs}}) \\
&= P\Big(\bigvee_{a \in \text{dom}(A)} A = a, \bigvee_{b \in \text{dom}(B)} B = b, C = c \,\Big|\, A = a_{\text{obs}} \Big) \\
&\stackrel{(1)}{=} \sum_{a \in \text{dom}(A)} \sum_{b \in \text{dom}(B)} P(A = a, B = b, C = c \mid A = a_{\text{obs}}) \\
&\stackrel{(2)}{=} \sum_{a \in \text{dom}(A)} \sum_{b \in \text{dom}(B)} P(A = a, B = b, C = c) \cdot \frac{P(A = a \mid A = a_{\text{obs}})}{P(A = a)} \\
&\stackrel{(3)}{=} \sum_{a \in \text{dom}(A)} \sum_{b \in \text{dom}(B)} \frac{P(A = a, B = b) P(B = b, C = c)}{P(B = b)} \\
&\qquad \cdot \frac{P(A = a \mid A = a_{\text{obs}})}{P(A = a)} \\
&= \sum_{b \in \text{dom}(B)} \frac{P(B = b, C = c)}{P(B = b)} \\
&\qquad \underbrace{\sum_{a \in \text{dom}(A)} P(A = a, B = b) \cdot \frac{R(A = a \mid A = a_{\text{obs}})}{P(A = a)}}_{= P(B = b \mid A = a_{\text{obs}})} \\
&= \sum_{b \in \text{dom}(B)} P(B = b, C = c) \cdot \frac{P(B = b \mid A = a_{\text{obs}})}{P(B = b)}.
\end{aligned}
$$

Here (1), (2), and (3) hold for the same reasons as above. It is obvious that the right part of Figure 3.13 on page 77 is only a graphical representation, for each value of attribute C, of the above formula. Note that, as above, this propagation formula is directly analogous to the formula for the relational case (cf. page 73 in Section 3.2.6).

In the same fashion as above, we can also compute the influence of observations of more than one attribute. Suppose, for example, that the values of attributes A and C have both been observed and found to be a_{obs} and c_{obs}, respectively. To compute the resulting conditional probabilities of the values of B given these observations, we have to compute $\forall b \in \text{dom}(B)$:

$$
\begin{aligned}
&P(B = b \mid A = a_{\text{obs}}, C = c_{\text{obs}}) \\
&\quad = P\Big(\bigvee_{a \in \text{dom}(A)} A = a, B = b, \bigvee_{c \in \text{dom}(C)} C = c \,\Big|\, A = a_{\text{obs}}, C = c_{\text{obs}} \Big) \\
&\quad \overset{(1)}{=} \sum_{a \in \text{dom}(A)} \sum_{c \in \text{dom}(C)} P(A = a, B = b, C = c \mid A = a_{\text{obs}}, C = c_{\text{obs}}) \\
&\quad \overset{(2)}{=} \alpha \sum_{a \in \text{dom}(A)} \sum_{c \in \text{dom}(C)} P(A = a, B = b, C = c) \cdot \frac{P(A = a \mid A = a_{\text{obs}})}{P(A = a)} \\
&\qquad\qquad \cdot \frac{P(C = c \mid C = c_{\text{obs}})}{P(C = c)} \\
&\quad \overset{(3)}{=} \alpha \sum_{a \in \text{dom}(A)} \sum_{c \in \text{dom}(C)} \frac{P(A = a, B = b) P(B = b, C = c)}{P(B = b)} \\
&\qquad\qquad \cdot \frac{P(A = a \mid A = a_{\text{obs}})}{P(A = a)} \frac{P(C = c \mid C = c_{\text{obs}})}{P(C = c)} \\
&\quad = \frac{\alpha}{P(B = b)} \cdot \Bigg(\sum_{a \in \text{dom}(A)} P(A = a, B = b) \cdot \frac{P(A = a \mid A = a_{\text{obs}})}{P(A = a)} \Bigg) \\
&\qquad\qquad \cdot \Bigg(\sum_{c \in \text{dom}(C)} P(B = b, C = c) \cdot \frac{P(C = c \mid C = c_{\text{obs}})}{P(C = c)} \Bigg),
\end{aligned}
$$

where $\alpha = \frac{P(A=a_{\text{obs}})P(C=c_{\text{obs}})}{P(A=a_{\text{obs}}, C=c_{\text{obs}})}$ is a normalization factor that enables us to have separate factors for the attributes A and C and thus to keep the propagation scheme uniform. (Note, however, the additional factor $\frac{1}{P(B=b)}$.) (1), (2), and (3) hold for similar/the same reasons as above. The evidence propagation process can easily be depicted in the style of Figure 3.13.

As in the relational case, the principle applied in the above derivation, namely shifting the sums so that terms independent of their index variable are moved out of their range, can easily be generalized to more attributes.

3.4 Possibilistic Decomposition

The method of decomposing a relation can be transferred to possibility distributions as easily as it could be transferred to probability distributions in Section 3.3. Again only the definitions of projection, cylindrical extension, and intersection have to be modified. Projection now consists in computing the maximal degrees of possibility over the dimensions removed by it. Extension and intersection are combined and consist in calculating the minimum of the prior joint and the posterior marginal possibility degrees.

3.4.1 Transfer from Relational Decomposition

Actually possibilistic decomposition is formally identical to relational decomposition in the possibility-based formalization studied in Section 3.2.5. The only difference is that instead of only 0 and 1 a (general) possibility measure can assume any value in the interval $[0, 1]$, thus quantifying the notion of a possibility. Therefore, in analogy to the treatment of the relational case in Section 3.2.5, we complement the semantical introduction of a possibility measure and a possibility distribution (cf. Section 2.4) by an axiomatic approach (compare also [Dubois and Prade 1988]):

Definition 3.4.1 *Let Ω be a (finite) sample space. A (general)* **possibility measure** Π *on Ω is a function* $\Pi : 2^\Omega \to [0, 1]$ *satisfying*

1. $\Pi(\emptyset) = 0$ *and*
2. $\forall E_1, E_2 \subseteq \Omega : \Pi(E_1 \cup E_2) = \max\{\Pi(E_1), \Pi(E_2)\}$.

Note that this definition differs from Definition 3.2.6 on page 67 only in the range of values of the measure. Note also that the measure of Definition 2.4.8 on page 46 satisfies the axioms of the above definition.

By the transition to a (general) possibility measure carried out above it is explained why there is no axiom $R(\Omega) = 1$ for binary possibility measures: It would have been necessary to revoke this axiom now. With *degrees* of possibility, $\Pi(\Omega) = \max_{\omega \in \Omega} \Pi(\{\omega\})$ need not be 1. Adding this constraint would introduce the normalization condition (cf. Definition 2.4.3 on page 37), which we rejected in Section 2.4.6.

Due to the close formal proximity of binary and general possibility measures there is not much left to be said. Everything developed following the definition of a binary possibility measure in Definition 3.2.6 on page 67 carries over directly to (general) possibility measures, since the fact that binary possibility measures can assume only the values 0 and 1 was not exploited.

3.4.2 A Simple Example

Although there is a direct transfer from the relational case, it is useful to illustrate the decomposition of possibility distributions with a simple example. Figure 3.14 shows a three-dimensional possibility distribution on the joint domain of the attributes A, B, and C and its marginal distributions (maxima over rows/columns). In analogy to the probabilistic example it is closely related to the relational example: Those value combinations that were possible have a high degree of possibility and those that were impossible have a low degree of possibility. This possibility distribution can be decomposed into the marginal distributions on the subspaces $\{A, B\}$ and $\{B, C\}$, because it can

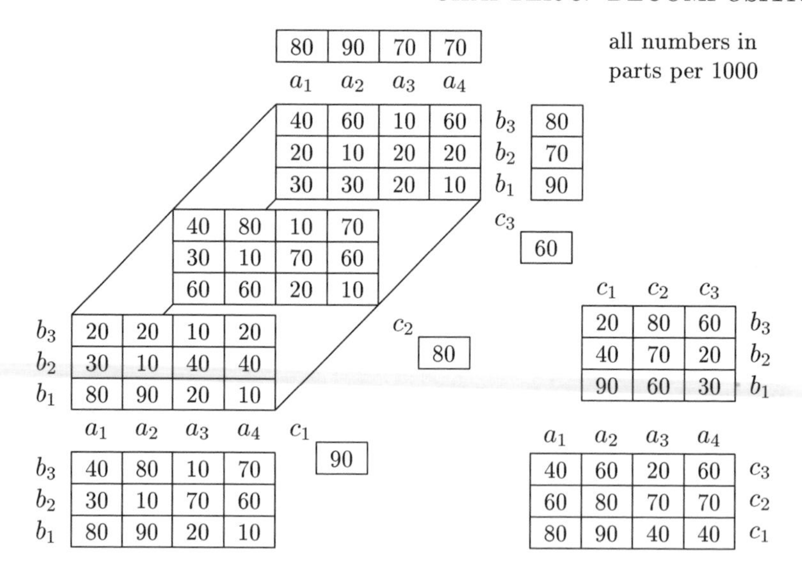

Figure 3.14 A three-dimensional possibility distribution with marginal distributions (maxima over rows/columns).

be reconstructed using the formula

$$
\begin{aligned}
&\forall a \in \mathrm{dom}(A) : \forall b \in \mathrm{dom}(B) : \forall c \in \mathrm{dom}(C) : \\
&\quad \Pi(A = a, B = b, C = c) = \min_{b \in \mathrm{dom}(B)} \{\Pi(A = a, B = b), \Pi(B = b, C = c)\} \\
&\qquad = \min_{b \in \mathrm{dom}(B)} \{ \max_{c \in \mathrm{dom}(C)} \Pi(A = a, B = b, C = c), \\
&\qquad\qquad \max_{a \in \mathrm{dom}(A)} \Pi(A = a, B = b, C = c).\}
\end{aligned}
$$

3.4.3 Reasoning in the Simple Example

Let us assume as usual that from an observation it is known that attribute A has value a_4. Obviously the corresponding (conditional) possibility distribution can be determined from the three-dimensional distribution by restricting it to the "slice" corresponding to $A = a_4$, i.e. by conditioning it on $A = a_4$, and computing the marginal distributions of that "slice". This is demonstrated in Figure 3.15. Note that the numbers in the "slices" corresponding to other values of attribute A have been set to zero, because these are known now to be impossible. Note also that—in contrast to the probabilistic case—the numbers in the "slice" corresponding to $A = a_4$ are unchanged, i.e. no renormalization takes place.

However, as in the probabilistic case studied in Section 3.3.2, distributions on two-dimensional subspaces are also sufficient to draw this inference; see

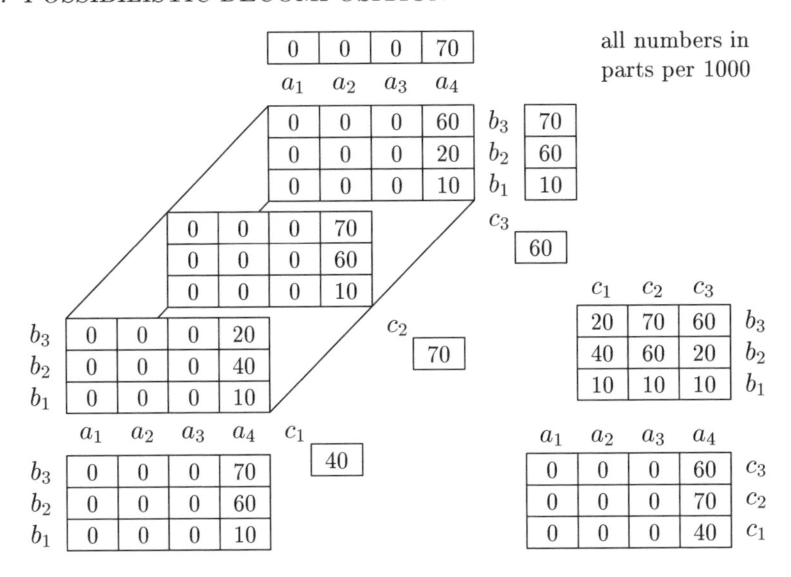

Figure 3.15 Reasoning in the domain as a whole.

Figure 3.16. The information that $A = a_4$ is extended to the subspace $\{A, B\}$ by computing the minimum of the prior joint degrees of possibility and the posterior degrees of possibility of $A = a_i$, $i = 1, 2, 3, 4$. Then the marginal distribution on B is determined by taking the maximum over the rows. In the same way the information of the new possibility distribution on B is propagated to C: The minimum of the prior joint distribution on $\{B, C\}$ and the posterior distribution on B is computed and projected to attribute C by taking the maximum over the columns. It is easy to check that the results obtained in this way are the same as those that follow from the computations on the three-dimensional domain (see above).

3.4.4 Conditional Degrees of Possibility and Independence

This reasoning scheme can be justified in the same way as in the relational and in the probabilistic case by drawing on the notion of a *conditional degree of possibility* (cf. Definition 3.2.9 on page 70 and Definition 2.4.9 on page 47). The derivation is formally identical to the one carried out in Section 3.2.6, pages 71ff, since for the relational case the fact that a binary possibility measure can assume only the values 0 and 1 was not exploited. This formal identity stresses that possibilistic networks can be seen as a “fuzzyfication” of relational networks, which is achieved in the usual way: A restriction to the values 0 and 1 is removed by considering instead all values in the interval $[0, 1]$.

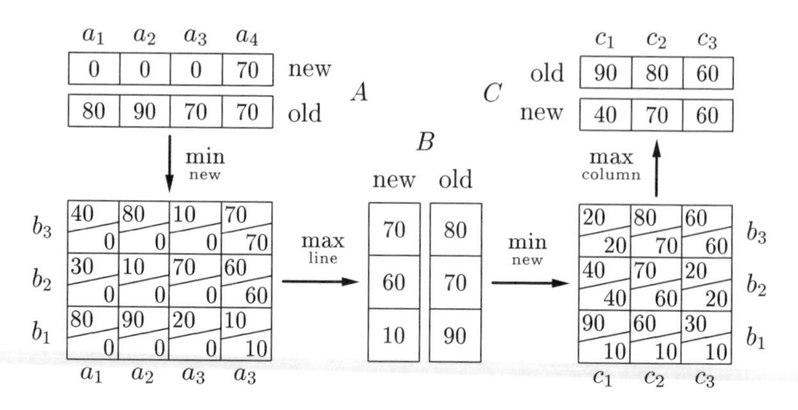

Figure 3.16 Propagation of the evidence that attribute A has value a_4 in the three-dimensional possibility distribution shown in Figure 3.14 using the projections to the subspaces $\{A, B\}$ and $\{B, C\}$.

Possibilistic decomposition as we study it here is based on a specific notion of conditional possibilistic independence, which is defined in direct analogy to the relational case (cf. Definition 3.2.10 on page 70).

Definition 3.4.2 *Let Ω be a (finite) sample space, Π a possibility measure on Ω, and A, B, and C attributes with respective domains* $\text{dom}(A)$, $\text{dom}(B)$, *and* $\text{dom}(C)$. *A and C are called* **conditionally possibilistically independent** *given B, written $A \perp\!\!\!\perp_\Pi C \mid B$, iff*

$$\begin{aligned} &\forall a \in \text{dom}(A) : \forall b \in \text{dom}(B) : \forall c \in \text{dom}(C) : \\ &\qquad \Pi(A = a, C = c \mid B = b) = \min\{\Pi(A = a \mid B = b), \Pi(C = c \mid B = b)\}. \end{aligned}$$

Of course, this definition is easily extended to sets of attributes. This specific notion of conditional possibilistic independence is usually called *possibilistic non-interactivity* [Dubois and Prade 1988]. However, in contrast to probability theory, for which there is unanimity about the notion of conditional probabilistic independence, for possibility theory several alternative notions have been suggested. Discussions can be found in, for example, [Farinas del Cerro and Herzig 1994, Fonck 1994]. The main problem seems to be that possibility theory is a calculus for uncertain *and* imprecise reasoning, the first of which is more closely related to probability theory, the latter more closely to relational algebra (cf. Section 2.4.11). Thus there are at least two ways to arrive at a definition of conditional possibilistic independence, namely either *uncertainty-based* by a derivation from *Dempster's rule of conditioning* [Shafer 1976], or *imprecision-based* by a derivation from the relational setting (which leads to possibilistic non-interactivity) [de Campos *et al.* 1995]. In this book we concentrate on the latter approach, because its semantical justification is much clearer and it has the advantage to be in accordance with the so-called *extension principle* [Zadeh 1975].

Note that conditional possibilistic independence can be used, in analogy to the probabilistic case, to derive a decomposition formula for a multivariate possibility distribution based on a chain rule like formula, namely

$$\forall a_1 \in \mathrm{dom}(A_1) : \ldots \forall a_n \in \mathrm{dom}(A_n) :$$
$$\Pi\Big(\bigwedge\nolimits_{i=1}^{n} A_i = a_i\Big) = \min\nolimits_{i=1}^{n} \Pi\Big(A_i = a_i \Bigm| \bigwedge\nolimits_{j=1}^{i-1} A_j = a_j\Big).$$

Obviously, this formula holds generally, since the term for $i = n$ in the minimum on the right is equal to the term on the left. However, in order to cancel conditions, we have to take some care, because even if $A \perp\!\!\!\perp_\Pi B \mid C$, it will be $\Pi(A = a \mid B = b, C = c) \neq \Pi(A = a \mid C = c)$ in general. Fortunately, there is a way of writing a conditional possibilistic independence statement that is equally useful, namely (for three attributes)

$$\forall a \in \mathrm{dom}(A) : \forall b \in \mathrm{dom}(B) : \forall c \in \mathrm{dom}(C) :$$
$$\Pi(A = a \mid B = b, C = c) = \min\{\Pi(A = a \mid C = c), \Pi(B = b, C = c)\}.$$

With such formulae we can cancel conditions in the terms of the formula above, if we proceed in the order of descending values of i. Then the unconditional possibility in the minimum can be neglected, because among the remaining, unprocessed terms there must be one that is equal to it or refers to more attributes and thus restricts the degree of possibility more.

3.5 Possibility versus Probability

From the simple examples of three-dimensional probability and possibility distributions discussed above it should be clear that the two approaches exploit entirely different properties to decompose distributions. This leads, of course, to substantial differences in the interpretation of the reasoning results. To make this clear, we consider in this section how, in the two calculi, the marginal distributions on single attributes relate to the joint distribution they are derived from. This is important, because the reasoning process, as it was outlined in this chapter, produces only marginal distributions on single attributes (conditioned on the observations). Since the relation of these marginal distributions to the underlying joint distribution is very different for probability distributions compared to possibility distributions, one has to examine whether it is actually the joint distribution one is interested in.

The difference is, of course, due to the way in which projections, i.e. marginal distributions, are computed in the two calculi. In probability theory the summation over the dimensions to be removed wipes out any reference

Σ	36	18	18	28	
28	0	0	0	28	28
18	18	0	0	0	18
18	18	0	0	0	18
36	0	18	18	0	18
	18	18	18	28	max

Figure 3.17 Possibility versus probability w.r.t. the interpretation of marginal distributions (all numbers are percent).

to these dimensions. In the resulting marginal distribution no trace of the attributes underlying these dimensions or their values is left: The marginal distribution refers exclusively to the attributes scaffolding the subspace projected to. The reason is, of course, that all values of the removed attributes contribute to the result of the projection w.r.t. their relative "importance", expressed in their relative probability.

In possibility theory this is different. Because the maximum is taken over the dimensions to be removed, not all values of the attributes underlying these dimensions contribute to the result of the projection. Only the values describing the elementary event or events having the highest degree of possibility determine the marginal degree of possibility. Thus not all information about the values of the removed attributes is wiped out. These attributes are implicitly fixed to those values describing the elementary event or events having the highest degree of possibility. It follows that—unlike marginal probabilities, which refer only to tuples over the attributes of the subspace projected to—marginal degrees of possibility always refer to *value vectors over all attributes of the universe of discourse*, although only the values of the attributes of the subspace are stated explicitly in the marginal distribution.

In other words, a marginal probability distribution states: "The probability that attribute A has value a is p." This probability is aggregated over all values of all other attributes and thus refers to a one element vector (a). A marginal possibility distribution states instead: "The degree of possibility of a value vector with the highest degree of possibility of all tuples in which attribute A has value a is p." That is, it refers to a value vector over all attributes of the universe of discourse, although the values of all attributes other than A are left implicit.

As a consequence of the difference just studied one has ask oneself whether one is interested in tuples instead of the value of only a single attribute. To understand this, reconsider the result of the probabilistic reasoning process shown in Figure 3.13 on page 77. It tells us that (given attribute A has value a_4, i.e. the object is grey) the most probable value of attribute B is b_3, i.e. that the object is most likely to be a triangle, and that the most probable value of attribute C is c_2, i.e. that the object is most likely to be of medium

0	40	0		40
40	0	0		40
0	0	20		20
40	40	20	max	

Figure 3.18 Maximum projections may also lead to an incorrect decision due to an "exclusive-or" effect.

size. However, from this we cannot conclude that the object is most likely to be a medium-sized grey triangle (the color we know from the observation). As can be seen from Figure 3.12 on page 76 or from the joint distribution shown on the bottom right in Figure 3.13 on page 77, the object is most likely to be *large* grey triangle, i.e. in the most probable tuple attribute C has the value c_3. The reason for this difference is, obviously, that grey triangles as well as grey squares of medium size, i.e. the tuples (a_4, b_2, c_2) and (a_4, b_3, c_2), respectively, have a relatively high probability, whereas of large grey objects $(A = a_4, C = c_3)$ only triangles $(B = b_2)$ have a high probability.

An even more extreme example is shown in Figure 3.17, which, in the center square, shows a probability distribution over the joint domain of two attributes having four values each. The marginal distributions are shown to the left and above this square. Here selecting the tuple containing the values with the highest marginal probabilities decides on an impossible tuple. It follows that in the probabilistic case we may decide incorrectly if we rely exclusively on the marginal distributions (and, indeed, this is not a rare situation). To make the correct decision, we have to compute the joint distribution first or must apply other specialized techniques [Pearl 1988].

For possibility distributions, however, the situation is different. If in each marginal distribution on a single attribute there is only one value having the highest degree of possibility, then the tuple containing these values is the one having the highest degree of possibility. This is illustrated in Figure 3.17, where marginal distributions computed by taking the maximum are shown below and to the right of the square (recall that, according to Chapter 2, a probability distribution is only a special possibility distribution and thus we may use maximum projection for probability distributions, too). These marginal distributions indicate the correct tuple.

It should be noted, though, that in the possibilistic setting we may also choose incorrectly, due to a kind of "exclusive-or" effect. This is illustrated in Figure 3.18. If we decide on the first value for both attributes (since for attributes we have to choose between the first and the second value), we decide on an impossible tuple. Therefore, in this case, we are also forced to compute the joint distribution to ensure a correct decision.

Note that the property of maximum projections just discussed provides the justification for using the maximum operation to compute the degree of possibility of sets of elementary events which we promised in Section 2.4.10.

Computing a marginal distribution can be seen as computing the degree of possibility of specific sets of elementary events, namely those that can be defined using a subset of all attributes (cf. Definition 3.2.7 and its extension to general possibility measures). Therefore in a multidimensional domain the maximum operation to compute the degree of possibility of sets of elementary events is useful, because it serves the task to identify—attribute by attribute—the values of the tuple or tuples of the joint domain that have the highest degree of possibility.

Note also that this property of maximum projections, which may appear as an unrestricted advantage at first sight, can also turn out to be disadvantageous, namely in the case, where we are not interested in a tuple over *all* unobserved attributes that has the highest degree of possibility. The reason is that—as indicated above—we cannot get rid of the implicitly fixed values of the attributes that were projected out. If we want to neglect an attribute entirely, we have to modify the universe of discourse and compute the possibility distribution and its decomposition on this modified universe.

Chapter 4

Graphical Representation

When we discussed the simple examples of decompositions in the previous chapter we already mentioned that the idea suggests itself to represent decompositions and evidence propagation in decompositions by *graphs* or *networks* (cf. Figure 3.7 on page 60). In these graphs there is a node for each attribute used to describe the underlying domain of interest. The edges indicate which projections are needed in the decomposition of a distribution and thus the paths along which evidence has to be propagated. In this chapter we study this connection to graphs in more detail, since it is a very intuitive and powerful way to handle decompositions of distributions.

Formally, decompositions of distributions are connected to graphs by the notion of *conditional independence*, which is closely related to the notion of *separation* in graphs. In Section 4.1 we study this relation w.r.t. both directed and undirected graphs based on a qualitative description of the properties of conditional independence by the so-called *graphoid* and *semi-graphoid axioms* [Dawid 1979, Pearl and Paz 1987, Geiger 1990], which are also satisfied by separation in graphs. This leads to natural definitions of *conditional independence graphs* based on the so-called *Markov properties* of graphs [Whittaker 1990, Lauritzen *et al.* 1990, Lauritzen 1996]. Finally, conditional independence graphs are shown to be direct descriptions of decompositions.

In Section 4.2 we turn to evidence propagation in graphs, the basic ideas of which were also indicated in the previous chapter already. We review briefly two well-known propagation methods, namely the polytree propagation method [Pearl 1986, Pearl 1988] and the join tree propagation method [Lauritzen and Spiegelhalter 1988]. For the former we provide a derivation that is based on the notion of *evidence factors*. Of course, since propagation has been an area of intensive research in the past years, there are also several other methods for drawing inferences with decompositions of distributions. However, discussing these in detail is beyond the scope of this book, which focuses on learning from data, and thus they are only mentioned.

4.1 Conditional Independence Graphs

In Chapter 3 we indicated that the decomposition of distributions can be based on a notion of conditional independence of (sets of) attributes. To study this notion independent of the imprecision or uncertainty calculus, it is convenient to have a qualitative characterization of its properties that does not refer to specific numerical equalities. Such a characterization is achieved by the so-called *graphoid* and *semi-graphoid axioms* (cf. Section 4.1.1). At least the latter are satisfied by probabilistic as well as possibilistic conditional independence, but also by separation in graphs (cf. Section 4.1.3). Hence separation in graphs can be used to represent conditional independence (although isomorphism cannot be had in general), which leads to the definition of a (minimal) conditional independence graph (cf. Section 4.1.4). However, this definition is based on a global criterion, which is inconvenient to test. To cope with this problem the so-called *Markov properties* of graphs are examined and shown to be equivalent under certain conditions (cf. Section 4.1.5). Finally, the connection of conditional independence graphs and decompositions is established by showing that the latter can be read from the former (cf. Section 4.1.6).

4.1.1 Axioms of Conditional Independence

Axioms for conditional independence were stated first by [Dawid 1979], but were also independently suggested later by [Pearl and Paz 1987].

Definition 4.1.1 *Let U be a set of (mathematical) objects and $(\cdot \perp\!\!\!\perp \cdot \mid \cdot)$ a three-place relation of subsets of U. Furthermore, let W, X, Y, and Z be four disjoint subsets of U. The four statements*

symmetry: $(X \perp\!\!\!\perp Y \mid Z) \Rightarrow (Y \perp\!\!\!\perp X \mid Z)$

decomposition: $(W \cup X \perp\!\!\!\perp Y \mid Z) \Rightarrow (W \perp\!\!\!\perp Y \mid Z) \wedge (X \perp\!\!\!\perp Y \mid Z)$

weak union: $(W \cup X \perp\!\!\!\perp Y \mid Z) \Rightarrow (X \perp\!\!\!\perp Y \mid Z \cup W)$

contraction: $(X \perp\!\!\!\perp Y \mid Z \cup W) \wedge (W \perp\!\!\!\perp Y \mid Z) \Rightarrow (W \cup X \perp\!\!\!\perp Y \mid Z)$

are called the **semi-graphoid axioms**. *A three-place relation $(\cdot \perp\!\!\!\perp \cdot \mid \cdot)$ that satisfies the semi-graphoid axioms for all W, X, Y, and Z is called a* **semi-graphoid**. *The above four statements together with*

intersection: $(W \perp\!\!\!\perp Y \mid Z \cup X) \wedge (X \perp\!\!\!\perp Y \mid Z \cup W) \Rightarrow (W \cup X \perp\!\!\!\perp Y \mid Z)$

are called the **graphoid axioms**. *A three-place relation $(\cdot \perp\!\!\!\perp \cdot \mid \cdot)$ that satisfies the graphoid axioms for all W, X, Y, and Z is called a* **graphoid**.

Of course, as can already be guessed from the notation, the set U is intended to be the set of attributes used to describe the domain under consideration and the relation $(\cdot \perp\!\!\!\perp \cdot \mid \cdot)$ is intended to denote a notion of conditional independence w.r.t. some imprecision or uncertainty calculus. With this interpretation these axioms can be read as follows [Pearl 1988]:

The *symmetry* axiom states that in any state of knowledge Z (i.e. for any instantiation of the attributes in Z), if X tells us nothing new about Y (i.e. if finding out the values of the attributes in X does not change our knowledge about the values of the attributes in Y), then Y tells us nothing new about X. The *decomposition* axiom asserts that if two combined items of information are irrelevant to X, then each separate item is irrelevant as well. The *weak union* axiom states that learning irrelevant information W cannot help the irrelevant information Y become relevant to X. The *contraction* axiom states that if X is irrelevant to Y after learning some irrelevant information W, then X must have been irrelevant before we learned W. Together the weak union and contraction properties mean that irrelevant information should not alter the relevance of other propositions in the system; what was relevant remains relevant, and what was irrelevant remains irrelevant. It is plausible that any reasonable notion of conditional independence should satisfy these axioms.

The *intersection* axiom states that unless W affects Y if X is held constant or X affects Y if W is held constant, neither W nor X nor their combination can affect Y. This axiom is less plausible than the other four. Two attributes can be relevant to a third, although each of them is irrelevant if the other is held constant. The reason may be a strong dependence between them, for instance, a 1-to-1 relationship of their values. In such a case either attribute is irrelevant if the other is known (since its value is implicitly fixed by the value of the other), but can be relevant if the other is unknown. Thus it is not surprising that, in general, the intersection axiom is satisfied neither for conditional probabilistic independence nor for conditional possibilistic independence (see below for an example).

Nevertheless, we have the following theorem:

Theorem 4.1.2 *Conditional probabilistic independence and conditional possibilistic independence satisfy the semi-graphoid axioms. If the considered joint probability distribution is strictly positive, conditional probabilistic independence satisfies the graphoid axioms.*

Proof. The proof of this theorem is rather simple and only exploits the definitions of conditional probabilistic and possibilistic independence, respectively. It can be found in Section A.1 in the appendix.

That in general neither probabilistic nor possibilistic conditional independence satisfies the intersection axiom can be seen from the simple relational example shown in Figure 4.1 (a probabilistic example can be derived by assigning a probability of 0.5 to each tuple). It is obvious that in the relation shown on the left in Figure 4.1 $A \perp\!\!\!\perp_R B \mid C$, $A \perp\!\!\!\perp_R C \mid B$, and $B \perp\!\!\!\perp_R C \mid A$ hold (cf. Figure 3.10 on page 66). From these statements $A \perp\!\!\!\perp_R BC$, $B \perp\!\!\!\perp_R AC$, and $C \perp\!\!\!\perp_R AB$ can be inferred with the intersection axiom. Applying the decomposition axiom to these statements yields $A \perp\!\!\!\perp_R B$, $A \perp\!\!\!\perp_R C$, and $B \perp\!\!\!\perp_R C$, but neither of these independences hold, as the projections on the right in Fig-

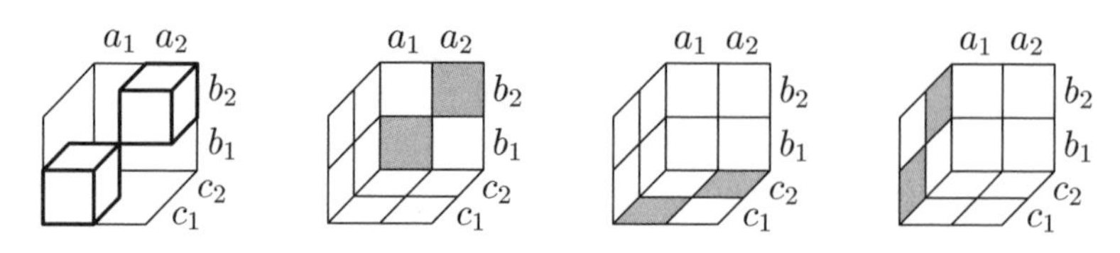

Figure 4.1 Conditional relational independence does not satisfy the intersection axiom. In the relation on the left, it is $A \perp\!\!\!\perp_R B \mid C$ and $A \perp\!\!\!\perp_R C \mid B$. However, the projections show that neither $A \perp\!\!\!\perp_R B$ nor $A \perp\!\!\!\perp_R C$.

r_{ABC}	$B = b_1$		$B = b_2$	
	$C = c_1$	$C = c_2$	$C = c_1$	$C = c_2$
$A = a_1$	$\bullet$	–	–	$\circ$
$A = a_2$	$\circ$	–	–	$\bullet$

Table 4.1 In the relation shown on the left in Figure 4.1 it is $A \not\perp\!\!\!\perp_R BC$: The relation contains only the tuples marked with $\bullet$, but for $A \perp\!\!\!\perp_R BC$ to hold, at least the tuples marked with $\circ$ have to be possible, too.

ure 4.1 demonstrate. Since the decomposition axiom holds for conditional possibilistic and thus for conditional relational independence (see Theorem 4.1.2), it must be the intersection axiom that is not satisfied. Alternatively, one can see directly from Table 4.1 that $A \not\perp\!\!\!\perp_R BC$ (the other two cases are analogous).

The main advantage of the graphoid and the semi-graphoid axioms is that they facilitate reasoning about conditional independence. If we have a set of conditional independence statements, we can easily find implied conditional independence statements by drawing inferences based on the graphoid and the semi-graphoid axioms.

This situation parallels the situation in symbolic logic, where we try to find inference rules that allow us to derive syntactically the semantical implications of a set of formulae. In the case of conditional independence the graphoid and semi-graphoid axioms are the syntactical inference rules. On the other hand, a conditional independence statement I is implied semantically by a set $\mathcal{I}$ of conditional independence statements if I holds in all distributions satisfying all statements in $\mathcal{I}$. Consequently, in analogy to symbolic logic, the question arises whether the syntactical rules are *sound* and *complete*, i.e. whether they yield only semantically correct consequences and whether all semantical consequences can be derived with them.

The soundness is ensured by Theorem 4.1.2. It has been conjectured by [Pearl 1988] that the semi-graphoid axioms are also complete for general (i.e.

not only strictly positive) probability distributions. However, this conjecture fails [Studený 1992]. Whether they are complete for conditional possibilistic independence seems to be an open problem.

4.1.2 Graph Terminology

Before we define separation in graphs in the next section, it is convenient to review some basic notions used in connection with graphs (although most of them are well known) and, more importantly, to introduce our notation.

Definition 4.1.3 *A* **graph** *is a pair $G = (V, E)$, where V is a (finite) set of* **vertices** *or* **nodes** *and $E \subseteq V \times V$ is a (finite) set of* **edges**. *It is understood that there are no loops, i.e. no edges (A, A) for any $A \in V$. G is called* **undirected** *iff*

$$\forall A, B \in V : \quad (A, B) \in E \ \Rightarrow \ (B, A) \in E.$$

Two ordered pairs (A, B) and (B, A) are identified and represent only one (undirected) edge.[1] *G is called* **directed** *iff*

$$\forall A, B \in V : \quad (A, B) \in E \ \Rightarrow \ (B, A) \notin E.$$

An edge (A, B) is considered to be directed from A towards B.

Note that the graphs defined above are *simple*, i.e. there are no multiple edges between two nodes and no loops. In order to distinguish between directed and undirected graphs, we write $\vec{G} = (V, \vec{E})$ for directed graphs.

The next four definitions introduce notions specific to undirected graphs.

Definition 4.1.4 *Let $G = (V, E)$ be an undirected graph. A node $B \in V$ is called* **adjacent to** *a node $A \in V$ or a* **neighbor** *of A iff there is an edge between them, i.e. iff $(A, B) \in E$. The set of all neighbors of A,*

$$\text{boundary}(A) = \{B \in V \mid (A, B) \in E\},$$

is called the **boundary** *of a node A and the set*

$$\text{closure}(A) = \text{boundary}(A) \cup \{A\}$$

is called the **closure** *of A.*

Of course, the notions of boundary and closure can easily be extended to sets of nodes, but we do not need this extension in this book. In the next definition the notion of adjacency of nodes is extended to paths.

[1] This way of expressing that edges are undirected, i.e. removing the "direction" of the ordered pairs by requiring both directions to be present, has the advantage that it can easily be extended to capture graphs with both directed and undirected edges: For a directed edge only one of the two possible pairs is in E.

Definition 4.1.5 *Let $G = (V, E)$ be an undirected graph. Two distinct nodes $A, B \in V$ are called* **connected** *in G, written $A \underset{G}{\sim} B$, iff there is a sequence $C_1, \ldots, C_k$, $k \geq 2$, of distinct nodes, called a* **path**, *with $C_1 = A$, $C_k = B$, and $\forall i, 1 \leq i < k : (C_i, C_{i+1}) \in E$.*

Note that in this definition a path is defined as a sequence of nodes (instead of a sequence of edges), because this is more convenient for our purposes. Note also that the nodes on the path must be distinct, i.e. the path must not lead back to a node already visited.

An important special case of an undirected graph is the *tree*, in which there is only a restricted set of paths.

Definition 4.1.6 *An undirected graph is called* **singly connected** *or a* **tree** *iff any pair of distinct nodes is connected by exactly one path.*

The following notions, especially the notion of a maximal clique, are important for the connection of decompositions and undirected graphs.

Definition 4.1.7 *Let $G = (V, E)$ be an undirected graph. An undirected graph $G_X = (X, E_X)$ is called a* **subgraph** *of G (induced by X) iff $X \subseteq V$ and $E_X = (X \times X) \cap E$, i.e. iff it contains a subset of the nodes in G and all corresponding edges.*

An undirected graph $G = (V, E)$ is called **complete** *iff its set of edges is complete, i.e. iff all possible edges are present, or formally iff*

$$E = V \times V - \{(A, A) \mid A \in V\}.$$

A complete subgraph is called a **clique**. *A clique is called* **maximal** *iff it is not a subgraph of a larger clique, i.e. a clique having more nodes.*

Note that in publications on graphical models the term *clique* is often used for what is called a *maximal clique* in the definition above, while what is called a *clique* in the above definition is merely called a *complete subgraph*. We chose the above definition, because it is the standard terminology used in graph theory [Bodendiek and Lang 1995].

The remaining definitions introduce notions specific to directed graphs.

Definition 4.1.8 *Let $\vec{G} = (V, \vec{E})$ be a directed graph. A node $B \in V$ is called a* **parent** *of a node $A \in V$ and, conversely, A is called the* **child** *of B iff there is a directed edge from B to A, i.e. iff $(B, A) \in E$. B is called* **adjacent to** *A iff it is either a parent or a child of A. The set of all* **parents** *of a node A is denoted*

$$\text{parents}(A) = \{B \in V \mid (B, A) \in \vec{E}\}$$

and the set of its **children** *is denoted*

$$\text{children}(A) = \{B \in V \mid (A, B) \in \vec{E}\}.$$

In the next definition the notion of adjacency of nodes is extended to paths.

Definition 4.1.9 *Let $\vec{G} = (V, \vec{E})$ be a directed graph.*

Two nodes $A, B \in V$ are called **d-connected** *in $\vec{G}$, written $A \underset{\vec{G}}{\rightsquigarrow} B$, iff there is a sequence $C_1, \ldots, C_k$, $k \geq 2$, of distinct nodes, called a* **directed path**, *with $C_1 = A$, $C_k = B$, and $\forall i, 1 \leq i < k : (C_i, C_{i+1}) \in \vec{E}$.*

Two nodes $A, B \in V$ are called **connected** *in $\vec{G}$, written $A \underset{\vec{G}}{\leftrightsquigarrow} B$, iff there is a sequence $C_1, \ldots, C_k$, $k \geq 2$, of distinct nodes, called a* **path**, *with $C_1 = A$, $C_k = B$, and $\forall i, 1 \leq i < k :\ (C_i, C_{i+1}) \in \vec{E} \ \vee\ (C_{i+1}, C_i) \in \vec{E}$.*

$\vec{G}$ is called **acyclic** *iff it does not contain a* **directed cycle**, *i.e. iff for all pairs of nodes A and B with $A \underset{\vec{G}}{\rightsquigarrow} B$ it is $(B, A) \notin \vec{E}$.*

Note that in a *path*, in contrast to a *directed path*, the edge directions are disregarded. Sometimes it is also called a *trail* in order to distinguish it from a directed path. With directed paths we can now easily define the notions of *ancestor* and *descendant* and the important set of all *non-descendants.*

Definition 4.1.10 *Let $\vec{G} = (V, \vec{E})$ be a directed acyclic graph. A node $A \in V$ is called an* **ancestor** *of a node $B \in V$ and, conversely, B is called a* **descendant** *of A iff there is a directed path from A to B. B is called a* **non-descendant** *of A iff it is distinct from A and not a descendant of A. The set of all* **ancestors** *of a node A is denoted*

$$\text{ancestors}(A) = \{B \in V \mid B \underset{\vec{G}}{\rightsquigarrow} A\},$$

the set of its **descendants** *is denoted*

$$\text{descendants}(A) = \{B \in V \mid A \underset{\vec{G}}{\rightsquigarrow} B\},$$

and the set of its **non-descendants** *is denoted*

$$\text{nondescs}(A) = V - \{A\} - \text{descendants}(A).$$

In analogy to undirected graphs there are the special cases of a tree and a polytree, in which the set of paths is severely restricted.

Definition 4.1.11 *A directed acyclic graph is called* **singly connected** *or a* **polytree** *iff each pair of distinct nodes is connected by exactly one path.*

A directed acyclic graph is called a (directed) **tree** *iff (1) it is a polytree and (2) exactly one node (the so-called root node) has no parents.*

An important concept for directed acyclic graphs is the notion of a *topological order* of the nodes of the graph. It can be used to test whether a directed graph is acyclic, since it only exists for acyclic graphs, and is often useful to fix the order in which the nodes of the graph are to be processed (cf. the proof of Theorem 4.1.20 in Section A.3 in the appendix).

Definition 4.1.12 *Let $\vec{G} = (V, \vec{E})$ be a directed acyclic graph. A numbering of the nodes of $\vec{G}$, i.e. a function $o : V \to \{1, \ldots, |V|\}$, satisfying*

$$\forall A, B \in V : \quad (A, B) \in \vec{E} \;\Rightarrow\; o(A) < o(B)$$

is called a **topological order** *of the nodes of $\vec{G}$.*

It is obvious that for any directed acyclic graph $\vec{G}$ a topological order can be constructed with the following simple recursive algorithm: Select an arbitrary childless node A in $\vec{G}$ and assign to it the number $|V|$. Then remove A from $\vec{G}$ and find a topological order for the reduced graph. It is also clear that for graphs with directed cycles there is no topological order, because a directed cycle cannot be reduced by the above algorithm: It must eventually reach a situation in which there is no childless node.

4.1.3 Separation in Graphs

As already indicated, the notion of conditional independence is strikingly similar to *separation* in graphs. What is to be understood by "separation" depends on whether the graph is directed or undirected. If it is undirected, separation is defined as follows:

Definition 4.1.13 *Let $G = (V, E)$ be an undirected graph and X, Y, and Z three disjoint subsets of nodes (vertices). Z* **u-separates** *X and Y in G, written $\langle X \mid Z \mid Y \rangle_G$, iff all paths from a node in X to a node in Y contain a node in Z. A path that contains a node in Z is called* **blocked** *(by Z), otherwise it is called* **active**.

Alternatively we may say that Z u-separates X and Y in G iff after removing the nodes in Z and their associated edges from G there is no path from a node in X to a node in Y. i.e. in the graph without the nodes in Z the nodes in X and Y are not connected.

If the graph is directed, a slightly more complicated criterion is used [Pearl 1988, Geiger *et al.* 1990, Verma and Pearl 1990]. It is less natural than u-separation and one can clearly tell that it was defined to capture conditional independence w.r.t. chain rule decompositions (cf. Section 4.1.6 below).

Definition 4.1.14 *Let $\vec{G} = (V, \vec{E})$ be a directed acyclic graph and X, Y, and Z three disjoint subsets of nodes (vertices). Z* **d-separates** *X and Y in $\vec{G}$, written $\langle X \mid Z \mid Y \rangle_{\vec{G}}$, iff there is no path from a node in X to a node in Y along which the following two conditions hold:*

1. *every node with converging edges (from its predecessor and its successor on the path) either is in Z or has a descendant in Z,*
2. *every other node is not in Z.*

A path satisfying the conditions above is said to be **active**, *otherwise it is said to be* **blocked** *(by Z).*

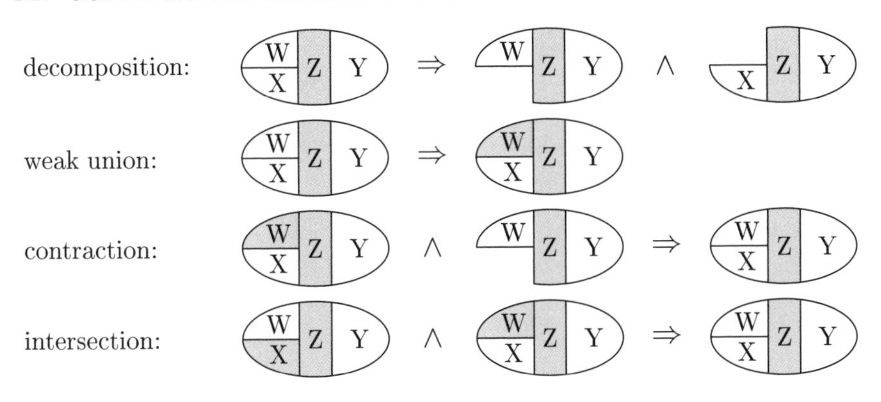

Figure 4.2 Illustration of the graphoid axioms and of separation in graphs.

Both u-separation and d-separation satisfy the graphoid axioms. For u-separation this is evident from the illustration shown in Figure 4.2 [Pearl 1988] (symmetry is trivial and thus neglected). Therefore the graphoid axioms are sound w.r.t. inferences about u-separation in graphs. However, they are not complete, because they are much weaker than u-separation.

To be more precise, the weak union axiom allows us only to extend the separating set Z by specific sets of nodes, namely those, of which it is already known that they are separated by Z from one of the sets X and Y (which are separated by Z), whereas u-separation is *monotonic*, i.e. *any* superset of a separating set is also a separating set. That is, u-separation cannot be destroyed by enlarging a separating set by *any* set of nodes, while the graphoid axioms do not exclude this. A set of axioms for u-separation that has been shown to be sound and complete [Pearl and Paz 1987] are the graphoid axioms with the weak union axiom replaced by the following two axioms:

strong union: $(X \perp\!\!\!\perp Y \mid Z) \;\Rightarrow\; (Y \perp\!\!\!\perp X \mid Z \cup W)$

transitivity: $(X \perp\!\!\!\perp Y \mid Z) \;\Rightarrow\; \forall A \in V - (X \cup Y \cup Z):$
$$(\{A\} \perp\!\!\!\perp Y \mid Z) \;\vee\; (X \perp\!\!\!\perp \{A\} \mid Z)$$

Obviously, the strong union axiom expresses the monotony of u-separation. The transitivity axiom is easily understood by recognizing that if a node A not in X, Y, or Z were not separated by Z from at least one of the sets X and Y, then there must be paths from A to a node in X and from A to a node in Y, both of which are not blocked by Z. But concatenating these two paths yields a path from X to Y that is not blocked by Z and thus X and Y could not have been u-separated by Z in the first place.

To verify that d-separation satisfies the graphoid axioms, the illustration in Figure 4.2 is also helpful. However, we have to take into account that d-separation is weaker than u-separation. d-separation does not satisfy the strong union axiom, because a path that is blocked by a separating set Z need not be blocked by a superset of Z: It may be blocked by Z, because a node with converging edges is not in Z and neither are any of its descendants. It may be active given a superset of Z, because this superset may contain the node with converging edges or any of its descendants, thus activating a path that was blocked before.

Nevertheless, the validity of the graphoid axioms is easily established: Decomposition obviously holds, since the set of paths connecting W and Y (or X and Y) are a subset of the paths connecting $W \cup X$ and Y and since the latter are all blocked, so must be the former. Weak union holds, because the nodes in W cannot activate any paths from X to Y. Even if W contained a node with converging edges of a path from X to Y or a descendant of such a node, the path would still be blocked, because any path from this node to Y is blocked by Z. (Note that the separation of W and Y by Z is essential.) Contraction is similar, only the other way round: Since all paths from W to Y are blocked by Z, we do not need the nodes in W to block the paths from X to Y, Z is sufficient. Basically the same reasoning shows that the intersection axiom holds. The only complication are paths from a node in $W \cup X$ that zig-zag between W and X before going to a node in Y. However, from the presuppositions of the intersection axiom it is clear that all such paths must be blocked by Z.

Since d-separation is much weaker than u-separation, the graphoid axioms could be complete w.r.t. inferences about d-separation. However, they are not. Consider, for instance, four nodes A, B, C, and D and the two separation statements $\langle A \mid B \mid C\rangle_{\vec{G}}$ and $\langle A \mid B \mid D\rangle_{\vec{G}}$. From these two statements it follows, simply by the definition of d-separation, that $\langle A \mid B \mid C, D\rangle_{\vec{G}}$. (To see this, one only has to consider the paths connecting A and C or A and D: All of these paths must be blocked because of the two separation statements we presupposed.) However, $\langle A \mid B \mid C, D\rangle_{\vec{G}}$ cannot be derived with the graphoid axioms, because only the symmetry axiom is applicable to $\langle A \mid B \mid C\rangle_{\vec{G}}$ and $\langle A \mid B \mid D\rangle_{\vec{G}}$ and it does not lead us anywhere.

4.1.4 Dependence and Independence Maps

Since separation in graphs is so similar to conditional independence, the idea suggests itself to represent a set of conditional independence statements—in particular, all conditional independence statements that hold in a given probability or possibility distribution—by a graph. At best we could check whether two sets are conditionally independent given a third or not by determining whether they are separated by the third in the graph.

However, this optimum, i.e. an isomorphism of conditional independence and separation, cannot be achieved in general. For undirected graphs this is immediately clear from the fact that u-separation satisfies axioms much stronger than the graphoid axioms (see above). In addition, probabilistic conditional independence for not strictly positive distributions and possibilistic conditional independence only satisfy the semi-graphoid axioms. Thus the validity of the intersection axiom for u- and d-separation already goes beyond what can be inferred about conditional independence statements. Finally, it is not immediately clear whether it is possible to represent simultaneously in a graph sets of conditional independence statements which may hold simultaneously in a distribution if these statements are *not* logical consequences of each other. Indeed, there are such sets of conditional independence statements for both undirected and directed acyclic graphs (examples are given below).

As a consequence we have to take refuge to a weaker way of defining conditional independence graphs than requiring isomorphism of conditional independence and separation [Pearl 1988]. It is sufficient, though, for the purpose of characterizing decompositions, because for this purpose isomorphism is certainly desirable, but not a *conditio sine qua non.*

Definition 4.1.15 *Let* $(\cdot \perp\!\!\!\perp_\delta \cdot \mid \cdot)$ *be a three-place relation representing the set of conditional independence statements that hold in a given distribution* δ *over a set* U *of attributes. An undirected graph* $G = (U, E)$ *over* U *is called a* **conditional dependence graph** *or a* **dependence map** *w.r.t.* δ *iff for all disjoint subsets* $X, Y, Z \subseteq U$ *of attributes*

$$X \perp\!\!\!\perp_\delta Y \mid Z \;\Rightarrow\; \langle X \mid Z \mid Y \rangle_G,$$

i.e. if G *captures by u-separation all (conditional) independences that hold in* δ *and thus represents only valid (conditional) dependences. Similarly,* G *is called a* **conditional independence graph** *or an* **independence map** *w.r.t.* δ *iff for all disjoint subsets* $X, Y, Z \subseteq U$ *of attributes*

$$\langle X \mid Z \mid Y \rangle_G \;\Rightarrow\; X \perp\!\!\!\perp_\delta Y \mid Z,$$

i.e. if G *captures by u-separation only (conditional) independences that are valid in* δ. G *is said to be a* **perfect map** *of the conditional (in)dependences in* δ *iff it is both a dependence map and an independence map.*

It is clear that the same notions can be defined for directed acyclic graphs $\vec{G} = (U, \vec{E})$ in exactly the same way, so we do not provide a separate definition.

A conditional dependence graph of a distribution guarantees that (sets of) attributes that are not separated in the graph are indeed conditionally dependent in the distribution. It may, however, display dependent (sets of) attributes as separated nodes (or node sets, respectively). Conversely, a conditional independence graph of a distribution guarantees that (sets of) attributes that are separated in the graph are indeed conditionally independent in the distribution. There may be, however, (sets of) conditionally independent attributes that are not separated in the graph.

It is obvious that a graph with no edges is a trivial conditional dependence graph and a complete graph is a trivial conditional independence graph, sim-

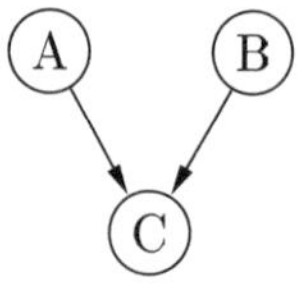

p_{ABC}	$A = a_1$		$A = a_2$	
	$B = b_1$	$B = b_2$	$B = b_1$	$B = b_2$
$C = c_1$	$^4/_{24}$	$^3/_{24}$	$^3/_{24}$	$^2/_{24}$
$C = c_2$	$^2/_{24}$	$^3/_{24}$	$^3/_{24}$	$^4/_{24}$

Figure 4.3 Marginal independence and conditional dependence can be represented by directed graphs but not by undirected graphs.

ply because the former represents no dependences and the latter no independences and thus obviously no false ones. However, it is also clear that these graphs are entirely useless. Therefore we need some restriction which ensures that a conditional dependence graph represents as many dependences as possible and a conditional independence graph represents as many independences as possible. This is achieved with the next definition.

Definition 4.1.16 *A conditional dependence graph is called* **maximal** *w.r.t. a distribution δ (or a maximal dependence map w.r.t. δ) iff no edge can be added to it so that the resulting graph is still a conditional dependence graph w.r.t. the distribution δ.*

A conditional independence graph is called **minimal** *w.r.t. a distribution δ (or a minimal independence map w.r.t. δ) iff no edge can be removed from it so that the resulting graph is still a conditional independence graph w.r.t. the distribution δ.*

The fact that decompositions depend on (conditional) independences makes it more important to truthfully record independences. If an invalid conditional independence can be read from a separation in the graph, we may arrive at an invalid decomposition formula (cf. Section 4.1.6 below). If, on the other hand, a valid conditional independence is not represented by separation in a graph, we only may not be able to exploit it and thus may not find the best decomposition. However, a suboptimal decomposition can never lead to incorrect inferences as an incorrect decomposition can. Therefore we neglect conditional *dependence* graphs in the following and concentrate on conditional *independence* graphs.

As already indicated above, the expressive power of conditional independence graphs is limited. For both directed acyclic graphs and undirected graphs there are sets of conditional independence statements that cannot be represented by separation, although they may hold simultaneously in a distribution. This is most easily demonstrated by considering a directed acyclic graph for which no equivalent undirected graph exists and an undirected graph for which no equivalent directed acyclic graph exists. An example of the former is shown in Figure 4.3. The directed acyclic graph on the left is a perfect

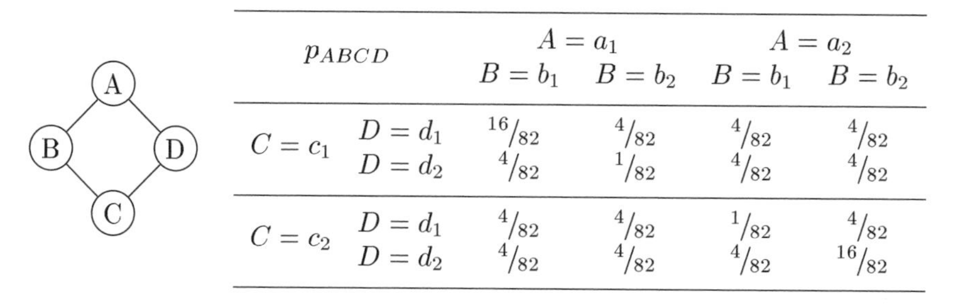

p_{ABCD}		$A = a_1$		$A = a_2$	
		$B = b_1$	$B = b_2$	$B = b_1$	$B = b_2$
$C = c_1$	$D = d_1$	$^{16}/_{82}$	$^{4}/_{82}$	$^{4}/_{82}$	$^{4}/_{82}$
	$D = d_2$	$^{4}/_{82}$	$^{1}/_{82}$	$^{4}/_{82}$	$^{4}/_{82}$
$C = c_2$	$D = d_1$	$^{4}/_{82}$	$^{4}/_{82}$	$^{1}/_{82}$	$^{4}/_{82}$
	$D = d_2$	$^{4}/_{82}$	$^{4}/_{82}$	$^{4}/_{82}$	$^{16}/_{82}$

Figure 4.4 Sets of conditional independence statements with certain symmetries can be represented by undirected graphs but not by directed graphs.

map w.r.t. the probability distribution p_{ABC} on the right: It is $A \perp\!\!\!\perp_P B$, but $A \not\perp\!\!\!\perp_P B \mid C$. That is, A and B are marginally independent, but conditionally dependent. It is immediately clear that there is no undirected perfect map for this probability distribution. The monotony of u-separation prevents us from representing $A \perp\!\!\!\perp_P B$, because this would entail the (invalid) conditional independence statement $A \perp\!\!\!\perp_P B \mid C$.

An example of the latter, i.e. of an undirected conditional independence graph for which there is no equivalent directed acyclic graph, is shown in Figure 4.4. As can easily be verified, the graph on the left is a perfect map w.r.t. the probability distribution p_{ABCD} on the right: It is $A \perp\!\!\!\perp_P C \mid BD$ and $B \perp\!\!\!\perp_P D \mid AC$, but no other conditional independence statements hold (except the symmetric counterparts of the above). It is clear that a directed acyclic graph must contain at least directed counterparts of the edges of the undirected graph. However, if we confine ourselves to these edges, an additional conditional independence statement is represented, independent of how the edges are directed. If another edge is added to exclude this statement, one of the two valid conditional independences is not represented.

4.1.5 Markov Properties of Graphs

In the previous section conditional independence graphs were defined based on a global criterion. This makes it hard to test whether a given graph is a conditional independence graph or not: One has to check for all separations of node sets in the graph whether the corresponding conditional independence statement holds. Thus the question arises whether there are simpler criteria. Fortunately, under certain conditions simpler criteria can be found by exploiting the equivalence of the so-called Markov properties of graphs.

For undirected graphs the Markov properties are defined as follows [Whittaker 1990, Frydenberg 1990, Lauritzen *et al.* 1990, Lauritzen 1996]:

Definition 4.1.17 *Let $(\cdot \perp\!\!\!\perp_\delta \cdot \mid \cdot)$ be a three-place relation representing the set of conditional independence statements that hold in a given joint distribution δ over a set U of attributes. An undirected graph $G = (U, E)$ is said to have (w.r.t. the distribution δ) the*

pairwise Markov property
iff in δ any pair of attributes which are nonadjacent in the graph are conditionally independent given all remaining attributes, i.e. iff

$$\forall A, B \in U, A \neq B: \quad (A, B) \notin E \Rightarrow A \perp\!\!\!\perp_\delta B \mid U - \{A, B\},$$

local Markov property
iff in δ any attribute is conditionally independent of all remaining attributes given its neighbors, i.e. iff

$$\forall A \in U: \quad A \perp\!\!\!\perp_\delta U - \text{closure}(A) \mid \text{boundary}(A),$$

global Markov property
iff in δ any two sets of attributes which are u-separated by a third[2] *are conditionally independent given the attributes in the third set, i.e. iff*

$$\forall X, Y, Z \subseteq U: \quad \langle X \mid Z \mid Y \rangle_G \Rightarrow X \perp\!\!\!\perp_\delta Y \mid Z.$$

Definition 4.1.15 on page 101 used the global Markov property to define conditional independence graphs. However, the pairwise or the local Markov property would be much more natural and convenient. Therefore it is pleasing to observe that, obviously, the boundary of an attribute u-separates it from the attributes in the remainder of the graph and thus the local Markov property is implied by the global. Similarly, the set of all other attributes u-separates two nonadjacent attributes and thus the pairwise Markov property is implied by the global, too. If the relation $(\cdot \perp\!\!\!\perp_\delta \cdot \mid \cdot)$ satisfies the semi-graphoid axioms, we also have that the pairwise Markov property is implied by the local, since it follows from an application of the weak union axiom. That is, for semi-graphoids we have

$$\mathcal{G}_{\text{global}}(\delta) \subseteq \mathcal{G}_{\text{local}}(\delta) \subseteq \mathcal{G}_{\text{pairwise}}(\delta),$$

where $\mathcal{G}_{prop}(\delta)$ is the set of undirected graphs having the Markov property *prop* w.r.t. the distribution δ.

Unfortunately, despite the above inclusions, the three Markov properties are not equivalent in general [Lauritzen 1996]. Consider, for example, five attributes A, B, C, D, and E with $\text{dom}(A) = \ldots = \text{dom}(E) = \{0, 1\}$. Let $A = B$, $D = E$, $C = B \cdot D$, and $P(A = 0) = P(E = 0) = \frac{1}{2}$. With these presuppositions it is easy to check that the graph shown in Figure 4.5 has the pairwise and the local Markov property w.r.t. to the joint probability

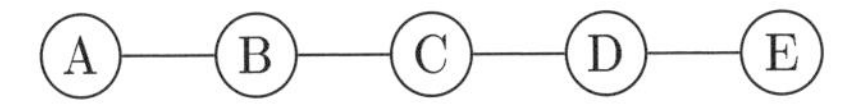

Figure 4.5 In general, the Markov properties are not equivalent.

distribution of A, B, C, D, and E. However, it does not have the global Markov property, because it is $A \not\perp\!\!\!\perp_P E \mid C$.

The equivalence of the three Markov properties can be established, though, if the relation that describes the set of conditional independence statements holding in a given distribution δ satisfies the graphoid axioms.

Theorem 4.1.18 *If a three-place relation* $(\cdot \perp\!\!\!\perp_\delta \cdot \mid \cdot)$ *representing the set of conditional independence statements that hold in a given joint distribution* δ *over a set* U *of attributes satisfies the graphoid axioms, then the pairwise, the local, and the global Markov property of an undirected graph* $G = (U, E)$ *are equivalent.*

Proof. From the observations made above we already know that the global Markov property implies the local and that the local Markov property implies the pairwise. So all that is left to show is that, given the graphoid axioms, the pairwise Markov property implies the global.

The idea of the proof is very simple. Consider three arbitrary nonempty disjoint subsets X, Y, and Z of nodes such that $\langle X \mid Z \mid Y \rangle_G$. We have to show that $X \perp\!\!\!\perp_\delta Y \mid Z$ follows from the pairwise conditional independence statements referring to attributes that are not adjacent in the graph. To do so we start from an arbitrary conditional independence statement $A \perp\!\!\!\perp_\delta B \mid U - \{A, B\}$ with $A \in X$ and $B \in Y$, and then shift nodes from the separating set to the separated sets, thus extending A to (a superset of) X and B to (a superset of) Y and shrinking $U - \{A, B\}$ to Z. The shifting is done by applying the intersection axiom, drawing on other pairwise conditional independence statements. Finally, any excess attributes in the separated sets are removed with the help of the decomposition axiom.

Formally, the proof is carried out by *backward* or *descending induction* [Pearl 1988, Lauritzen 1996]; see Section A.2 in the appendix.

If the above theorem applies, we can define a conditional independence graph in the following, very natural way: An undirected graph $G = (U, E)$ is a conditional independence graph w.r.t. to a joint distribution δ iff

$$\forall A, B \in V, A \neq B : \quad (A, B) \notin E \Rightarrow A \perp\!\!\!\perp_\delta B \mid U - \{A, B\}.$$

In addition, both the pairwise and the local Markov property are powerful criteria to *test* whether a given graph is a conditional independence graph.

[2]It is understood that the three sets are disjoint.

Of course, the three Markov properties can be defined not only for undirected, but also for directed graphs [Lauritzen 1996].

Definition 4.1.19 *Let* $(\cdot \perp\!\!\!\perp_\delta \cdot \mid \cdot)$ *be a three-place relation representing the set of conditional independence statements that hold in a given joint distribution* δ *over a set* U *of attributes. A directed acyclic graph* $\vec{G} = (U, \vec{E})$ *is said to have (w.r.t. the distribution* δ*) the*

pairwise Markov property
iff in δ *any attribute is conditionally independent of any non-descendant not among its parents given all remaining non-descendants, i.e. iff*

$$\forall A, B \in U : B \in \text{nondescs}(A) - \text{parents}(A) \Rightarrow A \perp\!\!\!\perp_\delta B \mid \text{nondescs}(A) - \{B\},$$

local Markov property
iff in δ *any attribute is conditionally independent of all remaining non-descendants given its parents, i.e. iff*

$$\forall A \in U : \quad A \perp\!\!\!\perp_\delta \text{nondescs}(A) - \text{parents}(A) \mid \text{parents}(A),$$

global Markov property
iff in δ *any two sets of attributes which are d-separated by a third*[3] *are conditionally independent given the attributes in the third set, i.e. iff*

$$\forall X, Y, Z \subseteq U : \quad \langle X \mid Z \mid Y \rangle_{\vec{G}} \Rightarrow X \perp\!\!\!\perp_\delta Y \mid Z.$$

As for undirected graphs, we can make some pleasing observations: It is clear that the parents of A d-separate A from all its non-descendants. The reason is that a path to a non-descendant must either pass through a parent—and then it is blocked by the set of parents—or it must pass through a descendant of A at which it has converging edges—and then it is blocked by the fact that neither this descendant nor any of its descendants are among the parents of A. Hence the global Markov property implies the local. For the same reasons the set of all remaining non-descendants d-separates a node A from a non-descendant node B that is not among its parents. Therefore the pairwise Markov property is also implied by the global. Finally, if the relation $(\cdot \perp\!\!\!\perp_\delta \cdot \mid \cdot)$ satisfies the semi-graphoid axioms, we also have that the pairwise Markov property is implied by the local, since it follows from an application of the weak union axiom. That is, for semi-graphoids we have

$$\vec{\mathcal{G}}_{\text{global}}(\delta) \subseteq \vec{\mathcal{G}}_{\text{local}}(\delta) \subseteq \vec{\mathcal{G}}_{\text{pairwise}}(\delta),$$

where $\vec{\mathcal{G}}_{prop}(\delta)$ is the set of directed acyclic graphs having the Markov property *prop* w.r.t. the distribution δ.

[3] It is understood that the three sets are disjoint.

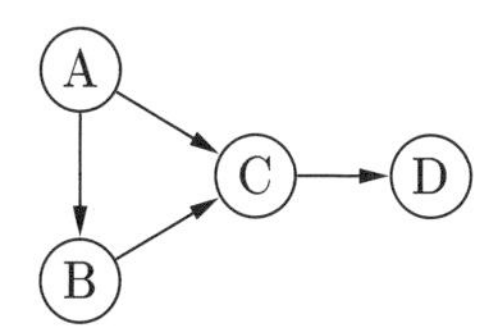

Figure 4.6 In general, the pairwise and the local Markov property are not equivalent.

Unfortunately, despite the above inclusions, the three Markov properties are again not equivalent in general [Lauritzen 1996]. Consider, for example, four attributes A, B, C, and D with $\text{dom}(A) = \ldots = \text{dom}(D) = \{0, 1\}$. Let $A = B = D$, C independent of A, and $P(A = 0) = P(C = 0) = \frac{1}{2}$. With these presuppositions it is easy to check that the graph shown in Figure 4.6 has the pairwise Markov property w.r.t. to the joint probability distribution of A, B, C, and D, but not the local.

The equivalence of the three Markov properties can be established, as for undirected graphs, if the graphoid axioms hold. However, one can also make a stronger assertion.

Theorem 4.1.20 *If a three-place relation $(\cdot \perp\!\!\!\perp_\delta \cdot \mid \cdot)$ representing the set of conditional independence statements that hold in a given joint distribution δ over a set U of attributes satisfies the semi-graphoid axioms, then the local and the global Markov property of a directed acyclic graph $\vec{G} = (U, \vec{E})$ are equivalent. If $(\cdot \perp\!\!\!\perp_\delta \cdot \mid \cdot)$ satisfies the graphoid axioms, then the pairwise, the local, and the global Markov property are equivalent.*

Proof. A proof of the first part of this theorem can be found, for example, in [Verma and Pearl 1990], although this may be a little difficult to recognize, because the definition of certain notions has changed since the publication of that paper. In Section A.3 in the appendix we provide our own proof (which is similar to the one in [Verma and Pearl 1990], though). Here we confine ourselves to making plausible why such a proof is possible for directed acyclic graphs, although it is clearly not for undirected graphs. Let A and B be two adjacent nodes that are separated by a set Z of nodes from a set Y of nodes. For undirected graphs, to derive the corresponding conditional independence statement from local conditional independence statements, we have to combine $A \perp\!\!\!\perp_\delta U - \text{closure}(A) \mid \text{boundary}(A)$ and $B \perp\!\!\!\perp_\delta U - \text{closure}(B) \mid \text{boundary}(B)$ in order to get a statement in which A and B appear on the same side. However, it is $A \in \text{boundary}(B)$ and $B \in \text{boundary}(A)$ and therefore the intersection axiom is needed. In contrast to this, in a directed acyclic graph it is either $A \in \text{parents}(B)$ or $B \in \text{parents}(A)$, but not both. Therefore the contraction axiom—which is similar to, but weaker than the intersection axiom—suffices to combine the local conditional independence statements.

Given the first part of the theorem the proof of the second part is rather simple. We already know that the local and the global Markov property are equivalent and, from the observations made above, that the local Markov property implies the pairwise. Therefore, all that is left to show is that the pairwise Markov property implies the local. However, this is easily demonstrated: We start from an arbitrary pairwise conditional independence statement for a node and combine it step by step, using the intersection axiom, with all other pairwise conditional independences for the same node and thus finally reach the local conditional independence statement for the node.

As for undirected graphs this theorem allows us to define a conditional independence graph in a more natural way based on the local or the pairwise Markov property. It is particularly convenient, though, that for a definition based on the local Markov property we only need to know that the semi-graphoid axioms hold (instead of the graphoid axioms, which often fail).

4.1.6 Graphs and Decompositions

The preceding sections were devoted to how conditional independence statements can be captured in a graphical representation. However, representing conditional independences is not a goal in itself, but only a pathway to finding a decomposition of a given distribution. To determine a conditional independence graph for a given distribution is equivalent to determining what terms are needed in a decomposition of the distribution, because they can be read directly from the graph, and finding a *minimal* conditional independence graph is tantamount to discovering the "best" decomposition, i.e. a decomposition into smallest terms. Formally, this connection is brought about by the theorems of this section. We study undirected graphs first.

Definition 4.1.21 *A probability distribution p_U over a set U of attributes is called* **decomposable** *or* **factorizable w.r.t. an undirected graph** *$G = (U, E)$ iff it can be written as a product of nonnegative functions on the maximal cliques of G. That is, let $\mathcal{M}$ be a family of subsets of attributes, such that the subgraphs of G induced by the sets $M \in \mathcal{M}$ are the maximal cliques of G. Then there must exist functions $\phi_M : \mathcal{E}_M \to \mathbb{R}_0^+$, $M \in \mathcal{M}$,*

$$\forall a_1 \in \text{dom}(A_1) : \ldots \forall a_n \in \text{dom}(A_n) :$$
$$p_U\Big(\bigwedge_{A_i \in U} A_i = a_i\Big) = \prod_{M \in \mathcal{M}} \phi_M\Big(\bigwedge_{A_i \in M} A_i = a_i\Big).$$

Similarly, a possibility distribution π_U over U is called **decomposable w.r.t. an undirected graph** *$G = (U, E)$ iff it can be written as the minimum of the marginal possibility distributions on the maximal cliques of G, i.e. iff*

$$\forall a_1 \in \text{dom}(A_1) : \ldots \forall a_n \in \text{dom}(A_n) :$$
$$\pi_U\Big(\bigwedge_{A_i \in U} A_i = a_i\Big) = \min_{M \in \mathcal{M}} \pi_M\Big(\bigwedge_{A_i \in M} A_i = a_i\Big).$$

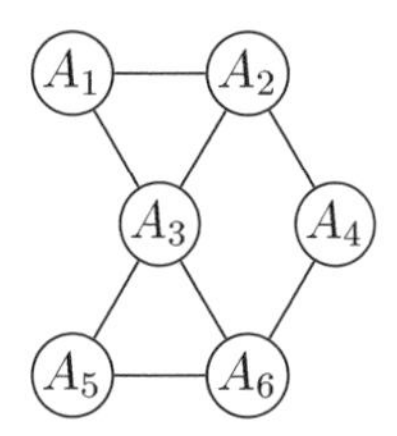

Figure 4.7 A simple undirected graph that represents a decomposition / factorization into four terms corresponding to the four maximal cliques.

Note that the decomposition formulae are the same as in Definition 3.3.1 on page 77 and in Definition 3.2.8 on page 69, respectively. The undirected graph G only fixes the set $\mathcal{M}$ of subsets of nodes in a specific way.

A simple example is shown in Figure 4.7. This graph has four maximal cliques, namely those induced by the four sets $\{A_1, A_2, A_3\}$, $\{A_3, A_5, A_6\}$, $\{A_2, A_4\}$, and $\{A_4, A_6\}$. Therefore, in the probabilistic case, this graph represents the factorization

$$\begin{aligned}
&\forall a_1 \in \mathrm{dom}(A_1) : \ldots \forall a_6 \in \mathrm{dom}(A_6) : \\
&\quad p_U(A_1 = a_1, \ldots, A_6 = a_6) = \phi_{A_1A_2A_3}(A_1 = a_1, A_2 = a_2, A_3 = a_3) \\
&\qquad\qquad \cdot \ \phi_{A_3A_5A_6}(A_3 = a_3, A_5 = a_5, A_6 = a_6) \\
&\qquad\qquad \cdot \ \phi_{A_2A_4}(A_2 = a_2, A_4 = a_4) \\
&\qquad\qquad \cdot \ \phi_{A_4A_6}(A_4 = a_4, A_6 = a_6).
\end{aligned}$$

The following theorem connects undirected conditional independence graphs to decompositions. It is usually attributed to [Hammersley and Clifford 1971], who proved it for the discrete case (to which we will confine ourselves, too), although (according to [Lauritzen 1996]) this result seems to have been discovered in various forms by several authors.

Theorem 4.1.22 *Let p_U be a strictly positive probability distribution on a set U of (discrete) attributes. An undirected graph $G = (U, E)$ is a conditional independence graph w.r.t. p_U iff p_U is factorizable w.r.t. G.*

Proof. The full proof, which is somewhat technical, can be found in Section A.4 in the appendix. In its first part it is shown that, if p_U is factorizable w.r.t. an undirected graph G, then G has the global Markov property w.r.t. conditional independence in p_U. This part exploits that two attributes which are u-separated in G cannot be in the same clique. Therefore, for any three disjoint subsets X, Y, and Z of attributes such that $\langle X \mid Z \mid Y \rangle_G$ the cliques can be divided into two sets and the functions ϕ_M can be combined into a corresponding product of two functions from which the desired conditional independence $X \perp\!\!\!\perp_{p_U} Y \mid Z$ follows. It is worth noting that the validity of this part of the proof is not restricted to strictly positive distributions p_U.

In the second part of the proof it is shown that p_U is factorizable w.r.t. a conditional independence graph G. This part is constructive, i.e. it provides a method to determine nonnegative functions ϕ_M from the joint distribution p_U, so that p_U can be written as a product of these functions.

The above theorem can be extended to more general distributions, for example, to distributions on real-valued attributes, provided they have a positive and continuous density [Lauritzen 1996]. However, since in this book we confine ourselves almost entirely to the discrete case, we do not discuss this extension. A possibilistic analog of the above theorem also holds and has been proven first in [Gebhardt 1997]. However, it is less general than its probabilistic counterpart. To show the desired equivalence of decomposition and representation of conditional independence, the permissible graphs have to be restricted to certain subset. The graphs in this subset are characterized by the so-called *running intersection property* of the family of attribute sets that induce their maximal cliques.

Definition 4.1.23 *Let $\mathcal{M}$ be a finite family of subsets of a finite set U and let $m = |\mathcal{M}|$. $\mathcal{M}$ is said to have the* **running intersection property** *iff there is an ordering $M_1, \ldots, M_m$ of the sets in $\mathcal{M}$, such that*

$$\forall i \in \{2, \ldots, m\} : \exists k \in \{1, \ldots, i-1\} : \quad M_i \cap \Big(\bigcup_{1 \leq j < i} M_j \Big) \subseteq M_k$$

If all pairs of nodes of an undirected graph G are connected in G and the family $\mathcal{M}$ of the node sets that induce the maximal cliques of G has the running intersection property, then G is said to have **hypertree structure**.

The idea underlying the notion of a hypertree structure is as follows: In normal graphs an edge can connect only two nodes. However, we may drop this restriction and introduce so-called *hypergraphs*, in which we have *hyperedges* that can connect any number of nodes. It is very natural to use a hyperedge to connect the nodes of a maximal clique, because by doing so we can make these cliques easier to recognize. (Note that the connectivity of the graph is unharmed if all edges of its maximal cliques are replaced by hyperedges, and therefore we do not loose anything by this operation.) If the sets of nodes that are connected by hyperedges have the running intersection property, then the hypergraph is, in a certain sense, *acyclic* (cf. Section 4.2.2). Since acyclic undirected normal graphs are usually called *trees*, the idea suggests itself to call such hypergraphs *hypertrees*. Therefore an undirected graph which becomes a hypertree, if the edges of its maximal cliques are replaced by hyperedges, is said to have *hypertree structure*.

For graphs with hypertree structure a possibilistic analog of the above theorem can be proven [Gebhardt 1997], although the hypertree structure of the conditional independence graph is only needed for one direction of the theorem, namely to guarantee the existence of a factorization.

A	B	C	D
a_1	b_1	c_1	d_2
a_1	b_1	c_2	d_1
a_2	b_2	c_1	d_1

Figure 4.8 The possibilistic decomposition theorem cannot be generalized to arbitrary graphs.

Theorem 4.1.24 *Let π_U be a possibility distribution on a set U of (discrete) attributes and let $G = (U, E)$ be an undirected graph over U. If π_U is decomposable w.r.t. G, then G is a conditional independence graph w.r.t. π_U. If G is a conditional independence graph w.r.t. π_U and if it has hypertree structure, then π_U is decomposable w.r.t. G.*

Proof. The full proof can be found in Section A.5 in the appendix. The first part of the theorem can be proven in direct analogy to the corresponding part of the proof of the probabilistic counterpart of the above theorem. Again it is exploited that two attributes that are u-separated cannot be in the same clique and therefore the cliques can be divided into two sets. Note that this part of the theorem does not require that the graph has hypertree structure and thus is valid for arbitrary undirected graphs.

For the second part of the theorem the hypertree structure of G is essential, since it allows an induction on a construction sequence for the graph G that exploits the global Markov property of G. The construction sequence is derived from the ordering of the cliques that results from the ordering underlying the running intersection property.

Unfortunately, the above theorem cannot be generalized to arbitrary graphs as the following example demonstrates (this is a slightly modified version of an example given in [Gebhardt 1997]). Consider the undirected graph and the simple relation shown in Figure 4.8 (and recall that a relation is only a special possibility distribution). It is easy to check that the graph is a conditional independence graph of the relation, since both conditional independence statements that can be read from it, namely $A \perp\!\!\!\perp_R C \mid \{B, D\}$ and $B \perp\!\!\!\perp_R D \mid \{A, C\}$, hold in the relation. However, it does not have hypertree structure, because the set of its four maximal cliques does not have the running intersection property, and, indeed, the relation is not decomposable w.r.t. the graph. If it were decomposable, then we would have

$$\forall a \in \text{dom}(A) : \forall b \in \text{dom}(B) : \forall C \in \text{dom}(C) : \forall d \in \text{dom}(D) :$$
$$R(a, b, c, d) = \min\{R(a, b), R(b, c), R(c, d), R(d, a)\},$$

where $R(a, b)$ is an abbreviation of $R(A = a, B = b)$ etc. However, it is $R(a_1, b_1, c_1, d_1) = 0$ (since this tuple is not contained in the relation), but $R(a_1, b_1) = 1$ (because of the first or the second tuple), $R(b_1, c_1) = 1$ (first tuple), $R(c_1, d_1) = 1$ (third tuple), and $R(d_1, a_1) = 1$ (second tuple) and therefore $\min\{R(a_1, b_1), R(b_1, c_1), R(c_1, d_1), R(d_1, a_1)\} = 1$. Note, however, that a multivariate possibility distribution *may* be decomposable w.r.t. a conditional independence graph that does not have hypertree structure. It is only that it cannot be *guaranteed* that it is decomposable.

In the following we turn to directed graphs. The connection of directed acyclic graphs and decompositions, in this case chain rule decompositions, is achieved in a similar way as for undirected graphs.

Definition 4.1.25 *A probability distribution p_U over a set U of attributes is called* **decomposable** *or* **factorizable w.r.t. a directed acyclic graph** $\vec{G} = (U, \vec{E})$ *iff it can be written as a product of the conditional probabilities of the attributes given their parents in $\vec{G}$, i.e. iff*

$$\forall a_1 \in \mathrm{dom}(A_1) : \ldots \forall a_n \in \mathrm{dom}(A_n) :$$
$$p_U\Big(\bigwedge_{A_i \in U} A_i = a_i\Big) = \prod_{A_i \in U} P\Big(A_i = a_i \;\Big|\; \bigwedge_{A_j \in \mathrm{parents}_{\vec{G}}(A_i)} A_j = a_j\Big).$$

Similarly, a possibility distribution π_U over U is called **decomposable w.r.t. a directed acyclic graph** $\vec{G} = (U, \vec{E})$ *iff it can be written as the minimum of conditional degrees of possibility of the attributes given their parents in $\vec{G}$. That is, iff*

$$\forall a_1 \in \mathrm{dom}(A_1) : \ldots \forall a_n \in \mathrm{dom}(A_n) :$$
$$\pi_U\Big(\bigwedge_{A_i \in U} A_i = a_i\Big) = \min_{A_i \in U} \Pi\Big(A_i = a_i \;\Big|\; \bigwedge_{A_j \in \mathrm{parents}_{\vec{G}} A_i} A_j = a_j\Big).$$

Note that the decomposition formulae are the same as the chain rule decomposition formulae on page 79 and page 87, respectively. The directed acyclic graph G only fixes the conditions of the conditional probabilities or conditional degrees of possibility in a specific way.

A simple example graph is shown in Figure 4.9. In the probabilistic case this graph represents the factorization

$$\begin{aligned}
&\forall a_1 \in \mathrm{dom}(A_1) : \ldots \forall a_7 \in \mathrm{dom}(A_7) : \\
&\quad p_U(A_1 = a_1, \ldots, A_7 = a_7) \\
&\quad = \; P(A_1 = a_1) \cdot P(A_2 = a_2 \mid A_1 = a_1) \cdot P(A_3 = a_3) \\
&\quad \cdot \; P(A_4 = a_4 \mid A_1 = a_1, A_2 = a_2) \cdot P(A_5 = a_5 \mid A_2 = a_2, A_3 = a_3) \\
&\quad \cdot \; P(A_6 = a_6 \mid A_4 = a_4, A_5 = a_5) \cdot P(A_7 = a_7 \mid A_5 = a_5).
\end{aligned}$$

We have similar theorems for directed graphs as for undirected graphs relating factorization and representation of conditional independence.

Theorem 4.1.26 *Let p_U be a probability distribution on a set U of (discrete) attributes. A directed acyclic graph $\vec{G} = (U, \vec{E})$ is a conditional independence graph w.r.t. p_U iff p_U is factorizable w.r.t. $\vec{G}$.*

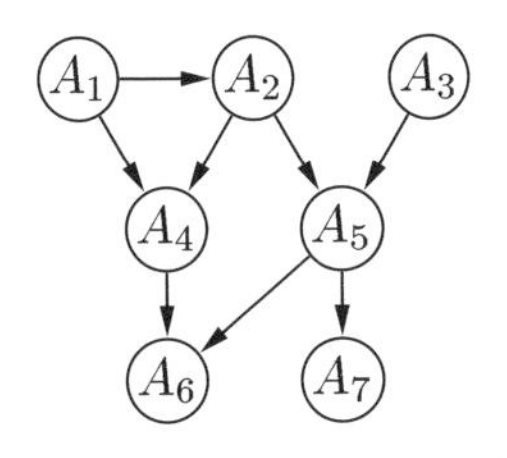

Figure 4.9 A simple directed acyclic graph that represents a decomposition/factorization into terms with at most two conditions.

Proof. The proof, which exploits the equivalence of the global and the local Markov property, the latter of which is directly connected to the factorization formula, can be found in Section A.6 in the appendix.

As for undirected graphs, there is only an incomplete possibilistic analog, only that here it is the other direction that does not hold in general.

Theorem 4.1.27 *Let π_U be a possibility distribution on a set U of (discrete) attributes. If a directed acyclic graph $\vec{G} = (U, \vec{E})$ is a conditional independence graph w.r.t. π_U, then π_U is decomposable w.r.t. $\vec{G}$.*

Proof. The proof, which can be found in Section A.7 in the appendix, is analogous to the proof of the corresponding part of the probabilistic case. As usual when going from probabilities to degrees of possibility, one has to replace the sum by the maximum and the product by the minimum.

Note that the converse of the above theorem, i.e. that a directed acyclic graph $\vec{G}$ is a conditional independence graph of a possibility distribution π_U if π_U is decomposable w.r.t. $\vec{G}$, does not hold. To see this, consider the simple relation r_{ABC} shown on the left in Figure 4.10 (and recall that a relation is a special possibility distribution). Trivially, r_{ABC} is decomposable w.r.t. the directed acyclic graph shown on the right in Figure 4.10, since

$$\begin{aligned}
&\forall a \in \text{dom}(A) : \forall b \in \text{dom}(B) : \forall c \in \text{dom}(C) : \\
&\quad r_{ABC}(A = a, B = b, C = c) \\
&\qquad = \min\{R(A = a), R(B = b), R(C = c \mid A = a, B = b)\} \\
&\qquad = \min\{R(A = a), R(B = b), R(C = c, A = a, B = b)\}
\end{aligned}$$

In the graph $\vec{G}$ the attributes A and B are d-separated given the empty set, i.e. $\langle A \mid \emptyset \mid B \rangle_{\vec{G}}$, and thus the global Markov property would imply $A \perp\!\!\!\perp_{r_{ABC}} B$. However, this is not the case, as the projection of r_{ABC} shown in the center of Figure 4.10 demonstrates.

Whether the set of graphs can be restricted to an easily characterizable subset (like the undirected graphs with hypertree structure) in order to make the converse of the above theorem hold, seems to be an open problem.

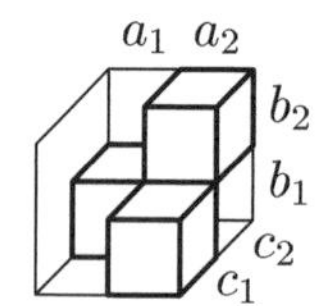

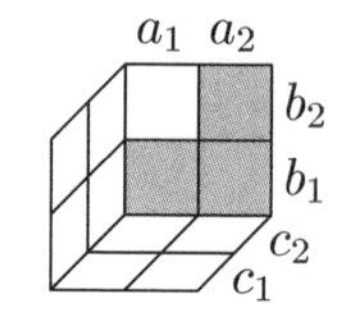

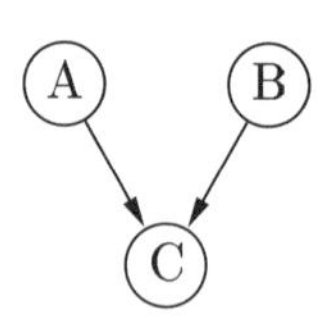

Figure 4.10 Relational and possibilistic decomposition w.r.t. a directed acyclic graph do not imply the global Markov property of the graph.

4.1.7 Markov Networks and Bayesian Networks

Since conditional independence graphs and decompositions of distributions are so intimately connected (as shown in the preceding section), the idea suggests itself to combine them in one structure. In such a structure *qualitative information* is available about the conditional (in)dependences between attributes in the form of a conditional independence graph, which indicates the paths along which evidence about the values of observed attributes has to be transferred to the remaining unobserved attributes. In addition, the terms of the decompositions provide *quantitative information* about the precise effects that different pieces of evidence have on the probability or degree of possibility of the unobserved attributes.

The combination of a conditional independence graph and the decomposition it describes finally leads us to the well-known notions of a *Markov network* and a *Bayesian network*.

Definition 4.1.28 *A* **Markov network** *is an undirected conditional independence graph of a probability distribution* p_U *together with the family of non-negative functions* ϕ_M *of the factorization induced by the graph.*

Definition 4.1.29 *A* **Bayesian network** *is a directed conditional independence graph of a probability distribution* p_U *together with the family of conditional probabilities of the factorization induced by the graph.*

We call both Markov networks and Bayesian networks *probabilistic networks*. Note that often only networks that are based on *minimal* conditional independence graphs are called *Markov networks* or *Bayesian networks* [Pearl 1988]. Although minimal conditional independence graphs are certainly desirable, in our opinion this is an unnecessary restriction, since, for example, the propagation algorithms for the two network types work just as well for networks based on non-minimal conditional independence graphs.

Notions analogous to *Markov network* or *Bayesian network* can, of course, be defined for the possibilistic case, too, although there are no special names for them. We call the analog of a Markov network an **undirected possibilistic network** and the analog of a Bayesian network a **directed possibilistic network**. Note that for undirected possibilistic networks we need not require that the conditional independence graph has hypertree structure, although this is needed to guarantee that a decomposition w.r.t. the graph exists. Clearly, even if it is not guaranteed to exist, it may exist for a given graph and a given distribution, and if it does, there is no reason why we should not use the corresponding network.

4.2 Evidence Propagation in Graphs

Conditional independence graphs not only provide a way to find a decomposition of a given multidimensional distribution as shown in the preceding section, they can also be used as a framework for the implementation of evidence propagation methods. The basic idea is that the edges of the graph indicate the paths along which evidence has to be transmitted. This is reasonable, because if two attributes are separated by a set S of other attributes, then there should not be any transfer of information from the one to the other if the attributes in S are instantiated. However, if all information is transmitted along the edges of the graph, this would necessarily be the result (at least for undirected graphs), provided we make sure that the information cannot permeate instantiated attributes. (For directed graphs, of course, we have to take special precautions due to the peculiar properties of nodes with converging edges, cf. the definition of d-separation.)

In this section we briefly review two of the best-known propagation methods that are based on the idea described above: The polytree propagation method developed by [Pearl 1986, Pearl 1988] (Section 4.2.1) and the join tree propagation method developed by [Lauritzen and Spiegelhalter 1988] (Section 4.2.2). Both algorithms have been developed for the probabilistic setting and therefore we confine ourselves to explaining them w.r.t. Bayesian and Markov networks. However, it is clear that the ideas underlying these methods can be transferred directly to the possibilistic setting. It is worth noting that join tree propagation underlies the evidence propagation in the commercial Bayesian network tool HUGIN [Andersen *et al.* 1989] and that its possibilistic counterpart has been implemented in POSSINFER [Gebhardt and Kruse 1996a, Kruse *et al.* 1994]. In addition, the approach has been generalized to other uncertainty calculi like belief functions [Shafer and Shenoy 1988, Shenoy 1992b, Shenoy 1993] in the so-called valuation-based networks [Shenoy 1992a]. This generalized version has been implemented in PULCINELLA [Saffiotti and Umkehrer 1991].

As its name already indicates, the polytree propagation method is restricted to *singly connected networks*. It exploits the fact that in a polytree there is only one path on which the information derived from an instantiated attribute can travel to another attribute. This makes it very simple to derive a message passing scheme for the evidence propagation, which can be implemented by locally communicating node processors.

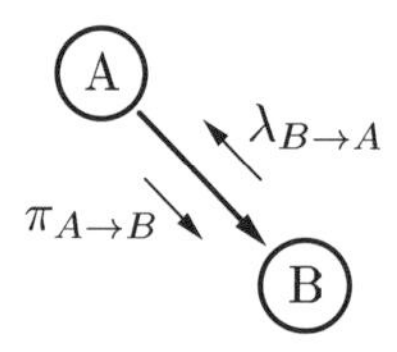

Figure 4.11 Node processors communicating by message passing: π-messages are sent from parent to child and λ-messages are sent from child to parent.

Multiply connected networks are much harder to handle, because there may be several paths on which evidence can travel from one node to another and thus it is difficult to ensure that it is used only once to update the probabilities of the values of an attribute. However, the join tree propagation method provides a general formal approach to deal with multiply connected graphs. Its key idea is to transform a given network into a singly connected structure, namely a join tree, for which a propagation scheme similar to the one for polytrees can be derived.

Since we look back on several years of research, it is clear that these methods are not the only possible ones. However, it is also clear that in this book we cannot provide an exhaustive treatment of evidence propagation. Therefore we only mention some other methods in Section 4.2.3.

4.2.1 Propagation in Polytrees

The polytree propagation method for Bayesian networks, which was suggested by [Pearl 1986], is the oldest exact probabilistic propagation method. It exploits that a polytree (cf. Definition 4.1.11 on page 97) is *singly connected*, i.e. that there is only one path from a node to another and thus there is no choice of how to transmit the evidence in the network.

The basic idea of the propagation method is to use node processors that exchange *messages* with their parents and their children. There are two types of messages: So-called π-messages[4] that are sent from parent to child, and so-called λ-messages that are sent from child to parent (cf. Figure 4.11). Intuitively, a π-message represents the (influence of the) information collected in the subgraph that can be reached from A if the way to B is barred, and a λ-message represents the (influence of the) information collected in the subgraph that can be reached from B if the way to A is barred. That the conditional independence graph is a polytree ensures that the two subgraphs are indeed disjoint and thus no multiple transmission of the same information occurs.

To derive the propagation formulae for the probabilistic case,[5] i.e. for a Bayesian network, we assume first that no evidence has been added, i.e. that

[4] We regret the equivocation of π, which also denotes a possibility distribution. However, we did not want to deviate from the standard notation of the polytree propagation algorithm either. So we decided to retain the letter π to denote a message from a parent.

[5] Note that the derivation given here differs considerably from the derivation given in [Pearl 1986] or in [Pearl 1988], although it leads, necessarily, to the same results.

no attributes have been instantiated. This simplifies the notation considerably. Later we will indicate how instantiated attributes change the results.

The idea of the derivation is the same as for the simple example in Chapter 3, cf. pages 79ff, although we momentarily neglect the evidence: We start from the definition of the marginal probability of an attribute. Then we exploit the factorization formula and move terms that are independent of a summation variable out of the corresponding sum. By comparing the results for attributes that are adjacent in the graph we finally arrive at the propagation formulae, i.e. at formulae that state how the outgoing messages are computed from the incoming messages.

In the discrete case, to which we confine ourselves here, the marginal probability of an attribute is computed from a joint distribution by

$$P(A_g = a_g) \quad = \sum_{\substack{\forall A_i \in U - \{A_g\}: \\ a_i \in \text{dom}(A_i)}} P\Big(\bigwedge_{A_j \in U} A_j = a_j\Big),$$

where the somewhat sloppy notation w.r.t. the sum is intended to indicate that the sum is to be taken over all values of all attributes in U except A_g. The index g was chosen to indicate that A_g is the current "goal" attribute. (In the following we need quite a lot of different indices, so it is convenient to choose at least some of them mnemonically.) In the first step it is exploited that the distribution is factorizable w.r.t. the conditional independence graph of the Bayesian network (cf. Definition 4.1.25 on page 112):

$$P(A_g = a_g) \quad = \sum_{\substack{\forall A_i \in U - \{A_g\}: \\ a_i \in \text{dom}(A_i)}} \prod_{A_k \in U} P\Big(A_k = a_k \;\Big|\; \bigwedge_{A_j \in \text{parents}(A_k)} A_j = a_j\Big).$$

In the second step it is exploited that the graph is a polytree, i.e. that it is singly connected. A convenient property of a polytree is that by removing an edge it is split into two unconnected subgraphs. Since we need to refer several times to certain subgraphs that result from such splits, it is helpful to define a notation for the sets of nodes underlying them. Let

$$U_B^A(C) = \{C\} \cup \{D \in U \mid D \underset{\vec{G}'}{\sim} C, \vec{G}' = (U, \vec{E} - \{(A, B)\})\},$$

i.e. let $U_B^A(C)$ be the set of those attributes that can still be reached from attribute C if the edge $A \rightarrow B$ is removed from the graph. With this set we can define the following sets which we need in the following:

$$U_+(A) = \bigcup_{C \in \text{parents}(A)} U_A^C(C), \qquad U_+(A, B) = \bigcup_{C \in \text{parents}(A) - \{B\}} U_A^C(C),$$

$$U_-(A) = \bigcup_{C \in \text{children}(A)} U_C^A(C), \qquad U_-(A, B) = \bigcup_{C \in \text{children}(A) - \{B\}} U_A^C(C),$$

Intuitively, $U_+(A)$ are the attributes in the graph G_+ "above" attribute A and $U_-(A)$ are the attributes in the graph G_- "below" it. Similarly, $U_+(A,B)$ are the attributes "above" A except those that can be reached via its parent B and $U_-(A,B)$ are the attributes "below" A except those that can be reached via its child B.

A split in the graph gives rise to a partitioning of the factors of the above product, since the structure of the graph is reflected in the conditions of the conditional probabilities. If we split the graph G w.r.t. the goal attribute A_g into an upper graph $G_+(A_g)$ and a lower graph $G_-(A_g)$, we get

$$\begin{aligned} P(A_g = a_g) \quad = \quad & \sum_{\substack{\forall A_i \in U - \{A_g\}: \\ a_i \in \text{dom}(A_i)}} \Bigg(P\Big(A_g = a_g \;\Big|\; \bigwedge_{A_j \in \text{parents}(A_g)} A_j = a_j\Big) \\ & \cdot \prod_{A_k \in U_+(A_g)} P\Big(A_k = a_k \;\Big|\; \bigwedge_{A_j \in \text{parents}(A_k)} A_j = a_j\Big) \\ & \cdot \prod_{A_k \in U_-(A_g)} P\Big(A_k = a_k \;\Big|\; \bigwedge_{A_j \in \text{parents}(A_k)} A_j = a_j\Big)\Bigg). \end{aligned}$$

Note that indeed the factors of the first product in the above formula refer only to attributes the upper part of the graph, while the factors of the second product refer only to attributes in the lower part of the graph (plus the goal attribute A_g), as can easily be seen from the structure of a polytree.

In the third step it is exploited that terms that are independent of a summation variable can be moved out of the corresponding sum. In addition we make use of

$$\sum_i \sum_j a_i b_j = \Big(\sum_i a_i\Big)\Big(\sum_j b_j\Big).$$

This yields a decomposition into two main factors, called the *π-value* and the *λ-value* of the attribute A_g:

$$\begin{aligned} & P(A_g = a_g) \\ = \; & \Bigg(\sum_{\substack{\forall A_i \in \text{parents}(A_g): \\ a_i \in \text{dom}(A_i)}} P\Big(A_g = a_g \;\Big|\; \bigwedge_{A_j \in \text{parents}(A_g)} A_j = a_j\Big) \\ & \cdot \Bigg[\sum_{\substack{\forall A_i \in U_+^*(A_g): \\ a_i \in \text{dom}(A_i)}} \prod_{A_k \in U_+(A_g)} P\Big(A_k = a_k \;\Big|\; \bigwedge_{A_j \in \text{parents}(A_k)} A_j = a_j\Big) \Bigg] \Bigg) \\ & \cdot \Bigg[\sum_{\substack{\forall A_i \in U_-(A_g): \\ a_i \in \text{dom}(A_i)}} \prod_{A_k \in U_-(A_g)} P\Big(A_k = a_k \;\Big|\; \bigwedge_{A_j \in \text{parents}(A_k)} A_j = a_j\Big) \Bigg] \\ = \; & \pi(A_g = a_g) \cdot \lambda(A_g = a_g), \end{aligned}$$

where $U_+^*(A_g) = U_+(A_g) - \text{parents}(A_g)$. The first factor represents the probabilistic influence of the upper part of the network, transmitted through the parents of A_g, while the second factor represents the probabilistic influence of the lower part of the network, transmitted through the children of A_g. It is clear that the second factor is equal to one if no evidence has been added (as we currently assume).[6] However, this will change as soon as evidence is added and therefore we retain this factor.

In the next step we consider the two factors more closely. We start with the parent side (assuming that A_g actually has parents). Neglecting for a moment the conditional probability of the values of A_g given the values of its parents, we can derive, exploiting again that we have a polytree, that

$$\begin{aligned}
&\sum_{\substack{\forall A_i \in U_+^*(A_g):\\ a_i \in \text{dom}(A_i)}} \prod_{A_k \in U_+(A_g)} P\Big(A_k = a_k \,\Big|\, \bigwedge_{A_j \in \text{parents}(A_k)} A_j = a_j\Big) \\
&= \prod_{A_p \in \text{parents}(A_g)} \\
&\quad \Bigg(\sum_{\substack{\forall A_i \in \text{parents}(A_p):\\ a_i \in \text{dom}(A_i)}} P\Big(A_p = a_p \,\Big|\, \bigwedge_{A_j \in \text{parents}(A_p)} A_j = a_j\Big) \\
&\quad \cdot \Bigg[\sum_{\substack{\forall A_i \in U_+^*(A_p):\\ a_i \in \text{dom}(A_i)}} \prod_{A_k \in U_+(A_p)} P\Big(A_k = a_k \,\Big|\, \bigwedge_{A_j \in \text{parents}(A_k)} A_j = a_j\Big)\Bigg]\Bigg) \\
&\quad \cdot \Bigg[\sum_{\substack{\forall A_i \in U_-(A_p, A_g):\\ a_i \in \text{dom}(A_i)}} \prod_{A_k \in U_-(A_p, A_g)} P\Big(A_k = a_k \,\Big|\, \bigwedge_{A_j \in \text{parents}(A_k)} A_j = a_j\Big)\Bigg] \\
&= \prod_{A_p \in \text{parents}(A_g)} \pi(A_p = a_p) \\
&\quad \cdot \Bigg[\sum_{\substack{\forall A_i \in U_-(A_p, A_g):\\ a_i \in \text{dom}(A_i)}} \prod_{A_k \in U_-(A_p, A_g)} P\Big(A_k = a_k \,\Big|\, \bigwedge_{A_j \in \text{parents}(A_k)} A_j = a_j\Big)\Bigg] \\
&= \prod_{A_p \in \text{parents}(A_g)} \pi_{A_p \to A_g}(A_p = a_p),
\end{aligned}$$

where $U^*(A_p) = U_+(A_p) - \text{parents}(A_p)$. That is, we can represent the influence of the parent side as a product with one factor for each parent A_p of A_g. (The

[6] This can be seen from the fact that for the sums over the values of the leaf attributes in $G_-(A_g)$ all factors except the conditional probability for this attribute can be moved out of the sum. Since we sum over a conditional probability distribution, the result must be 1. Working recursively upward in the network, we see that the whole sum must be 1.

index p was chosen to indicate that the A_p are the parents of the current goal attribute.) The notation $\pi_{A_p \to A_g}$ is intended to indicate that this is a message sent from parent attribute A_p to attribute A_g. This is justified, since the expression underlying it is very similar to the expression we could derive for $P(A_p = a_p)$ in analogy to the expression for $P(A_g = a_g)$. The first factor is identical and the second differs only slightly: From it all terms corresponding to the subgraph "rooted" at A_g are excluded. Therefore it is reasonable to compute $\pi_{A_p \to A_g}(A_p = a_p)$ not w.r.t. A_g but w.r.t. A_p, where it is needed anyway, and to send the result to A_g. $\pi_{A_p \to A_g}$ is parameterized only with $A_p = a_p$, because A_p is the only "free" attribute, i.e. the only attribute over the values of which it is not summed. (Note that the goal attribute A_g does not appear.) Combining the two preceding equations we have

$$\begin{aligned}\pi(A_g = a_g) &= \sum_{\substack{\forall A_i \in \text{parents}(A_g):\\ a_i \in \text{dom}(A_i)}} P\Big(A_g = a_g \,\Big|\, \bigwedge_{A_j \in \text{parents}(A_g)} A_j = a_j\Big) \\ &\qquad \cdot \prod_{A_p \in \text{parents}(A_g)} \pi_{A_p \to A_g}(A_p = a_p).\end{aligned}$$

Turning from the parents to the children of A_g, we consider next the second factor in the product for $P(A_g = a_g)$ (assuming that A_g actually has children). In a similar way as above, exploiting again that we have a polytree, we derive

$$\begin{aligned}&\lambda(A_g = a_g) \\ &= \sum_{\substack{\forall A_i \in U_-(A_g):\\ a_i \in \text{dom}(A_i)}} \prod_{A_k \in U_-(A_g)} P\Big(A_k = a_k \,\Big|\, \bigwedge_{A_j \in \text{parents}(A_k)} A_j = a_j\Big) \\ &= \prod_{A_c \in \text{children}(A_g)} \sum_{a_c \in \text{dom}(A_c)} \\ &\quad \Bigg(\sum_{\substack{\forall A_i \in \text{parents}(A_c) - \{A_g\}:\\ a_i \in \text{dom}(A_i)}} P\Big(A_c = a_c \,\Big|\, \bigwedge_{A_j \in \text{parents}(A_c)} A_j = a_j\Big) \\ &\quad \cdot \Bigg[\sum_{\substack{\forall A_i \in U_+^*(A_c, A_g):\\ a_i \in \text{dom}(A_i)}} \prod_{A_k \in U_+(A_c, A_g)} P\Big(A_k = a_k \,\Big|\, \bigwedge_{A_j \in \text{parents}(A_k)} A_j = a_j\Big)\Bigg]\Bigg) \\ &\quad \underbrace{\cdot \Bigg[\sum_{\substack{\forall A_i \in U_-(A_c):\\ a_i \in \text{dom}(A_i)}} \prod_{A_k \in U_-(A_c)} P\Big(A_k = a_k \,\Big|\, \bigwedge_{A_j \in \text{parents}(A_k)} A_j = a_j\Big)\Bigg]}_{= \lambda(A_c = a_c)} \\ &= \prod_{A_c \in \text{children}(A_g)} \lambda_{A_c \to A_g}(A_g = a_g),\end{aligned}$$

where $U_+^*(A_c, A_g) = U_+(A_c, A_g) - \text{parents}(A_c)$. That is, the influence of the lower part of the network can be represented as a product with one factor for each child A_c of A_g. (The index c was chosen to indicate that A_c is a child attribute of the current goal attribute.) The notation $\lambda_{A_c \to A_g}$ is intended to indicate that this is a message sent from child attribute A_c to attribute A_g. This is justified, since the expression underlying it is obviously very similar to the expression we could derive for $P(A_c = a_c)$ in analogy to the expression for $P(A_g = a_g)$. The only differences are that the factors are summed over all values of the attribute A_c and that from the first factor of the outer sum all terms are excluded that correspond to the subgraph "ending" in A_g. Therefore it is reasonable to compute $\lambda_{A_c \to A_g}(A_g = a_g)$ not w.r.t. A_g but w.r.t. A_c, where the necessary values are available anyway, and to send the result to A_g. $\lambda_{A_c \to A_g}$ is parameterized only with $A_g = a_g$, because A_g is the only "free" attribute of the expression, i.e. the only attribute over the values of which it is not summed.

From the above formulae the propagation formulae can easily be derived. To state them, we do no longer refer to a goal attribute A_g, but write the formulae w.r.t. a parent attribute A_p and a child attribute A_c, which are considered to be connected by an edge from A_p to A_c. With this presupposition we get for the π-message

$$\begin{aligned}
&\pi_{A_p \to A_c}(A_p = a_p) \\
&= \pi(A_p = a_p) \\
&\quad \cdot \left[\sum_{\substack{\forall A_i \in U_-(A_p, A_c): \\ a_i \in \text{dom}(A_i)}} \prod_{A_k \in U_-(A_p, A_c)} P\Big(A_k = a_k \,\Big|\, \bigwedge_{A_j \in \text{parents}(A_k)} A_j = a_j\Big) \right] \\
&= \frac{P(A_p = a_p)}{\lambda_{A_c \to A_p}(A_p = a_p)}.
\end{aligned}$$

This formula is very intuitive. The message $\lambda_{A_c \to A_p}(A_p = a_p)$, which is sent from child A_c to parent A_p, represents information gathered in the subgraph "rooted" in A_c. Obviously, this information should not be returned to A_c. However, all other information gathered at A_p should be passed to A_c. This is achieved by the above formula. It combines all incoming λ-messages except the one from A_c and all π-messages and passes the result to A_c. Since $P(A_p = a_p)$ represents the information gathered in the whole network and since the λ-message from A_c is only a factor in $P(A_p = a_p)$, we can derive the simple quotient shown above.

Similarly, the message $\pi_{A_p \to A_c}(A_p = a_p)$ sent from parent A_p to child A_c represents information gathered in the subgraph "ending" in A_p. Obviously, this information should not be returned to A_p, but all other information gathered at A_c should be passed to A_p. In analogy to the above, this can also be seen as removing the influence of $\pi_{A_p \to A_c}(A_p = a_p)$ from $P(A_c = a_c)$.

However, the influence of $\pi_{A_p \to A_c}(A_p = a_p)$ on $P(A_c = a_c)$ is a little more complex then the influence of $\lambda_{A_c \to A_p}(A_p = a_p)$ on $P(A_p = a_p)$ and thus we cannot write the message as a simple quotient. Instead we only have

$$\begin{aligned}
&\lambda_{A_c \to A_p}(A_p = a_p) \\
&= \sum_{a_c \in \mathrm{dom}(A_c)} \lambda(A_c = a_c) \\
&\qquad \cdot \sum_{\substack{\forall A_i \in \mathrm{parents}(A_c)-\{A_p\}: \\ a_i \in \mathrm{dom}(A_k)}} P\Big(A_c = a_c \,\Big|\, \bigwedge_{A_j \in \mathrm{parents}(A_c)} A_j = a_j\Big) \\
&\qquad\qquad \cdot \prod_{A_k \in \mathrm{parents}(A_c)-\{A_p\}} \pi_{A_k \to A_p}(A_k = a_k).
\end{aligned}$$

In analogy to the formula for the π-message examined above, this formula combines all incoming π-messages except the one from A_p and all λ-messages and passes the result to A_p.

Up to now we have assumed that no evidence has been added to the network, i.e. that no attributes have been instantiated. However, if attributes are instantiated, the formulae change only slightly. We have to add to the joint probability distribution a factor for each instantiated attribute: If X_{obs} is the set of observed (instantiated) attributes, we have to compute

$$\begin{aligned}
&P\Big(A_g = a_g \,\Big|\, \bigwedge_{A_k \in X_{\mathrm{obs}}} A_k = a_k^{(\mathrm{obs})}\Big) \\
&= \alpha \sum_{\substack{\forall A_i \in U-\{A_g\}: \\ a_i \in \mathrm{dom}(A_i)}} P\Big(\bigwedge_{A_j \in U} A_j = a_j\Big) \prod_{A_k \in X_{\mathrm{obs}}} P\Big(A_k = a_k \,\Big|\, A_k = a_k^{(\mathrm{obs})}\Big),
\end{aligned}$$

where the $a_k^{(\mathrm{obs})}$ are the observed values and α is a normalization constant,

$$\alpha = \frac{1}{P\Big(\bigwedge_{A_k \in X_{\mathrm{obs}}} A_k = a_k^{(\mathrm{obs})}\Big)}.$$

The justification for this formula is very similar to the justification of the introduction of similar *evidence factors* for the observed attributes in the simple three-attribute example discussed in Chapter 3 (compare page 80): Obviously,

$$\begin{aligned}
&P\Big(\bigwedge_{A_j \in U} A_j = a_j \,\Big|\, \bigwedge_{A_k \in X_{\mathrm{obs}}} A_k = a_k^{(\mathrm{obs})}\Big) \\
&= \alpha\, P\Big(\bigwedge_{A_j \in U} A_j = a_j, \bigwedge_{A_k \in X_{\mathrm{obs}}} A_k = a_k^{(\mathrm{obs})}\Big)
\end{aligned}$$

$$= \begin{cases} \alpha\, P\left(\bigwedge_{A_j \in U} A_j = a_j\right), & \text{if } \forall A_j \in X_{\text{obs}} : a_j = a_j^{(\text{obs})}, \\ 0, & \text{otherwise,} \end{cases}$$

with α defined as above. In addition it is clear that

$$\forall A_k \in X_{\text{obs}} : \quad P\left(A_k = a_k \,\middle|\, A_k = a_k^{(\text{obs})}\right) \;=\; \begin{cases} 1, & \text{if } a_k = a_k^{(\text{obs})}, \\ 0, & \text{otherwise,} \end{cases}$$

and therefore

$$\prod_{A_k \in X_{\text{obs}}} P\left(A_k = a_k \,\middle|\, A_k = a_k^{(\text{obs})}\right) \;=\; \begin{cases} 1, & \text{if } \forall A_k \in X_{\text{obs}} : a_k = a_k^{(\text{obs})}, \\ 0, & \text{otherwise.} \end{cases}$$

Combining the two equations, we arrive at the formula stated above. Note that we can neglect the normalization factor α (i.e. need not compute it explicitly), since it can always be recovered from the fact that a probability distribution, whether marginal or conditional, must be normalized.

It is easy to see that, if the derivation of the propagation formula is redone with the modified initial formula for the probability of a value of a goal attribute A_g, the *evidence factors* $P(A_k = a_k \mid A_k = a_k^{(\text{obs})})$ only influence the formulae for the messages that are sent out from the instantiated attributes: In the derivation each such factor accompanies the conditional probability for the same attribute. Therefore we get the following formula for the π-messages that are sent out from an instantiated attribute A_p:

$$\begin{aligned} &\pi_{A_p \to A_c}(A_p = a_p) \\ &\quad= P\left(A_p = a_p \,\middle|\, A_p = a_p^{(\text{obs})}\right) \cdot \pi(A_p = a_p) \\ &\qquad \cdot \left[\sum_{\substack{\forall A_i \in U_-(A_p, A_c): \\ a_i \in \text{dom}(A_i)}} \; \prod_{A_k \in U_-(A_p, A_c)} P\Big(A_k = a_k \,\Big|\, \bigwedge_{A_j \in \text{parents}(A_k)} A_j = a_j\Big) \right] \\ &\quad= \begin{cases} \beta, & \text{if } a_p = a_p^{(\text{obs})}, \\ 0, & \text{otherwise,} \end{cases} \end{aligned}$$

where β is some constant. It is clear that we can choose $\beta = 1$: We already have to determine the factor α by exploiting that the marginal probabilities must be normalized, so another factor does not matter.

This formula is again very intuitive. In a polytree, any attribute A_p d-separates all attributes in the subgraph "above" it from those in the subgraph "below" it. Consequently, if A_p is instantiated, no information from the upper network should be passed to the lower network, which explains why all π-messages from parents of an instantiated attribute A_p are discarded. Similarly, in a polytree any attribute A_p d-separates the subgraphs corresponding to any two of its children and therefore the underlying attribute sets are conditionally independent given A_p. Consequently, if A_p is instantiated, no information from one child should be passed to another, which explains why all λ-messages from the children of an instantiated attribute A_p are discarded.

For the λ-messages sent out from an instantiated attribute A_c we get

$$\begin{aligned}
&\lambda_{A_c \to A_p}(A_p = a_p) \\
&= \sum_{a_c \in \mathrm{dom}(A_c)} P\left(A_c = a_c \;\middle|\; A_c = a_c^{(\mathrm{obs})}\right) \cdot \lambda(A_c = a_c) \\
&\qquad \cdot \sum_{\substack{\forall A_i \in \mathrm{parents}(A_c) - \{A_p\}: \\ a_i \in \mathrm{dom}(A_k)}} P\Big(A_c = a_c \;\Big|\; \bigwedge_{A_j \in \mathrm{parents}(A_c)} A_j = a_j\Big) \\
&\qquad\qquad \cdot \prod_{A_k \in \mathrm{parents}(A_c) - \{A_p\}} \pi_{A_k \to A_c}(A_k = a_k) \\
&= \gamma \sum_{\substack{\forall A_i \in \mathrm{parents}(A_c) - \{A_p\}: \\ a_i \in \mathrm{dom}(A_k)}} P\Big(A_c = a_c^{(\mathrm{obs})} \;\Big|\; \bigwedge_{A_j \in \mathrm{parents}(A_c)} A_j = a_j\Big) \\
&\qquad\qquad \cdot \prod_{A_k \in \mathrm{parents}(A_c) - \{A_p\}} \pi_{A_k \to A_c}(A_k = a_k),
\end{aligned}$$

where γ is some constant. It is clear that we can choose $\gamma = 1$, for the same reason why we could use a value of 1 for the constant β above.

Again the formula is very intuitive. That the λ-messages from the children of A_c are discarded is due to the fact that in a polytree any attribute d-separates the attributes in the subgraph "above" it from the attributes in the subgraph "below" it (see above). However, the π-messages cannot be discarded, since no attribute d-separates any two of its parents. Therefore the π-messages from the parents of A_c are still processed, although their computation is, of course, restricted to the observed value $a_c^{(\mathrm{obs})}$.

The above formulae are all restricted to single values of attributes. However, we obviously need to determine the probability of all values of the goal attribute and we have to evaluate, even in the above formulae, the messages for all values of the message parameter attributes. Therefore it is convenient to write the above equations in vector form, with a vector for each attribute having as many elements as the attribute has values. The conditional probabilities can then be represented as matrices. Rewriting the formulae is straightforward, though, and therefore we do not restate them here.

With these formulae the propagation of evidence can be implemented by locally communicating node processors. Each node receives messages from and sends messages to its parents and its children. From these messages it also computes the marginal probability of the values of the corresponding attribute, conditioned on the available evidence. The node recomputes the marginal probabilities and recomputes and resends messages whenever it receives a new message from any of its parents or any of its children.

The network is initialized by setting all λ-messages to 1 for all values (reflecting the fact mentioned above that, without any evidence, the λ-value of an attribute is equal to 1 for all attribute values). The π-messages of the root attributes of the polytree, i.e. the parentless attributes, are initialized to the marginal probability of the values of these attributes (which are part of the factorization). This initiates an update process for all child attributes, which then, in turn, send out new messages. This triggers an update of their children and parents. Finally, when all attributes have been updated, the marginal probabilities of all attributes are established.

Whenever an attribute is instantiated, new λ- and π-messages are computed for this attribute. This triggers an update of the neighboring attributes, thus spreading the information in the network as in the initialization phase. Finally the marginal probabilities conditioned on the observed value(s) are computed for all attributes.

It is important to note that, as the above explanations should have made clear, evidence propagation consists in computing (conditioned) marginal probability distributions for *single* attributes. Usually the most probable *value vector* for several attributes cannot be found by selecting the most probable value for each attribute (recall the explanations of Section 3.5).

4.2.2 Join Tree Propagation

Directed acyclic graphs are not necessarily polytrees: There can be more than one path connecting two nodes. At first sight it may seem to be possible to apply the same method as for polytrees if a graph is multiply connected. However, such situations can be harmful, because evidence can travel on more than one route from one node to another, namely if more than one path is active given the set of instantiated attributes. Since probabilistic update is not idempotent—that is, incorporating the same evidence twice may invalidate the result—cycles must be avoided or dealt with in special ways. Possibilistic evidence propagation, by the way, is less sensitive to such situations, because the possibilistic update operation *is* idempotent (it does not matter how many times a degree of possibility is restricted to the same upper bound by the minimum operation) and thus the same evidence can be incorporated several times without invalidating the result. Nevertheless it is useful to avoid multiply connected graphs in possibilistic reasoning, because situations can arise where a cycle must be traversed many times to reach the reasoning result [Kruse *et al.* 1994]. It should be noted, though, that in possibilistic reasoning it is at most *desirable* to avoid cycles for reasons of efficiency, whereas in probabilistic reasoning they *must* be avoided in order to ensure the correctness of the inference results.

Multiply connected networks can be handled in several ways. One method is to temporarily fix selected unobserved attributes in order to "cut open" all cycles, so that the normal polytree propagation algorithm can be applied.

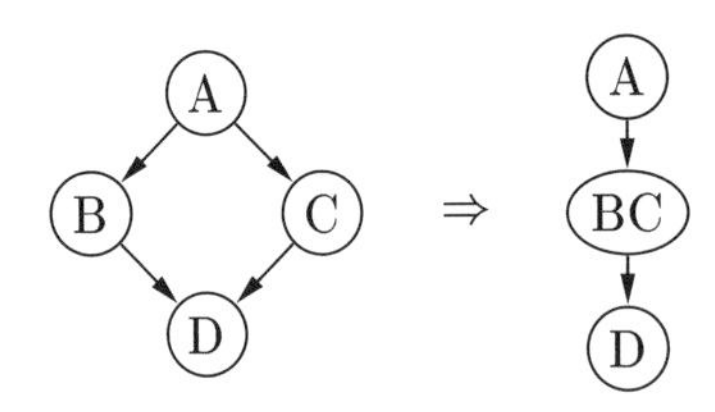

Figure 4.12 Merging attributes can make the polytree algorithm applicable in multiply connected networks.

The available evidence is then propagated for each combination of values of the fixed attributes and the respective results are averaged weighted with the probabilities of the value combinations [Pearl 1988]. This procedure can also be seen as introducing artificial evidence to make the propagation feasible and then to remove it again (by computing the weighted sum).

Another way to approach the problem is to merge attributes lying "opposite" to each other in a cycle into one pseudo-attribute in order to "flatten" the cycle to a string of attributes. A very simple example is shown in Figure 4.12. Combining the attributes B and C into one pseudo-attribute removes the cycle. In principle all cycles can be removed in this way and thus one finally reaches a situation in which the polytree propagation algorithm can be applied. Of course, if stated in this way, this is only a heuristic method, since it does not provide us with exact criteria which nodes should be merged. However, it indicates the key principle that underlies several methods to handle cycles, namely to transform a given network in such a way that a singly connected structure is obtained. For the resulting singly connected structure a propagation scheme can then easily be derived in analogy to the polytree propagation method described in the preceding section.

The *join tree propagation method* [Lauritzen and Spiegelhalter 1988], which we are going to discuss in this section, is based on a sophisticated version of the node merging approach. Its basic idea is to add edges to a given conditional independence graph so that it finally has hypertree structure.[7] As already mentioned in Section 4.1.6, a graph with hypertree structure is, in a certain sense, acyclic. As a consequence of this property each maximal clique of the resulting graph can be made a node of a so-called *join tree*, which is a singly connected structure and thus can be used to propagate evidence. We do not give this method a full formal treatment, though, but confine ourselves to explaining it w.r.t. a simple example, namely the application of a Bayesian networkfor blood group determination of Danish Jersey cattle in the F-blood group system, the primary purpose of which is parentage verification for pedigree registration [Rasmussen 1992]. This example also serves as an illustration of the more theoretical results of the first section of this chapter. A more detailed treatment of join tree propagation can be found in [Jensen 1996, Castillo *et al.* 1997].

[7] The notion of a hypertree structure was defined in Definition 4.1.23 on page 110.

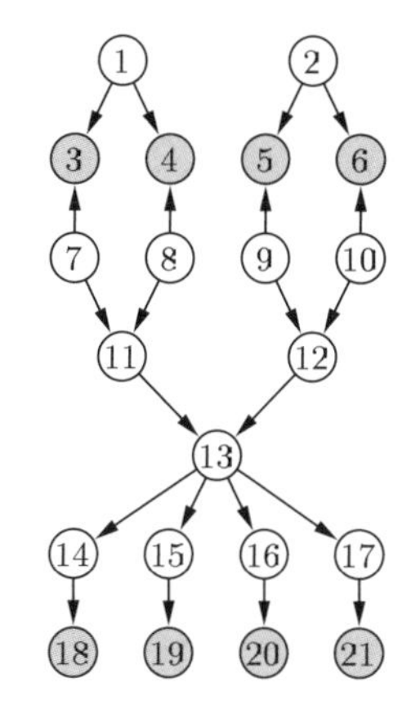

21 attributes:
1 – dam correct?
2 – sire correct?
3 – stated dam ph.gr. 1
4 – stated dam ph.gr. 2
5 – stated sire ph.gr. 1
6 – stated sire ph.gr. 2
7 – true dam ph.gr. 1
8 – true dam ph.gr. 2
9 – true sire ph.gr. 1
10 – true sire ph.gr. 2
11 – offspring ph.gr. 1
12 – offspring ph.gr. 2
13 – offspring genotype
14 – factor 40
15 – factor 41
16 – factor 42
17 – factor 43
18 – lysis 40
19 – lysis 41
20 – lysis 42
21 – lysis 43

The grey nodes correspond to observable attributes.

Figure 4.13 Domain expert designed network for the Danish Jersey cattle blood type determination example. ("ph.gr." stands for "phenogroup".)

sire	true sire	stated sire ph.gr. 1		
correct	ph.gr. 1	F1	V1	V2
yes	F1	1	0	0
yes	V1	0	1	0
yes	V2	0	0	1
no	F1	0.58	0.10	0.32
no	V1	0.58	0.10	0.32
no	V2	0.58	0.10	0.32

Table 4.2 A small fraction of the quantitative part of the Bayesian network, the conditional independence graph of which is shown in Figure 4.13: Conditional probability distributions for the phenogroup 1 of the stated sire.

The section of the world modeled in this example is described by 21 attributes, eight of which are observable. The size of the domains of these attributes ranges from two to eight values. The total frame of discernment has $2^6 \cdot 3^{10} \cdot 6 \cdot 8^4 = 92\,876\,046\,336$ possible states. This number makes it obvious that the knowledge about this world section must be decomposed in order to make reasoning feasible, since it is clearly impossible to store a probability for each state. Figure 4.13 lists the attributes and shows the conditional independence graph, which was designed by human domain experts. The grey nodes correspond to the observable attributes. This graph is the qualitative part of the Bayesian network.[8]

[8]Actually, the original Bayesian network has an additional attribute "parent correct?", which has "dam correct?" and "sire correct?" as its children, so that in all there are 22 attributes. We decided to discard this attribute, since it does not carry real information and without it the join tree construction is much simpler.

According to Theorem 4.1.22 (cf. page 109), a conditional independence graph enables us to factorize the joint probability distribution into a product of conditional probabilities with one factor for each attribute, in which it is conditioned on its parents. In the Danish Jersey cattle example, this factorization leads to a considerable simplification. Instead of having to determine the probability of each of the 92 876 046 336 elements of the 21-dimensional frame of discernment Ω, only 308 conditional probabilities need to be specified. An example of a conditional probability table, which is part of the factorization, is shown in Table 4.2. It states the conditional probabilities of the phenogroup 1 of the stated sire of a given calf conditioned on the phenogroup 1 of the true sire of the calf and whether the sire was correctly identified. The numbers in this table are derived from statistical data and the experience of human domain experts. The set of all 21 conditional probability tables is the quantitative part of the Bayesian network for this example.

After the Bayesian network is constructed, we want to draw inferences with it. In the Danish Jersey cattle example, for instance, the phenogroups of the stated dam and the stated sire can be determined and the lysis values of the calf can be measured. From the latter information we want to infer the genotype of the calf and thus wish to assess whether the stated parents of the calf are the true parents. However, the conditional independence graph of this example is no polytree and therefore the algorithm described in the previous section cannot be applied. Instead we preprocess the graph, so that it gets hypertree structure. This transformation is carried out in two steps. In the first step the so-called *moral graph* of the conditional independence graph is constructed and in the second the moral graph is *triangulated.*

A *moral graph* (the name was invented by [Lauritzen and Spiegelhalter 1988]) is constructed from a directed acyclic graph by "marrying" the parents of each attribute (hence the name "moral graph"). This is done by adding undirected edges between all pairs of parents and by discarding the directions of all other edges (the edges are kept, though). In general a moral graph represents only a subset of the independence relations of the underlying directed acyclic graph, so that this transformation may result in a loss of independence information. The reason for this was already explained w.r.t. the simple example of Figure 4.3 on page 102: In an undirected graph we cannot represent a situation of marginal independence, but conditional dependence, because u-separation is monotonous. Therefore edges must be added between the parents of an attribute, because these parents will become dependent if the child attribute (or any of its descendants) is instantiated. The moral graph for the Danish Jersey Cattle example is shown on the left in Figure 4.14. The edges that were added when parents were "married" are indicated by dotted lines.

Note that the moral graph can be chosen as the conditional independence graph of a Markov network for the same domain. Hence, if we have a Markov network, the join tree method can also be applied. We only have to leave out the first step of the transformation.

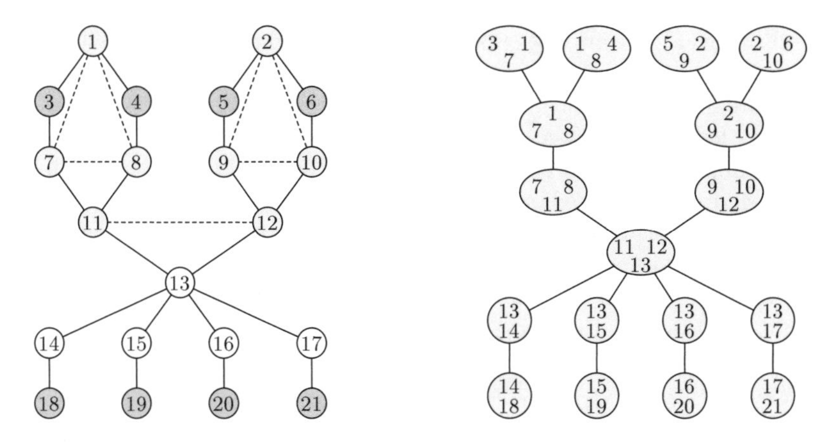

Figure 4.14 Triangulated moral graph (left) and join tree (right) for the conditional independence graph shown in Figure 4.13. The dotted lines are the edges added when parents were "married". The nodes of the join tree correspond to the maximal cliques of the triangulated moral graph.

In the second step, the moral graph is triangulated. An undirected graph is called *triangulated* or *chordal* if all cycles with four or more nodes have a *chord*, where a chord is an edge that connects two nodes that are nonadjacent w.r.t. the cycle. If this condition is satisfied, all cycles are build up from cycles consisting of three nodes (in other words: from triangles—hence the name "triangulation"). To achieve triangulation, it may be necessary to add edges, which may result in a (further) loss of independence information. A simple, though not optimal algorithm to test whether a given undirected graph is triangulated and to triangulate it, if it is not, is the following [Tarjan and Yannakakis 1984, Pearl 1988, Castillo *et al.* 1997]:

Algorithm 4.2.1 *(graph triangulation)*

Input: *An undirected graph* $G = (V, E)$.

Output: *A triangulated undirected graph* $G' = (V, E')$ *with* $E' \supseteq E$.

1. *Compute an ordering of the nodes of the graph using* maximum cardinality search, *i.e. number the nodes from 1 to* $n = |V|$, *in increasing order, always assigning the next number to the node having the largest set of previously numbered neighbors (breaking ties arbitrarily).*
2. *From* $i = n$ *to* $i = 1$ *recursively fill in edges between any nonadjacent neighbors of the node numbered* i *having lower ranks than* i *(including neighbors linked to the node numbered* i *in previous steps). If no edges are added, then the original graph is triangulated; otherwise the new graph is triangulated.*

Note that this algorithm does not necessarily add the smallest possible number of edges that are necessary to triangulate the graph. Note also that the triangulated graph is still a conditional independence graph, since it only represents fewer conditional independence statements. In the Danish Jersey cattle example the moral graph shown on the left in Figure 4.14 is already triangulated, so no new edges need to be introduced.

A triangulated graph is guaranteed to have hypertree structure[9] and hence it can be turned into a **join tree** [Lauritzen and Spiegelhalter 1988, Castillo *et al.* 1997]. In a join tree there is a node for each maximal clique of the triangulated graph and its edges connect nodes that represent cliques having attributes in common. In addition, any attribute that is contained in two nodes must also be contained in all nodes on the path between them. (Note that in general, despite this strong requirement, a join tree for a given triangulated graph is not unique.) A simple algorithm to construct a join tree from a triangulated graph is the following [Castillo *et al.* 1997]:

Algorithm 4.2.2 *(join tree construction)*

Input: *A triangulated undirected graph* $G = (V, E)$.

Output: *A join tree* $G' = (V', E')$ *for* G.

1. *Determine a numbering of the nodes of* G *using maximum cardinality search (see Algorithm 4.2.1).*
2. *Assign to each clique the maximum of the ranks of its nodes.*
3. *Sort the cliques in ascending order w.r.t. the numbers assigned to them.*
4. *Traverse the cliques in ascending order and connect each clique* C_i *to that clique of the preceding cliques* $C_1, \ldots, C_{i-1}$ *with which it has the largest number of nodes in common (breaking ties arbitrarily).*

With a join tree we finally have a singly connected structure, which can be used to derive propagation formulae. Note that the special property of a join tree, namely that any attribute contained in two nodes must also be contained in all nodes on the path between them, is important for evidence propagation, because it ensures that we have to incorporate evidence about the value of an attribute into only one node containing this attribute. Since all nodes containing the attribute are connected, the evidence is properly spread. Without this property, however, it may be necessary to incorporate the evidence into more than one node, which could lead to the same update anomalies as the cycles of the original graph: The same information could be used twice to update the probabilities of the values of some other attribute, thus invalidating the inference result. A join tree for the Danish Jersey cattle example is shown on the right in Figure 4.14.

Of course, a join tree is only the qualitative framework for a propagation method. The quantitative part of the original network has to be transformed,

[9]The notion of hypertree structure was defined in Definition 4.1.23 on page 110.

too. That is, from the family of conditional probability distributions of the original Bayesian network (or the factor potentials of the original Markov network), we have to compute appropriate functions, for instance, marginal distributions, on the maximal cliques of the triangulated graph. However, we do not consider this transformation here. Details can be found, for example, in [Jensen 1996, Castillo *et al.* 1997].

Having constructed the quantitative part, too, we can finally turn to evidence propagation itself. Evidence propagation in join trees is basically an iterative extension and projection process, as was already demonstrated in the simple examples of Chapter 3. (Note that we used in the probabilistic example what we can now call a Markov network and that its join tree has a node for each of the two subspaces of the decomposition.) When evidence about the value of an attribute becomes available, it is first extended to a join tree node the attribute is contained in. This is done by conditioning the associated marginal distribution. we call this an extension, because by this conditioning we go from restrictions on the values of a single attribute to restrictions on tuples of attribute values. Hence the information is extended from a single attribute to a subspace formed by several attributes (cf. also the relational example discussed in Chapter 3). Then the conditioned distribution is projected to all intersections of the join tree node with other nodes (often called *separator sets*). Through these projections the information can be transferred to other nodes of the join tree, where the process is repeated: First it is extended to the subspace represented by the node, then it is projected to the intersections connecting it to other nodes. The process stops when all nodes have been updated. It is clear that this process can be implemented, as the polytree propagation method of the previous section, by locally communicating node processors.

4.2.3 Other Evidence Propagation Methods

Evidence propagation in inference networks like Bayesian networks has been studied for quite some time now and thus it is not surprising that there is an abundance of propagation algorithms. It is clear that we cannot deal with all of them and therefore, in the preceding two section, we confined ourselves to studying two of them a little more closely. In the following paragraphs we only mention a few other approaches by outlining their basic ideas.

Stochastic simulation [Pearl 1988] consists in randomly generating a large number of instantiations of all attributes w.r.t. to the joint probability distribution that is represented by a probabilistic network. Of these instantiations those are discarded that are not compatible with the given evidence. From the remaining ones the relative frequency of values of the unobserved attributes is determined. The instantiation process is most easily explained w.r.t. a Bayesian network. First all parentless attributes are instantiated w.r.t. their marginal probability. This fixes the values of all parents of some other attributes, so that the probability of their values can be determined from the given conditional distributions. Therefore they can be instantiated next. The process is repeated until all attributes have been instantiated. Obviously, the attributes are most easily processed w.r.t. a topological order, since this ensures that all parents are instantiated.

The main drawback of stochastic simulation is that for large networks and non-trivial evidence a huge amount of instantiations has to be generated in order to retain enough of them after those incompatible with the given evidence have been discarded. It should also be noted that stochastic simulation can be used only in the probabilistic setting, because possibility distributions do not allow for a random instantiation of the attributes.

The basic idea of *iterative proportional fitting* [Whittaker 1990, von Hasseln 1998] is to traverse the given network, usually a Bayesian network, several times and to make on each traversal small changes to the probabilities of the values of unobserved attributes in order to fit them to the constraints that are imposed by the conditional distributions and by the values of the observed attributes. When equilibrium is reached, the probabilities of the unobserved attributes can be read from the corresponding nodes.

Bucket elimination [Dechter 1996, Zhang and Poole 1996] is an evidence propagation method that is not bound directly to graphs, although it can be supported by a conditional independence graph. It is based on the idea that an attribute can be eliminated (hence the name "bucket elimination") by summing the product of all factors in which it appears for all of the values of the attribute. By successive summations all attributes except a given goal attribute are eliminated, so that finally a single factor remains. It is obvious that the efficiency of the bucket elimination algorithm depends heavily on the order of the summations. If a wrong order is chosen, the intermediate distributions can get very large, thus rendering the process practically infeasible. To find a good order of the attributes a conditional independence graph can be helpful, because it can be constructed by following the edges of the graph.

It is clear that all algorithms mentioned in this section are applicable even if the conditional independence graph is not a polytree and none of them needs any preprocessing of the conditional independence graph. However, their drawbacks are that they are either inefficient (stochastic simulation) or that the time for propagating given evidence is hard to estimate in advance (iterative proportional fitting), whereas with join tree propagation this time can easily be computed from the size of the maximal cliques and the lengths of the paths in the join tree.

Chapter 5

Computing Projections

With this chapter we turn to learning graphical models from data. We start by considering the problem of computing projections of relations and of database-induced multivariate probability and possibility distributions to a given subspace of the frame of discernment. That is, we consider the problem of how to estimate a marginal distribution w.r.t. a given set of attributes from a database of sample cases.[1]

Computing such projections is obviously important, because they form the quantitative part of a graphical model, i.e. they are the components of the factorization or decomposition of a multivariate distribution. Without a method to determine them it is usually impossible to induce the structure (i.e. the qualitative part) of a graphical model, because all algorithms for this task presuppose such a method in one way or the other.

It turns out that computing projections is trivial in the relational and the probabilistic case—at least if the database is precise, i.e. does not contain missing values or set-valued information—which explains why this task is often not considered explicitly in these cases. If the database contains missing values estimating marginal probability distributions is more difficult, but can be handled very nicely with the *expectation maximization algorithm*, even though its application can turn out to be computationally expensive.

The possibilistic case, on the other hand, poses an unpleasant problem. This is—at least to some extent—counterintuitive, because computing the sum projections of the probabilistic case is so very simple. However, it is not possible to use an analogous method (like simply computing the maximum), as we are going to demonstrate with a simple example. Fortunately, the database to learn from can be preprocessed (by computing its *closure under tuple intersection* [Borgelt and Kruse 1998c]) so that computing maximum projections becomes simple and, for most practical problems, efficient.

[1] Recall the simple examples of Chapter 3, especially the relational example, in order to understand why we call the computation of a marginal distribution a *projection*.

5.1 Databases of Sample Cases

Before we can state clearly the problems underlying the computation of projections from a database of sample cases, it is helpful to define formally what we understand by "database". We distinguish two cases: databases with precise tuples and databases with imprecise tuples. However, since the latter are a generalization of the former, it suffices to consider the latter case.

To define the notion of a database of imprecise tuples, we start by extending the definitions of a tuple and of a relation (cf. Definition 3.2.1 on page 63) to capture imprecision w.r.t. the values of the attributes underlying them.

Definition 5.1.1 *Let* $U = \{A_1, \ldots, A_n\}$ *be a (finite) set of attributes with respective domains* $\mathrm{dom}(A_i)$, $i = 1, \ldots, n$. *A* **tuple** *over* U *is a mapping*

$$t_U : U \to \bigcup_{A \in U} 2^{\mathrm{dom}(A)}$$

satisfying $\forall A \in U : t_U(A) \subseteq \mathrm{dom}(A)$ *and* $t_U(A) \neq \emptyset$. *The set of all tuples over* U *is denoted* T_U. *A* **relation** R_U *over* U *is a set of tuples over* U, *i.e.* $R_U \subseteq T_U$.

We still write tuples similar to the usual vector notation. For example, a tuple t over $\{A, B, C\}$ which maps A to $\{a_1\}$, B to $\{b_2, b_4\}$ and C to $\{c_1, c_3\}$ is written $t = (A \mapsto \{a_1\}, B \mapsto \{b_2, b_4\}, C \mapsto \{c_1, c_3\})$. If an implicit order of the attributes is fixed, the attributes may be omitted. In addition, we still write $\mathrm{dom}(t) = X$ to indicate that t is a tuple over X.

With the above definition a tuple can represent *imprecise* (i.e. set-valued) information about the state of the modeled section of the world. It is, however, restricted in doing so. It cannot represent *arbitrary* sets of instantiations of the attributes, but only such sets that can be defined by stating a *set of values* for each attribute. We chose not to use a more general definition (which would define a tuple as an arbitrary set of instantiations of the attributes), because the above definition is usually much more convenient for practical purposes. It should be noted, though, that all results of this chapter can be transferred directly to the more general case, because the restriction of the above definition is not exploited.

We can now define the notions of a *precise* and of an *imprecise* tuple.

Definition 5.1.2 *A tuple* t_U *over a set* U *of attributes is called* **precise** *iff* $\forall A \in U : |t_U(A)| = 1$. *Otherwise it is called* **imprecise**. *The set of all precise tuples over* X *is denoted* $T_U^{(\mathrm{precise})}$.

Clearly, Definition 3.2.1 on page 63 was restricted to precise tuples. Projections of tuples and relations are defined in analogy to the precise case (cf. Definitions 3.2.2 and 3.2.3 on page 64).

Definition 5.1.3 *If t_X is a tuple over a set X of attributes and $Y \subseteq X$, then $t_X|_Y$ denotes the* **restriction** *or* **projection** *of the tuple t_X to Y. That is, the mapping $t_X|_Y$ assigns sets of values only to the attributes in Y. Hence* $\mathrm{dom}(t_X|_Y) = Y$, *i.e. $t_X|_Y$ is a tuple over Y.*

Definition 5.1.4 *Let R_X be a relation over a set X of attributes and $Y \subseteq X$. The* **projection** $\mathrm{proj}_Y^X(R_X)$ *of the relation R_X from X to Y is defined as*

$$\mathrm{proj}_Y^X(R_X) \stackrel{\text{def}}{=} \{t_Y \in T_Y \mid \exists t_X \in R_X : t_Y \equiv t_X|_Y\}.$$

It is clear that to describe a dataset of sample cases a simple relation does not suffice. In a relation, as it is a *set* of tuples, each tuple can appear only once. In contrast to this, in a dataset of sample cases a tuple may appear several times, reflecting the frequency of the occurrence of the corresponding case. Since we cannot dispense with this frequency information (we need it for both the probabilistic and the possibilistic setting), we need a mechanism to represent the number of occurrences of a tuple.

Definition 5.1.5 *A* **database** *D_U over a set U of attributes is a pair (R_U, w_{R_U}), where R_U is a relation over U and w_{R_U} is a function mapping each tuple in R_U to a natural number, i.e. $w_{R_U} : R_U \to \mathbb{N}$.*

If the set U of attributes is clear from the context, we drop the index U. The function w_{R_U} is intended to indicate the number of occurrences of a tuple $t \in R_U$ in a dataset of sample cases. We call $w_{R_U}(t)$ the **weight** of tuple t.

When dealing with imprecise tuples, it is helpful to be able to speak of a precise tuple being "contained" in an imprecise one or of one imprecise tuple being "contained" in another (w.r.t. the set of represented instantiations of the attributes). These terms are made formally precise by introducing the notion of a tuple being *at least as specific* as another.

Definition 5.1.6 *A tuple t_1 over an attribute set X is called* **at least as specific as** *a tuple t_2 over X, written $t_1 \sqsubseteq t_2$ iff $\forall A \in X : t_1(A) \subseteq t_2(A)$.*

Note that $\sqsubseteq$ is not a total ordering, since there are tuples that are incomparable. For example, $t_1 = (\{a_1\}, \{b_1, b_2\})$ and $t_2 = (\{a_1, a_2\}, \{b_1, b_3\})$ are incomparable, since neither $t_1 \sqsubseteq t_2$ nor $t_2 \sqsubseteq t_1$ holds. Note also that $\sqsubseteq$ is obviously transitive, i.e. if t_1, t_2, t_3 are three tuples over an attribute set X with $t_1 \sqsubseteq t_2$ and $t_2 \sqsubseteq t_3$, then also $t_1 \sqsubseteq t_3$. Finally, note that $\sqsubseteq$ is preserved by projection. That is, if t_1 and t_2 are two tuples over an attribute set X with $t_1 \sqsubseteq t_2$ and if $Y \subseteq X$, then $t_1|_Y \sqsubseteq t_2|_Y$.

5.2 Relational and Sum Projections

As already said above, computing projections from a database of sample cases is trivial in the relational and in the probabilistic setting: Computing the projection of a relation of precise tuples is an operation of relational algebra (cf. also Definition 3.2.3 on page 64). In order to deal with a relation R_U of imprecise tuples it suffices to note that such a relation can always (at least formally) be replaced by a relation R'_U of precise tuples, because in the relational case all we are interested in is whether a (precise) tuple is possible or not. This relation R'_U contains those precise tuples that are contained in a tuple in R_U, or formally

$$R'_U = \left\{ t' \in T_U^{(\text{precise})} \;\middle|\; \exists t \in R_U : t' \sqsubseteq t \right\}.$$

Note that this replacement need not be carried out explicitly, because $\sqsubseteq$ is maintained by projection (see above). Therefore we can work directly with the projection of a relation of imprecise tuples as defined in Definition 5.1.4, which indicates equally well which tuples of the subspace are possible. Note that in the relational case the tuple weights are disregarded.

In the probabilistic case it is usually assumed that the given database represents a sample of independent cases, generated by some random process that is governed by a multivariate probability distribution. With this presupposition, in order to estimate a marginal probability distribution from a database of precise tuples, the relational projection operation needs to be extended only slightly, because we have to take the tuple weights into account. This is done by summing for each precise tuple t_X of the subspace defined by the set X of attributes the weights of all tuples t_U in the database the projection of which is equal to t_X (hence the name "sum projection").

The result of this operation is an absolute frequency distribution on the subspace, which can be represented as a database on the subspace. From this distribution a marginal probability distribution is estimated using standard statistical techniques. We may, for example, use *maximum likelihood estimation*. That is, we may estimate the parameters in such a way that the *likelihood* of the data, i.e. the probability of the data given the model and its parameters, is maximized. This yields [Larsen and Marx 1986]

$$\forall a_1 \in \text{dom}(A_1) : \ldots \forall a_n \in \text{dom}(A_n) :$$
$$\hat{p}_X\Big(\bigwedge_{A_i \in X} A_i = a_i \Big) = \frac{w_X\Big(\bigwedge_{A_i \in X} A_i = a_i\Big)}{w_X(\epsilon)},$$

where $w_X\left(\bigwedge_{A_i \in X} A_i = a_i\right)$ is the weight of the tuples in the database satisfying $\bigwedge_{A_i \in X} A_i = a_i$ (this number can be read from the sum projection) and $w_X(\epsilon)$ is the total weight of all tuples.[2] Alternatively, we may use *Bayesian estimation*, which enables us to add a (uniform) prior expectation of the prob-

[2] We use the symbol ϵ in analogy to its use in the theory of formal languages, namely to denote an empty expression, i.e. an expression that does not restrict the set of tuples.

abilities by

$$\forall a_1 \in \mathrm{dom}(A_1) : \ldots \forall a_n \in \mathrm{dom}(A_n) :$$
$$\hat{p}_X\Big(\bigwedge_{A_i \in X} A_i = a_i \Big) = \frac{w_X\Big(\bigwedge_{A_i \in X} A_i = a_i \Big) + w_0}{w_X(\epsilon) + w_0 \prod_{A_i \in X} |\,\mathrm{dom}(A_i)|},$$

where the term w_0, which represents a uniform prior distribution, is most often chosen to be 1.[3] This uniform prior expectation often features also under the name of *Laplace correction*. It is clear that, by changing the above formula appropriately, arbitrary prior distributions can be incorporated.

5.3 Expectation Maximization

Unfortunately, in real-world applications, complete and precise databases, as we assumed them in the preceding section, are rare. To deal with imprecise tuples, we may apply the insufficient reason principle (cf. Section 2.4.3), which prescribes to distribute the weight of an imprecise tuple equally on all precise tuples contained in it. Formally, this enables us to work with a database of precise tuples if we remove the restriction that the weight function must assign natural numbers to the tuples. (Clearly, to represent a distributed weight, we need fractions.) Alternatively, we may preprocess the database and impute precise values (for example averages or most frequent values) or we may apply more sophisticated statistical methods like, for instance, expectation maximization (EM) [Dempster *et al.* 1977, Bauer *et al.* 1997] and gradient ascent [Russel *et al.* 1995]. In this section we study the basics of the expectation maximization algorithm using a simple Bayesian network as an example.

As its name already indicates, the expectation maximization algorithm consists of two steps: an expectation step and a maximization step. In the expectation step the parameters of the probability distribution to be estimated, which may be initialized randomly, are held constant and expected frequencies of the different precise tuples compatible with the imprecise tuples of the database are computed from it, thus completing the frequency information of the database. In the maximization step the frequencies of the tuples are held constant and the parameters of the distribution(s) are estimated from them, maximizing the likelihood of the data. (Due to this maximization of the likelihood of the data in the second step, the expectation maximization algorithm may be seen as a generalization of *maximum likelihood estimation*.

Of course, the two steps of the expectation maximization algorithm are executed not only once, but are iterated, until a stable state is reached, in which the parameters of the distribution to be estimated do not change any

[3]Compare, for example, the K2 (Bayesian Dirichlet uniform) metric discussed in Section 7.2.4. The likelihood equivalent Bayesian Dirichlet uniform metric, however, uses a different value (cf. the same section).

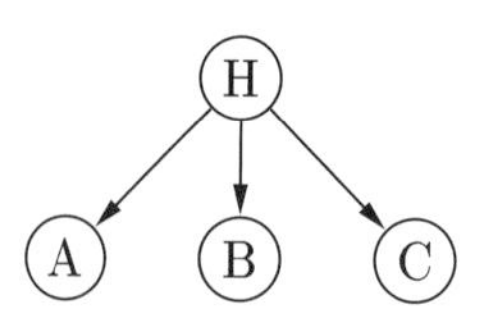

Figure 5.1 The structure of a Bayesian network and an imprecise database for this network.

H	A	B	C	w
*	a_0	b_0	c_0	14
*	a_0	b_0	c_1	11
*	a_0	b_1	c_0	20
*	a_0	b_1	c_1	20
*	a_1	b_0	c_0	5
*	a_1	b_0	c_1	5
*	a_1	b_1	c_0	11
*	a_1	b_1	c_1	14

more. It can be shown that, at least under certain conditions, the expectation maximization always converges to a stable state, although this state may be only a local optimum of the likelihood of the database [Dempster *et al.* 1977].

To illustrate how the expectation maximization algorithm works, we consider a simple example: Figure 5.1 shows on the left the graph of a Bayesian network over four attributes. All attributes are binary, i.e. each of them has only two possible values. In addition, we have the database shown on the right in Figure 5.1. In this database the value of H is missing for all tuples (indicated by a star). That is, attribute H is a *hidden* or *latent* attribute, i.e. an attribute that cannot be observed. The values of the other three attributes A, B, and C, however, are always known. The last column of the table states the weights of the different tuples, i.e. their absolute frequencies. In all, there are 100 sample cases (sum of the tuple weights).

We consider now one iteration of the expectation maximization algorithm, with which we want to estimate the parameters of the Bayesian network from the dataset shown on the right in Figure 5.1. To initialize the algorithm, we choose random initial values for the probabilities in the distributions of the Bayesian network. That is, we fix random marginal probabilities for the two values of the attribute H as well as random conditional probabilities for the values of the attributes A, B, and C given the values of attribute H. The random initialization we will use in this example is shown in Table 5.1.

The randomly chosen probabilities complete the Bayesian network and therefore it can be used to draw inferences (cf. Section 4.2). In particular, we can compute the conditional probabilities of the two values of the attribute H given an instantiation of the attributes A, B, and C. In this case, this is especially simple, because the decomposition formula of the Bayesian network is (cf. Section 4.1.6, especially Definition 4.1.25 on page 112)

$$\begin{aligned}
&\forall a \in \{a_0, a_1\} : \forall b \in \{b_0, b_1\} : \forall c \in \{c_0, c_1\} : \forall h \in \{h_0, h_1\} : \\
&\quad P(H = h, A = a, B = b, C = c) \\
&\quad = P(H = h)P(A = a \mid H = h)P(B = b \mid H = h)P(C = c \mid H = h),
\end{aligned}$$

Table 5.1 Random initial probabilities for the example execution of the expectation maximization algorithm.

p_H	h_0	h_1
	0.3	0.7

$p_{A\|H}$	h_0	h_1
a_0	0.4	0.6
a_1	0.6	0.4

$p_{B\|H}$	h_0	h_1
b_0	0.7	0.8
b_1	0.3	0.2

$p_{C\|H}$	h_0	h_1
c_0	0.8	0.5
c_1	0.2	0.5

and therefore the conditional probabilities of the values of the attribute H given values of the attributes A, B, and C can be computed as

$$\begin{aligned}
&\forall a \in \{a_0, a_1\} : \forall b \in \{b_0, b_1\} : \forall c \in \{c_0, c_1\} : \forall h \in \{h_0, h_1\} : \\
&\quad P(H = h \mid A = a, B = b, C = c) \\
&\quad = \alpha \cdot P(H = h)P(A = a \mid H = h)P(B = b \mid H = h)P(C = c \mid H = h),
\end{aligned}$$

where α is a normalization constant, which is to be determined so that

$$\sum_{h \in \{h_0, h_1\}} P(H = h \mid A = a, B = b, C = c) = 1.$$

(See the derivation of the evidence propagation formulae in Section 4.2.1, but also the discussion of naive Bayes classifiers in the next chapter.)

With the help of the above formula we can compute expected values for the absolute frequencies of the two values h_0 and h_1 of the attribute H for the different imprecise tuples in the database, namely by simply multiplying the conditional probabilities with the weights of the corresponding tuples. These expected values are shown in Table 5.2. Note that the frequencies of tuples with the same instantiation for the attributes A, B, and C sum to the weight of the corresponding tuple in Table 5.1.

With this computation we have executed the expectation step of the expectation maximization algorithm. We now have a completed database, from which we can estimate the probability parameters of the Bayesian network. For this estimation, which is the maximization step of the expectation maximization algorithm, we use maximum likelihood estimation (cf. the preceding section). That is, we compute, for instance, $P(A = a_0 \mid H = h_0)$ from the relative frequency of the value a_0 given $H = h_0$ by simply "counting" the corresponding tuples (or rather summing the tuple weights):

$$\begin{aligned}
&\hat{P}(A = a_0 \mid H = h_0) \\
&\approx \frac{1.27 + 3.14 + 2.93 + 8.14}{1.27 + 3.14 + 2.93 + 8.14 + 0.92 + 2.73 + 3.06 + 8.49} \approx 0.51.
\end{aligned}$$

Table 5.2 Expected values of the absolute tuple frequencies.

H	A	B	C	w	H	A	B	C	w
h_0	a_0	b_0	c_0	1.27	h_1	a_0	b_0	c_0	12.73
h_0	a_0	b_0	c_1	3.14	h_1	a_0	b_0	c_1	7.86
h_0	a_0	b_1	c_0	2.93	h_1	a_0	b_1	c_0	17.07
h_0	a_0	b_1	c_1	8.14	h_1	a_0	b_1	c_1	11.86
h_0	a_1	b_0	c_0	0.92	h_1	a_1	b_0	c_0	4.08
h_0	a_1	b_0	c_1	2.37	h_1	a_1	b_0	c_1	2.63
h_0	a_1	b_1	c_0	3.06	h_1	a_1	b_1	c_0	7.94
h_0	a_1	b_1	c_1	8.49	h_1	a_1	b_1	c_1	5.51

Table 5.3 Probabilities of the Bayesian network after the first step.

p_H	h_0	h_1
	0.3	0.7

$p_{A\|H}$	h_0	h_1
a_0	0.51	0.71
a_1	0.49	0.29

$p_{B\|H}$	h_0	h_1
b_0	0.25	0.39
b_1	0.75	0.61

$p_{C\|H}$	h_0	h_1
c_0	0.27	0.60
c_1	0.73	0.40

In this way we get the probabilities shown in Table 5.3 after one iteration of the expectation maximization algorithm, i.e. after one expectation and one maximization step. With these new probability distributions we redo the two steps, starting with the computation of new expected frequencies for the two values of the attribute H—exactly like we did it at the beginning with the randomly initialized probabilities—and then we reestimate the probabilities. We repeat this procedure until we reach a stable state. Technically we may test whether we have reached a stable state (or are at least sufficiently close to one) by checking whether the estimated probabilities changed no more than a predefined limit ε from one iteration of the expectation maximization algorithm to the next. In our example we eventually reach the probability distributions that are shown in Table 5.4.

Of course, the expectation maximization algorithm is not restricted to cases where one attribute is latent (as in this example). If the value of one or more attributes is missing only in some cases, it may be applied just as well. This algorithm is a very general and flexible method to estimate parameters of probability distributions from a database with incomplete tuples.

However, a serious drawback of the expectation maximization algorithm is that it can take very long to converge. Especially in the vicinity of a stable

Table 5.4 Probabilities of the Bayesian network after convergence.

p_H	h_0	h_1
	0.5	0.5

$p_{A\mid H}$	h_0	h_1
a_0	0.5	0.8
a_1	0.5	0.2

$p_{B\mid H}$	h_0	h_1
b_0	0.2	0.5
b_1	0.8	0.5

$p_{C\mid H}$	h_0	h_1
c_0	0.4	0.6
c_1	0.6	0.4

state convergence can be very slow. In our example, for instance, it takes 4210 iterations if $\varepsilon = 10^{-6}$ is chosen, i.e. if the algorithm is terminated if the probabilities of two consecutive iterations differ no more than 10^{-6}. In this case the computed probabilities differ less than $8 \cdot 10^{-4}$ from those shown in Table 5.4. For $\varepsilon = 10^{-4}$ the algorithm needs 687 iterations and the result differs less than 0.025 from Table 5.4.

In order to mitigate this disadvantage, several different acceleration methods have been suggested, among them (conjugate) gradient descent [Jamshidian and Jennrich 1993, Russel *et al.* 1995, Bauer *et al.* 1997] and a transformation to a root finding problem, so that the Newton–Raphson method can be applied [Jamshidian and Jennrich 1993]. Often standard expectation maximization is applied first to approach a convergence point and then it is switched to one of the mentioned methods to speed up convergence.

Here we do not discuss these methods in detail, but only study briefly an approach that is based on a simple idea from (artificial) neural network training [Haykin 1994, Zell 1994], namely the introduction of a momentum term. In this approach we view the change of the parameters from one iteration of the expectation maximization algorithm to the next as a generalized gradient.[4] To this generalized gradient we add a fraction of the change made in the previous iteration (this is the *momentum term*) to speed up the gradient ascent. Formally, we update a parameter θ using the rule

$$\theta_{t+1} = \theta_t + \Delta\theta_t \qquad \text{with} \qquad \Delta\theta_t = (\theta_t^{\mathrm{EM}} - \theta_t) + \beta \cdot \Delta\theta_{t-1},$$

where $\beta \in [0, 1)$ is a parameter of the method and t is the number of the iteration. That is, θ_t is the estimate for θ computed in the tth iteration and $\Delta\theta_t$ is the value by which θ_t is changed from the tth iteration to the next. θ_t^{EM} is the estimate computed from θ_t (and all other parameter estimates of step t) by an iteration of the expectation maximization algorithm.

Of course, with this scheme the resulting parameter estimate may not be valid. If, for instance, the parameter is a probability, the value may lie outside the interval [0,1], simply because the momentum term pushed it out of the

[4]This view is very intuitive, but there is also some formal justification for it [Jamshidian and Jennrich 1993].

acceptable range. In order to cope with this problem the new estimate θ_{t+1} is clamped to an interval $[\theta_{\min}, \theta_{\max}]$ of acceptable values. For our simple example we may use the interval $[0.01, 0.99]$, since all parameters are (conditional) probabilities. (Note that it may also be necessary to renormalize the distribution if there are more than two values per attribute.)

With this simple modification of the algorithm we achieve a considerable speedup in our simple example. With a limit of $\varepsilon = 10^{-6}$ the algorithm terminates after 624 iterations (compared to 4210 for pure expectation maximization) with the result differing less than $2 \cdot 10^{-5}$ from the values shown in Table 5.4 (compared to $8 \cdot 10^{-4}$ for pure expectation maximization). With a limit of $\varepsilon = 10^{-3}$ the algorithm terminates after 368 iterations with the result differing less than $8 \cdot 10^{-4}$ from Table 5.4.

5.4 Maximum Projections

In the possibilistic case, if we rely on the context model interpretation of a degree of possibility (cf. Section 2.4), a given database is interpreted as a description of a random set (cf. Definition 2.4.1 on page 35). Each tuple is identified with a context and thus the relative tuple weight is the context weight. The sample space Ω is assumed to be the set $T_U^{(\text{precise})}$ of all precise tuples over the set U of attributes of the database. With these presuppositions the possibility distribution $\pi_U^{(D)}$ that is induced by a database D over a set U of attributes can be defined as follows:

Definition 5.4.1 *Let $D = (R, w_R)$ be a non-empty database (i.e. $R \neq \emptyset$) over a set U of attributes. Then*

$$\pi_U^{(D)} : T_U^{(\text{precise})} \to [0, 1], \qquad \pi_U^{(D)}(t) \mapsto \frac{\sum_{s \in R, t \sqsubseteq s} w_R(s)}{\sum_{s \in R} w_R(s)},$$

is the **possibility distribution over U induced by D**.

That is, the degree of possibility of each precise tuple t is the relative weight of those (imprecise) tuples that contain it (cf. Definition 2.4.2 on page 36).

For a precise database computing maximum projections is equally simple as computing sum projections. The only difference is, as the names already indicate, that instead of summing the tuple weights we have to determine their maximum. That this simple procedure is possible can easily be seen from the fact that for a precise database the numerator of the fraction in the definition above is reduced to one term. Therefore we have

$$\forall t_X \in T_X : \quad \pi_X^{(D)}(t_X) = \max_{A \in U-X} \pi_U^{(D)}(t_U) = \frac{\max_{A \in U-X} w_R(t_U)}{\sum_{s \in R} w_R(s)}$$

(cf. Definition 2.4.8 on page 46 and Definition 3.2.7 on page 68).

Database:	$(\{a_1, a_2, a_3\}, \{b_3\})$	: 1
	$(\{a_1, a_2\}, \{b_2, b_3\})$	: 1
	$(\{a_3, a_4\}, \{b_1\})$	: 1

Table 5.5 A very simple imprecise database with three tuples (contexts), each having a weight of 1.

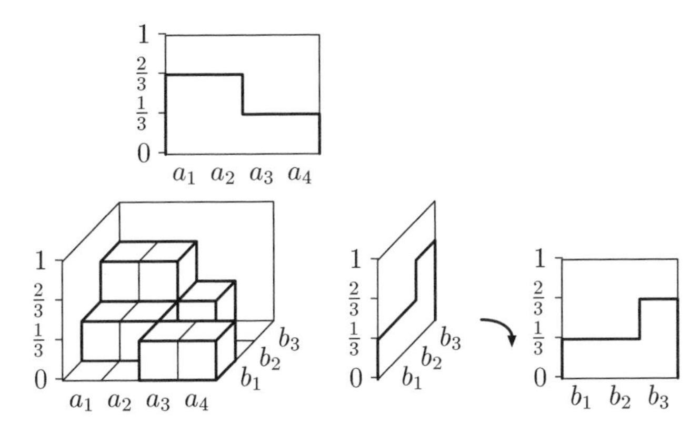

Figure 5.2 The possibility distribution induced by the three tuples of the database shown in Table 5.5.

Unfortunately this simple procedure cannot be transferred to databases with imprecise tuples, because in the presence of imprecise tuples the sum in the numerator has to be taken into account. This sum poses problems, because the computation of its terms can be very expensive.

5.4.1 A Simple Example

To understand the problems that result from databases of imprecise tuples it is helpful to study a simple example that clearly shows the difficulties that arise. Consider the very simple database shown in Table 5.5 that is defined over two attributes A and B. The possibility distribution on the joint domain of A and B that is induced by this database is shown graphically in Figure 5.2. This figure also shows the marginal possibility distributions (maximum projections) for each of the two attributes. Consider first the degree of possibility that attribute A has the value a_3, which is $\frac{1}{3}$. This degree of possibility *can* be computed by taking the maximum over all tuples in the database in which the value a_3 is possible: Both tuples in which it is possible have a weight of 1. On the other hand, consider the degree of possibility that attribute A has the value a_2, which is $\frac{2}{3}$. To get this value, we have to *sum* the weights of the tuples in which it is possible. Since both a_2 and a_3 are possible in two tuples of the database, we conclude that neither the sum nor the maximum of the tuple weights can, in general, yield the correct result.

Table 5.6 The maximum over tuples in the support equals the maximum over tuples in the closure.

Database	Support		Closure
$(\{a_1, a_2, a_3\}, \{b_3\}) : 1$	$(a_1, b_2) : 1$	$(a_3, b_1) : 1$	$(\{a_1, a_2, a_3\}, \{b_3\}) : 1$
$(\{a_1, a_2\}, \{b_2, b_3\}) : 1$	$(a_1, b_3) : 2$	$(a_3, b_3) : 1$	$(\{a_1, a_2\}, \{b_2, b_3\}) : 1$
$(\{a_3, a_4\}, \{b_1\}) : 1$	$(a_2, b_2) : 1$	$(a_4, b_1) : 1$	$(\{a_3, a_4\}, \{b_1\}) : 1$
	$(a_2, b_3) : 2$		$(\{a_1, a_2\}, \{b_3\}) : 2$
3 tuples	7 tuples		4 tuples

Note that this problem of computing maximum projections results from the fact that we consider unrestricted random sets, i.e. random sets that may have arbitrary focal sets. As can be seen from the detailed discussion in Section 2.4 and especially Sections 2.4.6 and 2.4.7, the problem vanishes immediately if the focal sets of the random set are required to be consonant. In this case, summing over the tuple weights always yields the correct result, because disjoint tuples (like the first and the third in the database), for which taking the maximum is necessary, are excluded. However, it is also clear that consonance of the focal sets is almost never to be had if random sets are used to interpret databases of sample cases.

Fortunately, the simple example shown in Figure 5.2 not only illustrates the problem that occurs w.r.t. computing maximum projections of database-induced possibility distribution, but also provides us with a hint how this problem may be solved. Obviously, the problem results from the fact that the first two tuples "intersect" on the precise tuples (a_1, b_3) and (a_2, b_3). If this intersection were explicitly represented—with a tuple weight of 2—we could always determine the correct projection by taking the maximum.

This is demonstrated in Table 5.6. The table on the left restates the database of Table 5.5. The table in the middle lists what we call the *support* of the database, which is itself a database. This database consists of all precise tuples that are contained in a tuple of the original database. The weights assigned to these tuples are the values of the numerator of the fraction in the definition of the database-induced possibility distribution. Obviously, the marginal degrees of possibility of a value of any of the two attributes A and B can be determined from this relation by computing the maximum over all tuples that contain this value (divided, of course, by the sum of the weights of all tuples in the original database), simply because this computation is a direct implementation of the definition. Therefore we can always fall back on this method of computing a maximum projection.

Note, however, that this method corresponds to the formal expansion of the database mentioned for the relational case, but that, in contrast to the relational case, we have to compute the expansion explicitly in order to determine the maximum projections. Unfortunately, this renders this method computationally infeasible in most cases, especially, if there are many attributes and several imprecise tuples (cf. the experimental results in Section 5.4.4). This problem is already indicated by the fact that even for this very simple example we need seven tuples in the support database, although the original database contains only three.

Consequently a better method than the computation via the support is needed. Such a method is suggested by the third column of Table 5.6. The first three tuples in this column are the tuples of the original database. In addition, this column contains an imprecise tuple that corresponds to the "intersection" of the first two tuples. Since this tuple is at least as specific as both the first and the second, it is assigned a weight of 2, the sum of the weights of the first and the second tuple. By adding this tuple to the database, the set of tuples becomes *closed under tuple intersection*, which explains the label *closure* of this column. That is, for any two tuples s and t in this database, if we construct the (imprecise) tuple that represents the set of precise tuples that are represented by both s and t, then this tuple is also contained in the database. It is easily verified that, in this example, the marginal degrees of possibility of a value of any of the two attributes A and B can be determined from this database by computing the maximum over all tuples that contain this value.

Hence, if we can establish this equality in general, preprocessing the database so that it is closed under tuple intersection provides an alternative to a computation of maximum projections via the support database. This is especially desirable, since it can be expected that in general only few tuples have to be added in order to achieve closure under tuple intersection. In the example, for instance, only one tuple needs to be added (cf. also the experimental results in Section 5.4.4).

5.4.2 Computation via the Support

The remainder of this chapter is devoted to introducing the technical notions needed to prove, in a final theorem, that a computation of a maximum projection via the closure under tuple intersection is always equal to a computation via the support of a possibility distribution (which, by definition, yields the correct value—see above). We start by making formally precise the notions of the support of a relation and the support of a database.

Definition 5.4.2 *Let R be a relation over a set U of attributes. The* **support of** R, *written* $\mathrm{support}(R)$, *is the set of all precise tuples that are at least as* *specific as a tuple in R, i.e.*

$$\text{support}(R) = \left\{ t \in T_U^{(\text{precise})} \;\middle|\; \exists r \in R : t \sqsubseteq r \right\}.$$

Obviously, $\text{support}(R)$ is also a relation over U. Using this definition we can define the support of a database.

Definition 5.4.3 *Let $D = (R, w_R)$ be a database over a set U of attributes. The* **support of** D *is the pair* $\text{support}(D) = (\text{support}(R), w_{\text{support}(R)})$, *where* $\text{support}(R)$ *is the support of the relation R and*

$$w_{\text{support}(R)} : \text{support}(R) \to \mathbb{N}, \qquad w_{\text{support}(R)}(t) \mapsto \sum_{s \in R, t \sqsubseteq s} w_R(s).$$

Obviously, $\text{support}(D)$ is also a database over U. Comparing this definition to Definition 5.4.1, we see that

$$\pi_U^{(D)}(t) = \begin{cases} \frac{1}{w_0} w_{\text{support}(R)}(t), & \text{if } t \in \text{support}(R), \\ 0, & \text{otherwise}, \end{cases}$$

where $w_0 = \sum_{s \in R} w_R(s)$. It follows that any maximum projection of a database-induced possibility distribution $\pi_U^{(D)}$ over a set U of attributes to a set $X \subseteq U$ can be computed from $w_{\text{support}(R)}$ as follows (although the two projections are identical, we write $\pi_X^{(\text{support}(D))}$ instead of $\pi_X^{(D)}$ to indicate that the projection is computed via the support of D):

$$\pi_X^{(\text{support}(D))} : T_X^{(\text{precise})} \to [0, 1],$$

$$\pi_X^{(\text{support}(D))}(t) \mapsto \begin{cases} \frac{1}{w_0} \max_{s \in S(t)} w_{\text{support}(R)}(s), & \text{if } S(t) \neq \emptyset, \\ 0, & \text{otherwise}, \end{cases}$$

where $S(t) = \{s \in \text{support}(R) \mid t \sqsubseteq s|_X\}$ and $w_0 = \sum_{s \in R} w_R(s)$.

It should be noted that, as already mentioned above, the computation of maximum projections via the support of a database is, in general, very inefficient, because of the usually huge number of tuples in $\text{support}(R)$.

5.4.3 Computation via the Closure

In this section we turn to the computation of a maximum projection via the closure of a database under tuple intersection. Clearly, we must begin by defining the notion of the *intersection* of two tuples.

Definition 5.4.4 *Let U be a set of attributes. A tuple s over U is called the* **intersection** *of two tuples t_1 and t_2 over U, written $s = t_1 \sqcap t_2$ iff* $\forall A \in U : s(A) = t_1(A) \cap t_2(A)$.

Note that the intersection of two given tuples need not exist. For example, $t_1 = (A \mapsto \{a_1\}, B \mapsto \{b_1, b_2\})$ and $t_2 = (A \mapsto \{a_2\}, B \mapsto \{b_1, b_3\})$ do not have an intersection, since $t_1(A) \cap t_2(A) = \emptyset$, but a tuple may not map an attribute to the empty set (cf. Definition 5.1.1 on page 134).

Note also that the intersection s of two tuples t_1 and t_2 is at least as specific as both of them, i.e. it is $s \sqsubseteq t_1$ and $s \sqsubseteq t_2$. In addition, s is the least specific of all tuples s' for which $s' \sqsubseteq t_1$ and $s' \sqsubseteq t_2$, i.e.

$$\forall s' \in T_U : \quad (s' \sqsubseteq t_1 \wedge s' \sqsubseteq t_2) \;\Rightarrow\; (s' \sqsubseteq s \equiv t_1 \sqcap t_2).$$

This is important, since it also says that any tuple that is at least as specific as each of two given tuples is at least as specific as their intersection. (This property is needed in the proof of Theorem 5.4.8.) Furthermore, note that intersection is idempotent, i.e. $t \sqcap t \equiv t$. (This is needed below, where some properties of closures are collected.) Finally, note that the above definition can easily be extended to the more general definition of an imprecise tuple, in which it is defined as an arbitrary set of instantiations of the attributes. Clearly, in this case tuple intersection reduces to simple set intersection.

From the intersection of two tuples we can proceed directly to the notions of *closed under tuple intersection* and *closure of a relation.*

Definition 5.4.5 *Let R be a relation over a set U of attributes. R is called* **closed under tuple intersection** *iff*

$$\forall t_1, t_2 \in R : \quad (\exists s \in T_U : s \equiv t_1 \sqcap t_2) \;\Rightarrow\; s \in R,$$

i.e. iff for any two tuples in R their intersection is also contained in R (provided it exists).

Definition 5.4.6 *Let R be a relation over a set U of attributes. The* **closure of** *R, written* closure(R), *is the set*

$$\text{closure}(R) = \left\{ t \in T_U \;\middle|\; \exists S \subseteq R : t \equiv \bigsqcap_{s \in S} s \right\},$$

i.e. the relation R together with all possible intersections of tuples from R.

Note that closure(R) is, obviously, also a relation and that it is closed under tuple intersection: If $t_1, t_2 \in \text{closure}(R)$, then, due to the construction,

$$\exists S_1 \subseteq R : \quad t_1 = \bigsqcap_{s \in S_1} s \qquad \text{and} \qquad \exists S_2 \subseteq R : \quad t_2 = \bigsqcap_{s \in S_2} s.$$

If now $\exists t \in T_U : t = t_1 \sqcap t_2$, then

$$t = t_1 \sqcap t_2 = \bigsqcap_{s \in S_1} s \sqcap \bigsqcap_{s \in S_2} s = \bigsqcap_{s \in S_1 \sqcup S_2} s \in \text{closure}(R).$$

(The last equality in this sequence holds, since $\sqcap$ is idempotent, see above.)

Note also that a direct implementation of the above definition is not the best way to compute closure(R). A better, because much more efficient way, is to start with a relation $R' = R$, to compute only intersections of pairs of tuples taken from R', and to add the results to R' until no new tuples can be added. The final relation R' is the closure of R.

As for the support, the notion of a closure is extended to databases.

Definition 5.4.7 *Let $D = (R, w_R)$ be a database over a set U of attributes. The* **closure of** *D is the pair* $\text{closure}(D) = (\text{closure}(R), w_{\text{closure}(R)})$, *where* closure($R$) *is the closure of the relation R and*

$$w_{\text{closure}(R)} : \text{closure}(R) \to \mathbb{N}, \qquad w_{\text{closure}(R)}(t) \mapsto \sum_{s \in R, t \sqsubseteq s} w_R(s).$$

We assert (and prove in the theorem below) that any maximum projection of $\pi_U^{(D)}$ to a set $X \subseteq U$ can be computed from $w_{\text{closure}(R)}$ as follows (we write $\pi_X^{(\text{closure}(D))}$ to indicate that the projection is computed via the closure of D):

$$\pi_X^{(\text{closure}(D))} : T_X^{(\text{precise})} \to [0, 1],$$

$$\pi_X^{(\text{closure}(D))}(t) \mapsto \begin{cases} \frac{1}{w_0} \max\limits_{c \in C(t)} w_{\text{closure}(R)}(c), & \text{if } C(t) \neq \emptyset, \\ 0, & \text{otherwise}, \end{cases}$$

where $C(t) = \{c \in \text{closure}(R) \mid t \sqsubseteq c|_X\}$ and $w_0 = \sum_{s \in R} w_R(s)$.

Since, as already mentioned, closure(R) usually contains much fewer tuples than support(R), a computation based on the above formula is much more efficient. We verify our assertion that any maximum projection can be computed in this way by the following theorem [Borgelt and Kruse 1998c]:

Theorem 5.4.8 *Let $D = (R, w_R)$ be a database over a set U of attributes and let $X \subseteq U$. Furthermore, let* $\text{support}(D) = (\text{support}(R), w_{\text{support}(R)})$ *and* $\text{closure}(D) = (\text{closure}(R), w_{\text{closure}(R)})$ *as well as $\pi_X^{(\text{support}(D))}$ and $\pi_X^{(\text{closure}(D))}$ be defined as above. Then*

$$\forall t \in T_X^{(\text{precise})} : \qquad \pi_X^{(\text{closure}(D))}(t) = \pi_X^{(\text{support}(D))}(t),$$

i.e. computing the maximum projection of the possibility distribution $\pi_U^{(D)}$ induced by D to the attributes in X via the closure of D is equivalent to computing it via the support of D.

Proof. The assertion of the theorem is proven in two steps. In the first, it is shown that, for an arbitrary tuple $t \in T_X^{(\text{precise})}$,

$$\pi_X^{(\text{closure}(D))}(t) \geq \pi_X^{(\text{support}(D))}(t),$$

dataset	cases	tuples in R	tuples in support(R)	tuples in closure(R)
djc	500	283	712957	291
soybean	683	631	unknown	631
vote	435	342	98934	400

Table 5.7 The number of tuples in support and closure of three databases.

and in the second that

$$\pi_X^{(\text{closure}(D))}(t) \leq \pi_X^{(\text{support}(D))}(t).$$

Both parts together obviously prove the theorem. The first part is carried out by showing that for the (precise) tuple $\hat{s}$ in support(D), which determines the value of $\pi_X^{(\text{support}(D))}(t)$, there must be a corresponding (imprecise) tuple in closure(D) with a weight no less than that of $\hat{s}$. The second part is analogous. The full proof can be found in Section A.8 in the appendix.

5.4.4 Experimental Results

We tested the method suggested in the preceding sections on three datasets, namely the Danish Jersey cattle blood type determination dataset (djc, 500 cases), the soybean diseases dataset (soybean, 683 cases), and the congress voting dataset (vote, 435 cases). (The latter two datasets are well known from the UCI Machine Learning Repository [Murphy and Aha 1994].) Each of these datasets contains a lot of missing values, which we treated as an imprecise attribute value. That is, for a missing value of an attribute A we assumed dom(A) as the set of values to which the corresponding tuple maps A. Unfortunately we could not get hold of any real-world dataset containing "true" imprecise attribute values, i.e. datasets with cases in which for an attribute A a set $S \subset \text{dom}(A)$ with $|S| > 1$ and $S \neq \text{dom}(A)$ was possible. If anyone can direct us to such a dataset, we would be very grateful.

For each of the mentioned datasets we compared the reduction to a relation (keeping the number of occurrences in the tuple weight), the expansion to the support of this relation, and the closure of the relation w.r.t. tuple intersection. The results are as shown in Table 5.7. The entry "unknown" means that the resulting relation is too large to be computed and hence we could not determine its size. It is obvious that using the closure instead of the support of a relation to compute the maximum projections leads to a considerable reduction in complexity, or, in some cases, makes it possible to compute a maximum projection in the first place.

A_1	A_2	$\cdots$	A_n	$: w$
$a_{1,1}$	?	$\cdots$	?	: 1
$\vdots$	$\vdots$		$\vdots$	$\vdots$
a_{1,m_1}	?	$\cdots$	?	: 1
?	$a_{2,1}$	$\cdots$	?	: 1
$\vdots$	$\vdots$		$\vdots$	$\vdots$
?	a_{2,m_2}	$\cdots$	?	: 1
$\vdots$	$\vdots$		$\vdots$	$\vdots$
?	?	$\cdots$	$a_{n,1}$	: 1
$\vdots$	$\vdots$		$\vdots$	$\vdots$
?	?	$\cdots$	a_{n,m_n}	: 1

Table 5.8 A pathological example for the computation of the closure under tuple intersection. Although there are only $\sum_{i=1}^{n} m_i$ tuples in this table, the closure under tuple intersection contains $(\prod_{i=1}^{n}(m_i + 1)) - 1$ tuples.

5.4.5 Limitations

It should be noted that, despite the promising results of the preceding section, computing the closure under tuple intersection of a relation with imprecise tuples does *not guarantee* that computing maximum projections is efficient. To see this, consider the pathological example shown in Table 5.8. $a_{i,j}$ is the j-th value in the domain of attribute i and m_i is the number of values of attribute A_i, i.e. $m_i = |\operatorname{dom}(A_i)|$. Question marks indicate missing values.

Although this table has only $\sum_{i=1}^{n} |\operatorname{dom}(A_i)|$ tuples, computing its closure under tuple intersection constructs all $\prod_{i=1}^{n} |\operatorname{dom}(A_i)|$ possible precise tuples over $U = \{A_1, \ldots, A_n\}$ and all possible imprecise tuples with a precise value for some and a missing value for the remaining attributes. In all there are $(\prod_{i=1}^{n}(|\operatorname{dom}(A_i)| + 1)) - 1$ tuples in the closure, because for the i-th element of a tuple there are $|\operatorname{dom}(A_i)| + 1$ possible entries: $|\operatorname{dom}(A_i)|$ attribute values and the star to indicate a missing value. The only tuple that does not occur is the one having only missing values.

To handle this database properly, an operation to merge tuples—for instance, tuples that differ in the (set of) value(s) for only one attribute—is needed. With such an operation, the above table can be reduced to a single tuple, having unknown values for all attributes. This shows that there is some potential for future improvements of this preprocessing method.

Chapter 6

Naive Classifiers

In this chapter we consider an important type of classifiers, which we call *naive classifiers*, because they naively make very strong independence assumptions. These classifiers can be seen as a special type of graphical model, the structure of which is fixed by the classification task.

The best-known naive classifier is, of course, the naive Bayes classifier, which is discussed in Section 6.1. It can be seen as a Bayesian network with a star-like structure. Due to the similarity of Bayesian networks and directed possibilistic networks, the idea suggests itself to construct a possibilistic counterpart of the naive Bayes classifier [Borgelt and Gebhardt 1999]. This classifier, which also has a star-like structure, is discussed in Section 6.2. For both naive Bayes classifiers and naive possibilistic classifiers there is a straightforward method to simplify them, i.e. to reduce the number of attributes used to predict the class. This method, which is based on a greedy approach, is reviewed in Section 6.3. Finally, Section 6.4 presents experimental results, in which both naive classifiers are compared to a decision tree classifier.

6.1 Naive Bayes Classifiers

Naive Bayes classifiers [Good 1965, Duda and Hart 1973, Langley *et al.* 1992, Langley and Sage 1994] are an old and well-known type of classifiers, i.e. of programs that assign a class from a predefined set to an object or case under consideration based on the values of attributes used to describe this object or case. They use a probabilistic approach, i.e. they try to compute conditional class probabilities and then predict the most probable class.

6.1.1 The Basic Formula

We start our discussion of naive Bayes classifiers by deriving the basic formula underlying them. Let C be a class attribute with a finite domain of m classes, i.e. $\text{dom}(C) = \{c_1, \ldots, c_m\}$, and let $U = \{A_1, \ldots, A_n\}$ be a set of other attributes used to describe a case or an object. These other attributes may be symbolic, i.e. $\text{dom}(A_k) = \{a_{k,1}, \ldots, a_{k,m_k}\}$, or numeric, i.e. $\text{dom}(A_k) = \mathbb{R}$.[1] If the second index of an attribute value does not matter, it is dropped and we simply write a_k for a value of an attribute A_k. With this notation a case or an object can be described, as usual, by an instantiation $(a_1, \ldots, a_n)$ of the attributes $A_1, \ldots, A_n$.

For a given instantiation $(a_1, \ldots, a_n)$ a naive Bayes classifier tries to compute the conditional probability $P(C = c_i \mid A_1 = a_1, \ldots, A_n = a_n)$ for all c_i and then predicts the class c_i for which this probability is highest. Of course, it is usually impossible to store all of these probabilities explicitly, so that the most probable class can be found by a simple lookup. If there are numeric attributes, this is obvious (in this case some parameterized function is needed). But even if all attributes are symbolic, we have to store a class (or a class probability distribution) for each point of the Cartesian product of the attribute domains, the size of which grows exponentially with the number of attributes. To cope with this problem, naive Bayes classifiers exploit—as their name already indicates—Bayes' rule and a set of (naive) conditional independence assumptions. With Bayes' rule the conditional probabilities are inverted. That is, naive Bayes classifiers consider[2]

$$\begin{aligned} &P(C = c_i \mid A_1 = a_1, \ldots, A_n = a_n) \\ &= \frac{f(A_1 = a_1, \ldots, A_n = a_n \mid C = c_i) \cdot P(C = c_i)}{f(A_1 = a_1, \ldots, A_n = a_n)}. \end{aligned}$$

Of course, for this inversion to be always possible, the probability density function $f(A_1 = a_1, \ldots, A_n = a_n)$ must be strictly positive.

There are two observations to be made about this inversion. In the first place, the denominator of the fraction on the right can be neglected, since for a given case or object to be classified it is fixed and therefore does not have any influence on the class ranking (which is all we are interested in). In addition, its influence can always be restored by normalizing the distribution on the classes, i.e. we can exploit

$$\begin{aligned} &f(A_1 = a_1, \ldots, A_n = a_n) \\ &= \sum_{j=1}^{m} f(A_1 = a_1, \ldots, A_n = a_n \mid C = c_j) \cdot P(C = c_j). \end{aligned}$$

Secondly, we can see that merely inverting the probabilities does not give us any advantage, since the probability space is equally large as it was before.

[1] In this chapter we temporarily deviate from the restriction to finite domains, because naive Bayes classifiers can be illustrated very well with numerical examples (see below).

[2] For simplicity we always use a probability density function f, although this is strictly correct only if there is at least one numeric attribute (otherwise it should be a distribution P). The only exception is the class attribute, which must always be symbolic.

However, here the (naive) conditional independence assumptions come in. To exploit them, we first apply the chain rule of probability to obtain

$$\begin{aligned} &P(C = c_i \mid A_1 = a_1, \ldots, A_n = a_n) \\ &\quad = \frac{P(C = c_i)}{p_0} \cdot \prod_{j=1}^{n} f\Big(A_j = a_j \;\Big|\; \bigwedge_{k=1}^{j-1} A_k = a_k, C = c_i\Big), \end{aligned}$$

where p_0 is an abbreviation for $f(A_1 = a_1, \ldots, A_n = a_n)$. Then we make the crucial assumption that, given the value of the class attribute, any attribute A_j is independent of any other. That is, we assume that knowing the class is enough to determine the probability (density) for a value a_j, i.e. that we need not know the values of any other attributes. Of course, this is a fairly strong assumption, which is truly "naive". However, it considerably simplifies the formula stated above, since with it we can cancel all attributes A_k appearing in the conditions. Thus we get

$$P(C = c_i \mid A_1 = a_1, \ldots, A_n = a_n) = \frac{P(C = c_i)}{p_0} \cdot \prod_{j=1}^{n} f(A_j = a_j \mid C = c_i).$$

This is the basic formula underlying naive Bayes classifiers. For a symbolic attribute A_j the conditional probabilities $P(A_j = a_j \mid C = c_i)$ are stored as a simple table. This is feasible now, since there is only one condition and hence only $m \cdot m_j$ probabilities have to be stored.[3] For numeric attributes it is often assumed that the probability density is a normal distribution and hence only the expected values $\mu_j(c_i)$ and the variances $\sigma_j^2(c_i)$ need to be stored in this case. Alternatively, numeric attributes may be discretized [Dougherty *et al.* 1995] and then handled like symbolic attributes.

Naive Bayes classifiers can easily be induced from a dataset of preclassified sample cases. All one has to do is to estimate the conditional probabilities/probability densities $f(A_j = a_j \mid C = c_i)$ using, for instance, maximum likelihood estimation (cf. Section 5.2). For symbolic attributes this yields

$$\hat{P}(A_j = a_j \mid C = c_i) = \frac{N(A_j = a_j, C = c_i)}{N(C = c_i)},$$

where $N(C = c_i)$ is the number of sample cases that belong to class c_i and $N(A_j = a_j, C = c_i)$ is the number of sample cases which in addition have the value a_j for the attribute A_j. To ensure that the probability is strictly positive (see above), unrepresented classes are deleted. If an attribute value does not occur given a class, its probability is either set to $\frac{1}{2N}$, where N is the total number of sample cases, or a uniform prior of $\frac{1}{N}$ is added to the estimated

[3]Actually only $m \cdot (m_j - 1)$ probabilities are really necessary. Since the probabilities have to add up to one, one value can be discarded from each conditional distribution. However, in implementations it is usually much more convenient to store all probabilities.

distribution, which is then renormalized (cf. the notion of Laplace correction mentioned in Section 5.2). For a numeric attribute A_j the standard maximum likelihood estimation functions for the parameters of a normal distribution may be used, namely

$$\hat{\mu}_j(c_i) = \frac{1}{N(C = c_i)} \sum_{k=1}^{N(C=c_i)} a_j(k)$$

for the expected value, where $a_j(k)$ is the value of the attribute A_j in the k-th sample case belonging to class c_i, and for the variance

$$\hat{\sigma}_j^2(c_i) = \frac{1}{N(C = c_i)} \sum_{k=1}^{N(C=c_i)} \left(a_j(k) - \hat{\mu}_j(c_i)\right)^2.$$

6.1.2 Relation to Bayesian Networks

As already mentioned above, a naive Bayes classifier can be seen as a special Bayesian network. This becomes immediately clear if we write the basic formula of a naive Bayes classifier as

$$\begin{aligned} P(C = c_i, A_1 = a_1, \ldots, A_n = a_n) &= P(C = c_i \mid A_1 = a_1, \ldots, A_n = a_n) \cdot p_0 \\ &= P(C = c_i) \cdot \prod_{j=1}^{n} f(A_j = a_j \mid C = c_i), \end{aligned}$$

which results from a simple multiplication by p_0. Obviously, this is the factorization formula of a Bayesian network with a star-like structure as shown on the left in Figure 6.1. That is, in this Bayesian network there is a distinguished attribute, namely the class attribute. It is the only unconditioned attribute (the only one without parents). All other attributes are conditioned on the class attribute and on the class attribute only. It is easy to verify that evidence propagation in this Bayesian network (cf. Section 4.2.1), if all attributes A_j are instantiated, coincides with the computation of the conditional class probabilities of a naive Bayes classifier.

Seeing a naive Bayes classifier as a special Bayesian network has the advantage that the strong independence assumptions underlying the derivation of its basic formula can be mitigated. If there are attributes that are conditionally dependent given the class, we may add edges between these attributes to capture this dependence (as indicated on the right in Figure 6.1).

Formally, this corresponds to not canceling *all* attributes A_k, after the chain rule of probability has been applied to attain

$$\begin{aligned} &P(C = c_i \mid A_1 = a_1, \ldots, A_n = a_n) \\ = \; &\frac{P(C = c_i)}{p_0} \cdot \prod_{j=1}^{n} f\Big(A_j = a_j \,\Big|\, \textstyle\bigwedge_{k=1}^{j-1} A_k = a_k, C = c_i\Big), \end{aligned}$$

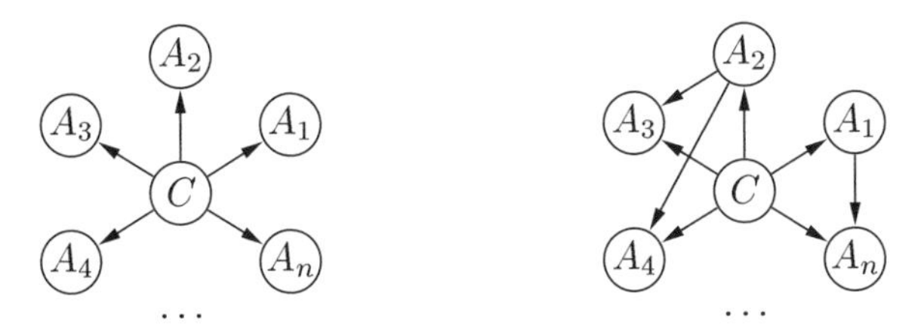

Figure 6.1 A naive Bayes classifier is a Bayesian network with a star-like structure. Additional edges can mitigate the independence assumptions.

but only those of which the attribute A_j is conditionally independent given the class and the remaining attributes. That is, we exploit a weaker set of conditional independence statements. Naive Bayes classifiers that had been improved in this way where used by [Geiger 1992], [Friedman and Goldszmidt 1996], and [Sahami 1996]. The former two restrict the additional edges to a tree, the latter studies a generalization in which each descriptive attribute has at most k parents (where k is a parameter of the method). However, these approaches bring us already into the realm of structure learning, because the additional edges have to be selected somehow (cf. Section 7.3.4).

6.1.3 A Simple Example

As an illustrative example of a naive Bayes classifier we consider the well-known iris data [Anderson 1935, Fisher 1936, Murphy and Aha 1994]. The classification problem is to predict the iris type (iris setosa, iris versicolor, or iris virginica) from measurements of the sepal length and width and the petal length and width. However, we confine ourselves to the latter two measures, which are the most informative w.r.t. a prediction of the iris type. (In addition, we cannot visualize a four-dimensional space.) The naive Bayes classifier induced from these two measures and all 150 cases (50 cases of each iris type) is shown in Table 6.1. The conditional probability density functions used to predict the iris type are shown graphically in Figure 6.2 on the left. The ellipses are the 2σ-boundaries of the (bivariate) normal distribution. These ellipses are axis-parallel, which is a consequence of the strong conditional independence assumptions underlying a naive Bayes classifier: The normal distributions are estimated separately for each dimension and no covariance is taken into account. However, even a superficial glance at the data points reveals that the two measures are far from independent given the iris type. Especially for iris versicolor the density function is a rather bad estimate. Nevertheless, the naive Bayes classifier is successful: It misclassifies only six cases (which can easily be made out in Figure 6.2: They are among the versicolor and virginica cases close to the intersection of the upper two ellipses).

Table 6.1 A naive Bayes classifier for the iris data. The normal distributions are described by $\hat{\mu} \pm \hat{\sigma}$ (i.e. expected value $\pm$ standard deviation).

iris type	iris setosa	iris versicolor	iris virginica
prior probability	0.333	0.333	0.333
petal length	1.46 ± 0.17	4.26 ± 0.46	5.55 ± 0.55
petal width	0.24 ± 0.11	1.33 ± 0.20	2.03 ± 0.27

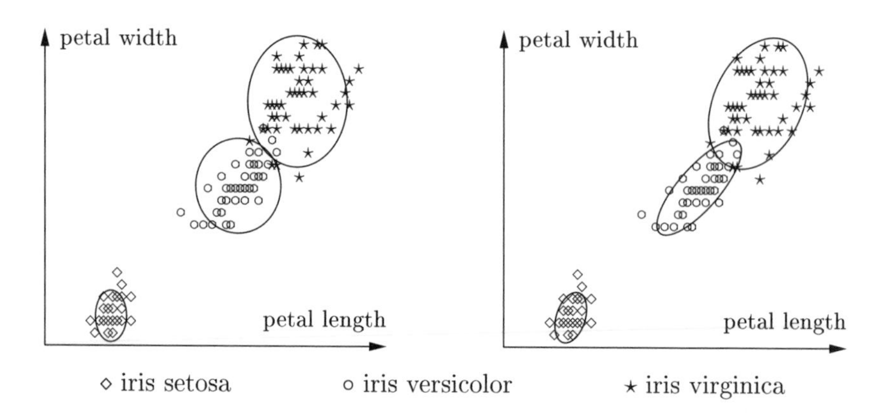

Figure 6.2 Naive Bayes density functions for the iris data (axis-parallel ellipses, left) and density functions that take into account the covariance of the two measures (general ellipses, right). The ellipses are the 2σ-boundaries.

However, if we allow for an additional edge between petal length and petal width, which is most easily implemented by estimating the covariance matrix of the two measures, a much better fit to the data can be achieved (see Figure 6.2 on the right, again the ellipses are the 2σ-boundaries of the probability density function). As a consequence the number of misclassifications drops from six to three (which can easily be made out in Figure 6.2).

6.2 A Naive Possibilistic Classifier

Due to the structural equivalence of probabilistic and possibilistic networks, which we pointed out in the Chapters 3 and 4, the idea suggests itself to construct a naive possibilistic classifier in strict analogy to the probabilistic case [Borgelt and Gebhardt 1999]: Let π be a possibility distribution over the attributes $A_1, \ldots, A_n$ and C. Because of the symmetry in the definition of a conditional degree of possibility (cf. Definition 2.4.9 on page 47), we have

$$\pi(C = c_i \mid A_1 = a_1, \ldots, A_n = a_n) \;=\; \pi(A_1 = a_1, \ldots, A_n = a_n \mid C = c_i).$$

This equation takes the place of Bayes' rule. It has the advantage of being much simpler than Bayes' rule and thus we need no normalization constant.

In the next step we apply the possibilistic analog of the chain rule of probability (cf. page 87 in Section 3.4.4) to obtain

$$\begin{aligned}
&\pi(C = c_i \mid A_1 = a_1, \ldots, A_n = a_n) \\
&= \min\nolimits_{j=1}^{n} \pi\Big(A_j = a_j \;\Big|\; \bigwedge\nolimits_{k=1}^{j-1} A_k = a_k, C = c_i\Big).
\end{aligned}$$

Finally we assume, in analogy to the probabilistic conditional independence assumptions, that given the value of the class attribute all other attributes are independent. With this assumption we arrive at

$$\pi(C = c_i \mid A_1 = a_1, \ldots, A_n = a_n) = \min\nolimits_{j=1}^{n} \pi(A_j = a_j \mid C = c_i).$$

This is the fundamental equation underlying a naive possibilistic classifier. Given an instantiation $(a_1, \ldots, a_n)$ it predicts the class c_i for which this equation yields the highest conditional degree of possibility. It is obvious that, as a naive Bayes classifier is a special Bayesian network, this possibilistic classifier is a special possibilistic network and, as a naive Bayes classifier, it has a star-like structure (cf. the left part of Figure 6.1). It is also clear that a naive possibilistic classifier may be improved in the same way as a naive Bayes classifier by adding edges between attributes that are conditionally dependent given the class (cf. the right part of Figure 6.1).

To induce a possibilistic classifier from data, we must estimate the conditional possibility distributions of the above equation. To do so, we can rely on the database preprocessing described in Section 5.4, by exploiting

$$\pi(A_j = a_j \mid C = c_i) \;=\; \pi(A_j = a_j, C = c_i).$$

However, this method works only for possibility distributions over attributes with a finite domain. If there are numeric attributes, it is not even completely clear how to define the joint possibility distribution that is induced by a database of sample cases. The main problem is that it is difficult to determine a possibility distribution on a continuous domain if some sample cases have precise values for the attribute under consideration (strangely enough, for possibilistic approaches *precision* can pose a problem). A simple solution would be to fix the size of a small interval to be used in such cases. However, such considerations are beyond the scope of this book and therefore, in the possibilistic case, we confine ourselves to attributes with finite domains.

6.3 Classifier Simplification

Both a naive Bayes classifier and a naive possibilistic classifier make strong independence assumptions. It is not surprising that these assumptions are likely to fail. If they fail—and they are the more likely to fail, the more attributes there are—the classifier may be worse than necessary. To cope with this problem, one may try to simplify the classifiers, naive Bayes as well as possibilistic, using a simple greedy attribute selection. With this procedure it can be hoped that a subset of attributes is found for which the strong assumptions hold at least approximately. The experimental results reported below indicate that this approach seems to be successful.

The attribute selection methods we used are the following: In the first method we start with a classifier that simply predicts the majority class. That is, we start with a classifier that does not use any attribute information. Then we add attributes one by one. In each step we select the attribute which, if added, leads to the smallest number of misclassifications on the training data (breaking ties arbitrarily). We stop adding attributes when adding any of the remaining attributes does not reduce the number of errors. This greedy approach to simplification was also used in [Langley and Sage 1994].

The second method is a reversal of the first. We start with a classifier that takes into account all available attributes and then we remove attributes step by step. In each step we select the attribute which, if removed, leads to the smallest number of misclassifications on the training data (breaking ties arbitrarily). We stop removing attributes when removing any of the remaining attributes leads to a higher number of misclassifications.

6.4 Experimental Results

We implemented the suggested possibilistic classifier along with a normal naive Bayes classifier [Borgelt 1999, Borgelt and Timm 2000] and tested both on four datasets from the UCI machine learning repository [Murphy and Aha 1994]. In both cases we used the simplification procedures described in the preceding section. The results are shown in Table 6.2, together with the results obtained with a decision tree classifier. The columns "add att." contain the results obtained by stepwise adding attributes, the columns "rem. att." the results obtained by removing attributes. The decision tree classifier[4] is similar to the well-known decision tree induction program C4.5 [Quinlan 1993]. The attribute selection measure used for these experiments was information gain ratio and the pruning method was confidence level pruning with a confidence level of 50% (these are the default values also used in C4.5).

[4]This decision tree classifier, which was implemented by the first author of this book, was already mentioned in a footnote in the introduction (cf. page 9, see also [Borgelt 1998, Borgelt and Timm 2000]).

Table 6.2 Experimental results on four datasets from the UCI machine learning repository.

dataset			possibilistic classifier		naive Bayes classifier		decision tree	
		tuples	add. att.	rem. att.	add. att.	rem. att.	unpruned	pruned
audio	train	113	0(0.0%)	1(0.9%)	0(0.0%)	0(0.0%)	12(10.6%)	13(11.5%)
	test	113	27(23.9%)	34(30.1%)	28(24.8%)	32(28.3%)	30(26.5%)	28(24.8%)
69 atts.	selected		14	17	14	63	12	10
horse	train	184	62(33.7%)	63(34.2%)	45(24.5%)	41(22.3%)	30(16.3%)	38(20.7%)
	test	184	63(34.2%)	69(37.5%)	64(34.8%)	65(35.3%)	66(35.9%)	59(32.1%)
20 atts.	selected		6	10	4	18	14	10
soybean	train	342	18(5.3%)	20(5.8%)	13(3.8%)	12(3.5%)	16(4.7%)	22(6.4%)
	test	341	59(17.3%)	57(16.7%)	45(13.2%)	45(13.2%)	48(14.1%)	42(12.3%)
36 atts.	selected		15	17	14	14	20	18
vote	train	300	9(3.0%)	8(2.7%)	9(3.0%)	9(2.7%)	6(2.0%)	7(2.3%)
	test	135	11(8.1%)	10(7.4%)	11(8.1%)	11(8.1%)	11(8.1%)	8(5.9%)
16 atts.	selected		2	3	2	2	6	4

It can be seen that the possibilistic classifier performs equally well as or only slightly worse than the naive Bayes classifier. This is encouraging, since none of the datasets is well suited to demonstrate the strengths of a possibilistic approach. Although all of the datasets contain missing values (which can be seen as imprecise information; cf. Section 5.4.4), the relative frequency of these missing values is rather low. None of the datasets contains true set valued information, which to treat possibility theory was designed.

Chapter 7

Learning Global Structure

In this chapter we study methods to learn the global structure of a graphical model from data. By *global structure* we mean the structure of the graph underlying the model. As discussed in Section 4.1.6 this graph indicates which conditional or marginal distributions constitute the represented decomposition. In contrast to this the term *local structure* refers to regularities in the individual conditional or marginal distributions of the decomposition, which cannot be exploited to achieve a finer-grained decomposition, but may at least be used to simplify the structures in which the distributions are stored. Learning local structure, although restricted to directed graphs (i.e. conditional distributions), is studied in the next chapter.

In analogy to Chapter 3 we introduce in Section 7.1 the general principles of learning the global structure of a graphical model based on some simple examples (actually the same examples as in Chapter 3 in order to achieve a coherent exposition), which are intended to provide an intuitive and comprehensible background. We discuss these examples in the same order as in Chapter 3, i.e. we start with the relational case and proceed with the probabilistic and the possibilistic case. The rationale is, again, to emphasize with the relational case the simplicity of the ideas, which can be disguised by the numbers in the probabilistic or the possibilistic case.

Although three major approaches can be distinguished, the structure of all learning algorithms for the global structure of a graphical model is very similar. Usually they consist of an *evaluation measure* (also called *scoring function*), by which a candidate model is assessed, and a (heuristic) *search method*, which determines the candidate models to be inspected. Often the search is guided by the value of the evaluation measure, but nevertheless the two components are relatively independent and therefore we discuss them in two separate sections, namely 7.2 and 7.3. Only the former distinguishes between the three cases (relational, probabilistic, and possibilistic), since the search methods are independent of the underlying calculus.

7.1 Principles of Learning Global Structure

As already indicated above, there are three main approaches to learn the global structure of a graphical model:

- Test whether a distribution is decomposable w.r.t. a given graph.

  This is the most direct approach. It is not bound to a graphical representation, but can also be carried out w.r.t. other representations of the set of subspaces to be used to compute the (candidate) decomposition of the given distribution.

- Find an independence map by conditional independence tests.

  This approach exploits the theorems of Section 4.1.6, which connect conditional independence graphs and graphs that represent decompositions. It has the advantage that by a single conditional independence test, if it fails, several candidate graphs can be excluded.

- Find a suitable graph by measuring the strength of dependences.

  This is a heuristic, but often highly successful approach, which is based on the frequently valid assumption that in a conditional independence graph an attribute is more strongly dependent on adjacent attributes than on attributes that are not directly connected to it.

Note that none of these methods is perfect. The first approach suffers from the usually huge number of candidate graphs (cf. Section 7.3.1). The second often needs the strong assumption that there is a perfect map w.r.t. the considered distribution. In addition, if it is not restricted to certain types of graphs (for example, polytrees), one has to test conditional independences of high order (i.e. with a large number of conditioning attributes), which tend to be unreliable unless the amount of data is enormous. The heuristic character of the third approach is obvious. Examples in which it fails can easily be found, since under certain conditions attributes that are not adjacent in a conditional independence graph can exhibit a strong dependence.

Note also that one may argue that the Bayesian approaches to learn Bayesian networks from data are not represented by the above list. However, we include them in the third approach, since in our view they only use a special dependence measure, although the statistical foundations of this measure may be somewhat stronger than those of other measures.

7.1.1 Learning Relational Networks

For the relational case the first approach to learn the global structure of a graphical model from data consists in computing, for the given relation, the intersection of the cylindrical extensions of the projections to the subspaces indicated by the graph to be tested (or, equivalently, the natural join of the projections; cf. Section 3.2.4). The resulting relation is compared to the original one and if it is identical, an appropriate graph has been found.

This is exactly what we did in Section 3.2.2 to illustrate what it means to decompose a relation (cf. Figure 3.5 on page 59). In that section, however, we confined ourself to the graph that actually represents a decomposition (and one other to show that decomposition is not a trivial property; cf. Figure 3.8 on page 61). Of course, when learning the graph structure from data, we do not know which graph is the correct one. We do not even know whether there is a graph (other than the complete one[1]) that represents a decomposition of the relation. Therefore we have to search the space of all possible graphs in order to find out whether there is a suitable graph.

For the simple relation used in Section 3.2.1 (cf. table 3.1 on page 56) this search is illustrated in Figure 7.1. It shows all eight possible undirected graphs over the three attributes A, B, and C together with the intersections of the cylindrical extensions of the projections to the subspaces indicated by the graphs. Clearly, graph 5 is the only graph (apart from the complete graph 8)[2] for which the corresponding relation is identical to the original one and therefore this graph is selected as the search result.

As discussed in Section 3.2.3, the relation R_{ABC} is no longer decomposable w.r.t. a graph other than the complete graph if the tuple (a_4, b_3, c_2) is removed. In such situations one either has to work with the relation as a whole (i.e. the complete graph) or be contented with an approximation that contains some additional tuples. Usually the latter alternative is chosen, since in applications the former is most often impossible because of the high number of dimensions of the world section to be modeled. Often an approximation has to be accepted even if there is an exact decomposition, because the exact decomposition contains one or more very large maximal cliques (i.e. maximal cliques with a large number of attributes), which cannot be handled efficiently. Therefore the problem of decomposing a relation can be stated in the following more general way: Given a relation and a maximal size for the maximal cliques, find an exact decomposition with maximal cliques as small as possible, or, if there is no suitable exact decomposition, find a "good" approximate decomposition of the relation, again with maximal cliques as small as possible.

Obviously, the simplest criterion for what constitutes a "good" approximation is the number of additional tuples in the relation corresponding to a given graph [Dechter 1990]: This relation cannot contain fewer tuples than

[1] Since a complete undirected graph consists of only one maximal clique, there is no decomposition and thus, trivially, a complete graph can represent any relation. This also follows from the fact that a complete graph is a trivial independence map. However, it is clear that a complete graph is useless, since it leads to no simplification.

[2] Note, again, that the complete graph 8 trivially represents the relation, since it consists of only one maximal clique. This graph does *not* represent the intersection of the cylindrical extensions of the projections to all three two-dimensional subspaces (cf. Definition 4.1.21 on page 108). As demonstrated in Section 3.2.3, this intersection need not be identical to the original relation.

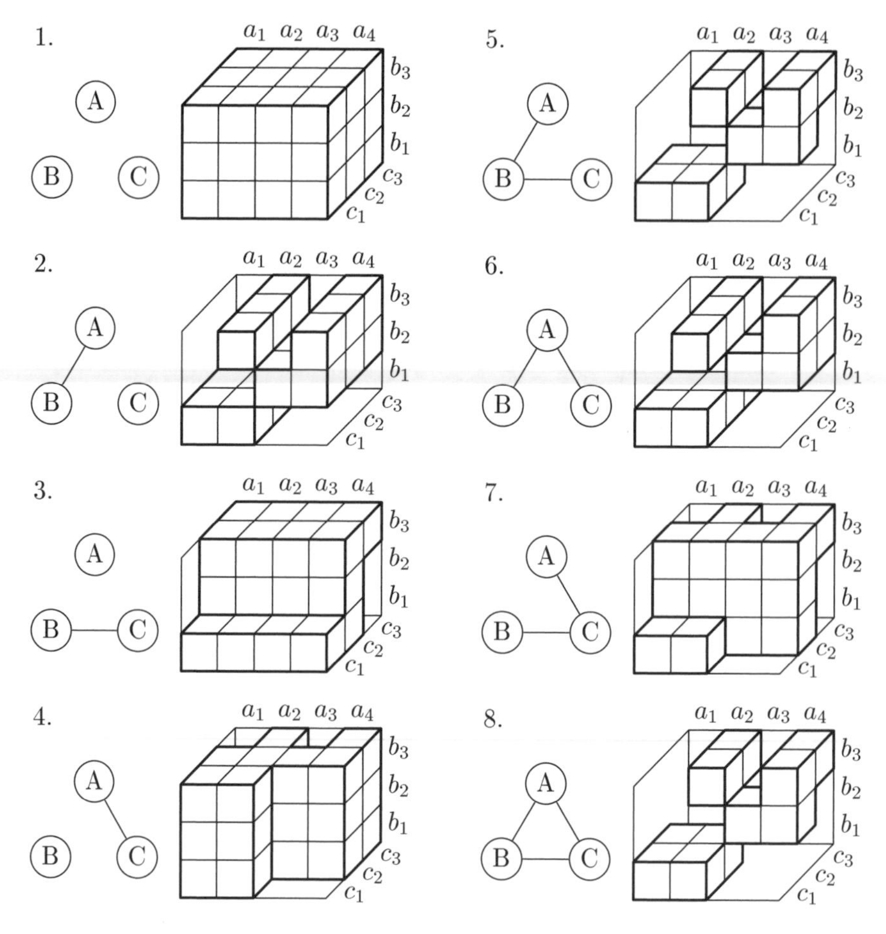

Figure 7.1 All eight possible graphs and the corresponding relations.

the original relation and thus the number of additional tuples is an intuitive measure of the "closeness" of this relation to the original one. Based on this criterion graph 5 of Figure 7.1 would be chosen for the relation R_{ABC} with the tuple (a_4, b_3, c_2) removed, because the relation corresponding to this graph contains only one additional tuple (namely the tuple (a_4, b_3, c_2)), whereas any of the other graphs (except the complete graph 8, of course) contains at least four additional tuples.

Note that in order to rank two graphs we need not compute the number of additional tuples, but can directly compare the number of tuples in the corresponding relations, since the number of tuples of the original relation is, obviously, the same in both cases. This simplifies the assessment: Find the graph among those of acceptable complexity (w.r.t. maximal clique size), for which the corresponding relation is the smallest among all such graphs, i.e. for which it has the least number of tuples.

Let us now turn to the second approach to learn a graphical model from data, namely finding an independence map by conditional independence tests. As already mentioned above, this approach draws on the theorems of Section 4.1.6 which state that a distribution is decomposable w.r.t. its conditional independence graph (although for possibility distributions and thus for relations one has to confine oneself to a restricted subset of all graphs if undirected graphs are considered). For the simple relational example this approach leads, as the approach discussed above, to a selection of graph 5, since the only conditional independence that can be read from it, namely $A \perp\!\!\!\perp C \mid B$, holds in the relation R_{ABC} (as demonstrated in Section 3.2.6) and therefore this graph is an independence map. All other graphs, however, are no independence maps (expect the complete graph 8, which is a trivial independence map). For instance, graph 3 indicates that A and B are marginally independent, but clearly they are not, since, for example, the values a_1 and b_3 are both possible, but cannot occur together. Note that the fact that A and B are not marginally independent also excludes, without further testing, graphs 1 and 4. That such a transfer of results to other graphs is possible is an important advantage of this approach.

To find approximate decompositions with the conditional independence graph approach we need a measure for the strength of (conditional) dependences. With such a measure we may decide to treat two attributes (or sets of attributes) as conditionally independent if the measure indicates only a weak dependence. This enables us to choose a sparser graph. It should be noted, though, that in this case the approximation is decided on "locally" (i.e. w.r.t. a conditional independence that may involve only a subset of all attributes) and thus need not be the "globally" best approximation.

In order to find a measure for the strength of the relational dependence of a set of attributes, recall that two attributes are relationally independent if their values are freely combinable (cf. Definition 3.2.10 on page 70) and that they are dependent if there is at least one combination of values which cannot occur (although both values of this combination can occur in combination with at least one other value). Therefore it is plausible to measure the strength of the relational dependence of two attributes by the number of possible value combinations: The fewer there are, the more strongly dependent the two attributes are.

Of course, we should take into account the number of value combinations that *could* be possible, i.e. the size of the (sub)space scaffolded by the two attributes. Otherwise the measure would tend to assess that attributes with only few values are more strongly dependent than attributes with many values. Therefore we define as a measure of the relational dependence of a set of attributes the *relative* number of possible value combinations, i.e. the quotient of the number of possible value combinations and the size of the subspace

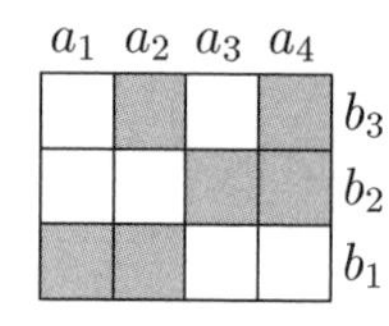

Hartley information needed to determine

coordinates:	$\log_2 4 + \log_2 3 = \log_2 12$	≈ 3.58
coordinate pair:	$\log_2 6$	≈ 2.58
gain:	$\log_2 12 - \log_2 6 = \log_2 2 = 1$	

Figure 7.2 Computation of Hartley information gain.

scaffolded by the attributes. Clearly, the value of this measure is one if and only if the attributes are independent, and it is the smaller, the more strongly dependent the attributes are.

This measure is closely related to the *Hartley information gain*, which is based on the *Hartley entropy* or *Hartley information* [Hartley 1928] of a set of alternatives (for example, the set of values of an attribute).

Definition 7.1.1 *Let S be a finite set of alternatives. The* **Hartley entropy** *or* **Hartley information** *of S, denoted $H^{(\text{Hartley})}(S)$, is the binary logarithm of the number of alternatives in S, i.e. $H^{(\text{Hartley})}(S) = \log_2 |S|$.*

The idea underlying this measure is the following: Suppose there is an oracle that knows the "correct" or "obtaining" alternative, but which accepts only questions that can be answered with "yes" or "no". How many questions do we have to ask? If we proceed in the following manner, we have to ask at most $\lceil \log_2 n \rceil$ questions, where n is the number of alternatives: We divide the set of alternatives into two subsets of about equal size (equal sizes if n is even, sizes differing by one if n is odd). Then we choose arbitrarily one of the sets and ask whether the correct alternative is contained in it. Independent of the answer, one half of the alternatives can be excluded. The process is repeated with the set of alternatives that were not excluded until only one alternative remains. Obviously, the information received from the oracle is at most $\lceil \log_2 n \rceil$ bits: one bit per question. By a more detailed analysis and by averaging over all alternatives, we find that the expected number of questions is about $\log_2 n$.[3]

The *Hartley information gain* is computed from the Hartley information as demonstrated in Figure 7.2. Suppose we want to find out the values assumed by two attributes A and B. To do so, we could determine first the value of attribute A using the question scheme indicated above, and then the value of attribute B. That is, we could determine the "coordinates" of the value combination in the joint domain of A and B. If we apply this method to the example shown in Figure 7.2, we need $\log_2 12 \approx 3.58$ questions on average.

[3]Note that $\log_2 n$ is not a precise expected value for all n. A better justification for using $\log_2 n$ is obtained by considering the problem of determining sequences of correct alternatives (cf. the discussion of the more general *Shannon entropy* or *Shannon information* in Section 7.2.4), where we will elaborate this view in more detail.

Table 7.1 The number of possible combinations relative to the size of the subspace and the Hartley information gain for the three attribute pairs.

attributes	relative number of possible value combinations	Hartley information gain
A, B	$\frac{6}{3\cdot 4} = \frac{1}{2} = 50\%$	$\log_2 3 + \log_2 4 - \log_2 6 = 1$
A, C	$\frac{8}{3\cdot 4} = \frac{2}{3} \approx 67\%$	$\log_2 3 + \log_2 4 - \log_2 8 \approx 0.58$
B, C	$\frac{5}{3\cdot 3} = \frac{5}{9} \approx 56\%$	$\log_2 3 + \log_2 3 - \log_2 5 \approx 0.85$

However, since not all value combinations are possible, we can save questions by determining the value combination directly, i.e. the "coordinate pair". There are six possible combinations, hence we need $\log_2 6 \approx 2.58$ questions on average and thus gain one question.

Formally Hartley information gain is defined as follows:

Definition 7.1.2 *Let A and B be two attributes and R a binary possibility measure with $\exists a \in \mathrm{dom}(A) : \exists b \in \mathrm{dom}(B) : R(A = a, B = b) = 1$. Then*

$$
\begin{aligned}
I_{\mathrm{gain}}^{(\mathrm{Hartley})}(A, B) &= \log_2 \Big(\sum_{a \in \mathrm{dom}(A)} R(A = a) \Big) + \log_2 \Big(\sum_{b \in \mathrm{dom}(B)} R(B = b) \Big) \\
&\quad - \log_2 \Big(\sum_{a \in \mathrm{dom}(A)} \sum_{b \in \mathrm{dom}(B)} R(A = a, B = b) \Big) \\
&= \log_2 \frac{\Big(\sum_{a \in \mathrm{dom}(A)} R(A = a) \Big) \Big(\sum_{b \in \mathrm{dom}(B)} R(B = b) \Big)}{\sum_{a \in \mathrm{dom}(A)} \sum_{b \in \mathrm{dom}(B)} R(A = a, B = b)},
\end{aligned}
$$

is called the **Hartley information gain** *of A and B w.r.t. R.*

With this definition the connection of Hartley information gain to the relative number of possible value combinations becomes obvious: It is simply the binary logarithm of the reciprocal of this relative number. It is also clear that this definition can easily be extended to more than two attributes by adding factors to the product in the numerator.

Note that Hartley information gain is zero if and only if the relative number of value combinations is one. Therefore the Hartley information gain is zero if and only if the considered attributes are relationally independent. If the attributes are dependent, it is the greater, the more strongly dependent the attributes are. Consequently, the Hartley information gain (or, equivalently, the relative number of possible value combinations) can be used directly to test for (approximate) marginal independence. This is demonstrated

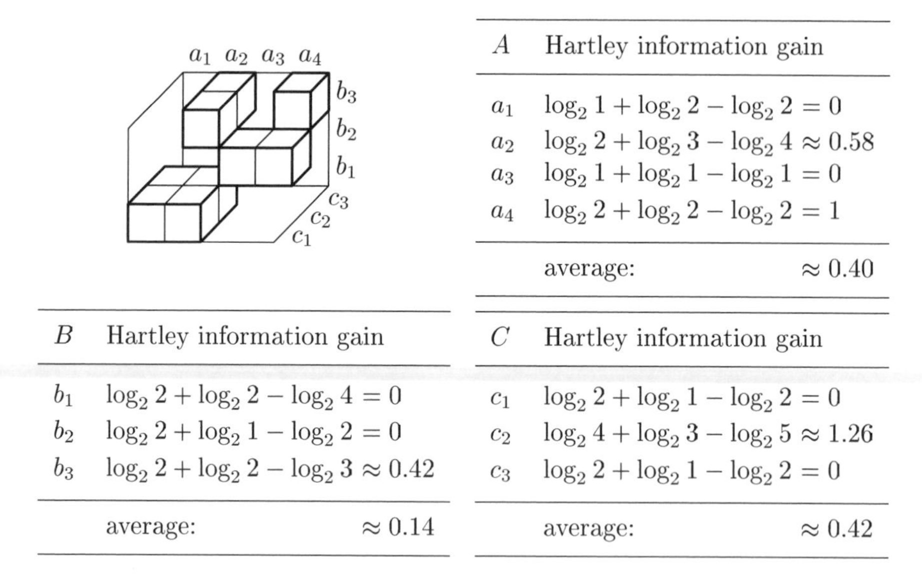

A	Hartley information gain
a_1	$\log_2 1 + \log_2 2 - \log_2 2 = 0$
a_2	$\log_2 2 + \log_2 3 - \log_2 4 \approx 0.58$
a_3	$\log_2 1 + \log_2 1 - \log_2 1 = 0$
a_4	$\log_2 2 + \log_2 2 - \log_2 2 = 1$
	average: ≈ 0.40

B	Hartley information gain
b_1	$\log_2 2 + \log_2 2 - \log_2 4 = 0$
b_2	$\log_2 2 + \log_2 1 - \log_2 2 = 0$
b_3	$\log_2 2 + \log_2 2 - \log_2 3 \approx 0.42$
	average: ≈ 0.14

C	Hartley information gain
c_1	$\log_2 2 + \log_2 1 - \log_2 2 = 0$
c_2	$\log_2 4 + \log_2 3 - \log_2 5 \approx 1.26$
c_3	$\log_2 2 + \log_2 1 - \log_2 2 = 0$
	average: ≈ 0.42

Figure 7.3 Using the Hartley information gain to test for approximate conditional independence.

in Table 7.1, which shows the relative number of possible value combinations and the Hartley information gain for the three attribute pairs of the simple relational example discussed above, namely the relation R_{ABC} (with or without the tuple (a_4, b_3, c_2), since removing this tuple does not change the two-dimensional projections). Clearly no pair of attributes is marginally independent, not even approximately, and hence all graphs with less than two edges (cf. Figure 7.1) can be excluded.

In order to use Hartley information gain (or, equivalently, the relative number of possible value combinations) to test for (approximate) conditional independence, one may proceed as follows: For each possible instantiation of the conditioning attributes the value of this measure is computed. (Note that in this case for the size of the subspace only the values that are possible *given the instantiation of the conditions* have to be considered—in contrast to a test for marginal independence, where usually all values have to be taken into account; cf. Definition 7.1.2.) The results are aggregated over all instantiations of the conditions, for instance, by simply averaging them.

As an example consider the relation R_{ABC} without the tuple (a_4, b_3, c_2), which is shown in the top left of Figure 7.3. This figure also shows the averaged Hartley information gain for B and C given the value of A (top right), for A and C given the value of B (bottom left) and for A and B given the value of C (bottom right). Obviously, since none of these averages is zero, neither $B \perp\!\!\!\perp C \mid A$ nor $A \perp\!\!\!\perp C \mid B$ nor $A \perp\!\!\!\perp C \mid B$ holds. But since A and C exhibit a rather weak conditional dependence given B, we may decide to treat them as conditionally independent and thus may choose graph 5 of Figure 7.1 to approximate the relation.

The notion of a measure for the strength of the dependence of two attributes brings us directly to the third method to learn a graphical model from data, namely to determine a suitable graph by measuring only the strength of (marginal) dependences. That such an approach is feasible can be made plausible as follows: Suppose that we choose the number of additional tuples in the intersection of the cylindrical extensions of the projections of a relation to the selected subspaces as a measure of the overall quality of an (approximate) decomposition (see above). In this case, to find a "good" approximate decomposition it is plausible to choose the subspaces in such a way that the number of possible value combinations in the cylindrical extensions of the projections to these subspaces is as small as possible, because the smaller the cylindrical extensions, the smaller their intersection. Obviously, the number of tuples in a cylindrical extension depends directly on the (relative) number of possible value combinations in the projection. Therefore it seems to be a good heuristic method to select projections in which the ratio of the number of possible value combinations to the size of the subspace is small (or, equivalently, for which the Hartley information gain—generalized to more than two attributes, if necessary—is large).

Another way to justify this approach is the following: If two attributes are conditionally independent given a third, then their marginal dependence is "mediated" through other attributes (not necessarily in a causal sense). But mediation usually weakens a dependence (cf. Section 7.3.4, where this is discussed in detail for the probabilistic case w.r.t. to trees). Therefore attributes that are conditionally independent given some (set of) attribute(s) are often less strongly dependent than attributes that are conditionally dependent given any (set of) other attribute(s). Consequently, it seems to be a good heuristic method to choose those subspaces, for which the underlying sets of attributes are as strongly dependent as possible.

It is clear that with this approach the search method is especially important, since it determines which graphs are considered. However, for the simple relational example, i.e. the relation R_{ABC} (with or without the tuple (a_4, b_3, c_2)), we may simply choose to construct an optimum weight spanning tree using, for example, the well-known Kruskal algorithm [Kruskal 1956]. As edge weights we may choose the relative number of possible value combinations (and determine a minimum weight spanning tree) or the Hartley information gain (and determine a maximum weight spanning tree). These weights are shown in Table 7.1. In this way graph 5 of Figure 7.1 is constructed, in agreement with the results of the other two learning methods.

Note that the Kruskal algorithm always yields a spanning tree. Therefore this search method excludes the graphs 1 to 4 of Figure 7.1. However, with a slight modification of the algorithm these graphs can also be reached (al-

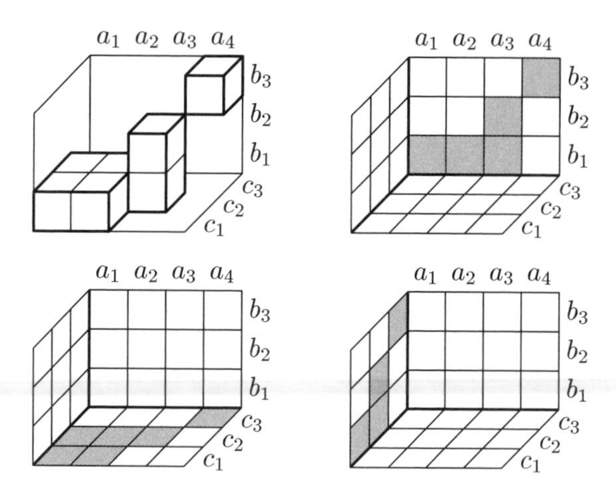

Figure 7.4 The Hartley information gain is only a heuristic criterion.

though this is of no importance for the example discussed). As is well known, the Kruskal algorithm proceeds by arranging the edges in the order of descending or ascending edge weights (depending on whether a maximum or a minimum weight spanning tree is desired) and then adds edges in this order, skipping those edges that would introduce a cycle. If we fix a lower (upper) bound for the edge weight and terminate the algorithm if all remaining edges have a weight less than (greater than) this bound, the algorithm may stop with a graph having fewer edges than a spanning tree.

For the discussed example the third method is obviously the most efficient. It involves the fewest number of computations, because one only has to measure the strength of the marginal dependences, whereas the conditional independence method also needs to assess the strength of conditional dependences. Furthermore, we need not search all graphs, but can *construct* a graph with the Kruskal algorithm (although this can be seen as a special kind of—very restricted—search). However, this efficiency is bought at a price. With this method we may not find a graph that represents an exact decomposition, although there is one. This is demonstrated with the simple relational example shown in Figure 7.4. It is easy to verify that the relation shown in the top left of this figure can be decomposed into the projections to the subspaces $\{A, B\}$ and $\{A, C\}$, i.e. w.r.t. the graph

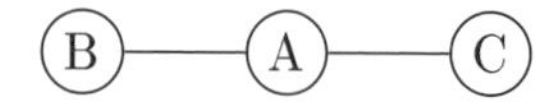

No other selection of subspaces yields a decomposition. However, if an optimum spanning tree is constructed with Hartley information gain providing the edge weights, the subspaces $\{A, B\}$ and $\{B, C\}$ are selected, i.e. the graph

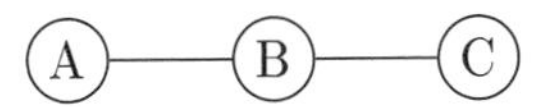

is constructed. Since the corresponding relation contains an additional tuple, namely the tuple (a_3, b_1, c_1), a suboptimal graph is chosen. But since the approximation of the relation is rather good, one may decide to accept this drawback of this learning method.

7.1.2 Learning Probabilistic Networks

Learning probabilistic networks is basically the same as learning relational networks. The main differences are in the criteria used to assess the quality of an approximate decomposition and in the measures used to assess the strength of the dependence of two or more attributes.

For the first method, i.e. the direct test for decomposability w.r.t. a given graph, suppose that we are given a multidimensional probability distribution, for which we desire to find an exact decomposition. In this case we can proceed in the same manner as in the relational setting, i.e. we compute the distribution that is represented by the graph and compare it to the given distribution. If it is identical, a suitable graph has been found. To compute the distribution represented by the graph we proceed as follows: For a directed graph we determine from the given distribution the marginal and conditional distributions indicated by the graph—their product is the probability distribution represented by the graph. For an undirected graph we compute functions on the maximal cliques as indicated in the proof of Theorem 4.1.22 (cf. Section A.4 in the appendix). Again their product is the probability distribution represented by the graph. For the simple example discussed in Section 3.3 (Figure 3.11 on page 75) this approach leads to a selection of graph 5 of Figure 7.1 on page 164. We refrain from illustrating the search in detail, because the figures would consume a lot of space, but are not very instructive.

Of course, as in the relational case, there need not be an exact decomposition of a given multidimensional probability distribution. In such a case, or if the maximal cliques of a graph that represents an exact decomposition are too large, we may decide to be contented with an approximation. A standard measure for the quality of a given approximation of a probability distribution, which corresponds to the number of additional tuples in the relational case, is the *Kullback–Leibler information divergence* [Kullback and Leibler 1951, Chow and Liu 1968, Whittaker 1990]. This measure is defined as follows:

Definition 7.1.3 *Let p_1 and p_2 be two strictly positive probability distributions on the same set $\mathcal{E}$ of events. Then*

$$I_{\text{KLdiv}}(p_1, p_2) = \sum_{E \in \mathcal{E}} p_1(E) \log_2 \frac{p_1(E)}{p_2(E)}$$

is called the **Kullback–Leibler information divergence** *of p_1 and p_2.*

It can be shown that the Kullback–Leibler information divergence is non-negative and that it is zero only if $p_1 \equiv p_2$ (cf. the proof of Lemma A.15.1 in Section A.15 in the appendix). Therefore it is plausible that this measure can be used to assess the approximation of a given multidimensional distribution and the distribution that is represented by a given graph: The smaller the value of this measure, the better the approximation.

Note that this measure does not treat the two distributions equally, but that it uses the values of the distribution p_1 as weights in the sum. Therefore this measure is not symmetric in general, i.e. in general it is $I_{\text{KLdiv}}(p_1, p_2) \neq I_{\text{KLdiv}}(p_2, p_1)$. Consequently, one has to decide which of p_1 and p_2 should be the actual distribution and which the approximation. We may argue as follows: The quotient of the probabilities captures the difference of the distributions w.r.t. single events E. The influence of such a "local" difference depends on the probability of the occurrence of the corresponding event E and therefore it should be weighted with the (actual) probability of this event. Hence p_1 should be the actual distribution.

As an illustration of this approach Figure 7.5 shows, for the simple three-dimensional probability distribution depicted in Figure 3.11 on page 75, all eight candidate graphs together with the Kullback–Leibler information divergence of the original distribution and its approximation by each of the graphs (upper numbers). This figure is the probabilistic counterpart of Figure 7.1. Clearly, graph 5 represents an exact decomposition of the distribution, since the Kullback–Leibler information divergence is zero for this graph. Note that, as in the relational case, the complete graph 8 always receives or shares the best assessment, since it consists of only one maximal clique and thus represents no real decomposition.

Up to now we have assumed that the multidimensional probability distribution, for which a decomposition is desired, is already given. However, in applications this is usually not the case. Instead we are given a database of sample cases. Clearly, the direct approach to handle this situation, namely to estimate the joint probability distribution from this database, so that we can proceed as indicated above, is, in general, infeasible. The reason is that the available data is rarely sufficient to make the estimation of the joint probabilities reliable, simply because the number of dimensions is usually large and thus the number of probabilities that have to be estimated is enormous (cf., for example, the Danish Jersey cattle example discussed in Section 4.2.2, which is, in fact, a rather simple application). For a reliable estimation the number of sample cases must be a multiple of this number.

Fortunately, if we confine to a maximum likelihood estimation approach, there is a feasible indirect method: The idea of maximum likelihood estimation is to choose that model or that (set of) probability parameter(s) that makes the observed data most likely. Yet, if we have a model, the probability of the database can easily be computed. Hence the idea suggests itself to estimate from the data only the marginal or conditional distributions that are indicated

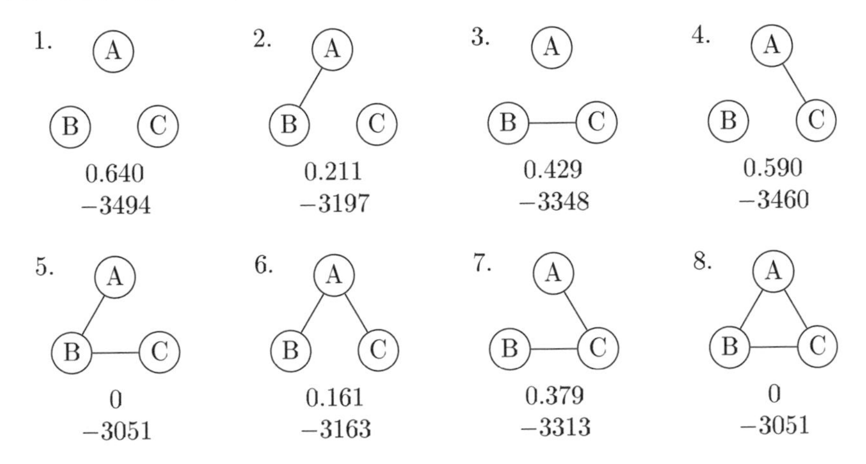

Figure 7.5 The Kullback–Leibler information divergence of the original distribution and its approximation (upper numbers) and the natural logarithms of the probability of an example database (log-likelihood of the data, lower numbers) for the eight possible candidate graphical models.

by a given graph. This is feasible, since these distributions are usually of limited size and therefore even a moderate amount of data is sufficient to estimate them reliably. With these estimated distributions we have a fully specified graphical model, from which we can compute the probability of each sample case (i.e. its *likelihood* given the model). Assuming that the sample cases are independent, we may then compute the probability of the database by simply multiplying the probabilities of the individual sample cases, which provides us with an assessment of the quality of the candidate model.

As an illustration reconsider the three-dimensional probability distribution shown in Figure 3.11 on page 75 and assume that the numbers are the (absolute) frequencies of sample cases with the corresponding value combination in a given database. The natural logarithms of the probabilities of this database given the eight possible graphical models, computed as outlined above, are shown in Figure 7.5 (lower numbers). Obviously, graph 5 would be selected.

Let us now turn to the second approach to learn a graphical model from data, namely finding an independence map by conditional independence tests. If we assume that we are given a probability distribution and that we desire to find an exact decomposition, it is immediately clear that the only difference to the relational case is that we have to test for conditional *probabilistic* independence instead of conditional *relational* independence. Thus it is not surprising that for the example of a probability distribution shown in Figure 3.11 on page 75, graph 5 of Figure 7.5 is selected, since the only conditional independence that can be read from it, namely $A \perp\!\!\!\perp B \mid C$, holds in this distribution (as demonstrated in Section 3.3.1).

	a_1	a_2	a_3	a_4
b_3	40	180	20	160
b_2	12	6	120	102
b_1	168	144	30	18

$I_{\text{mut}}(A, B) = 0.429$

a_1	a_2	a_3	a_4	
88	132	68	112	b_3
53	79	41	67	b_2
79	119	61	101	b_1

	a_1	a_2	a_3	a_4
c_3	50	115	35	100
c_2	82	133	99	146
c_1	88	82	36	34

$I_{\text{mut}}(A, C) = 0.050$

a_1	a_2	a_3	a_4	
66	99	51	84	c_3
101	152	78	129	c_2
53	79	41	67	c_1

	b_1	b_2	b_3
c_3	20	180	200
c_2	40	160	40
c_1	180	120	60

$I_{\text{mut}}(B, C) = 0.211$

b_1	b_2	b_3	
96	184	120	c_3
58	110	72	c_2
86	166	108	c_1

Figure 7.6 Mutual information/cross entropy in the simple example.

To find approximate decompositions we need, as in the relational case, a measure for the strength of dependences. For a strictly positive probability distribution, such a measure can easily be derived from the Kullback–Leibler information divergence by comparing two specific distributions, namely the joint distribution over a set of attributes and the distribution that can be computed from its marginal distributions under the assumption that the attributes are independent. For two attributes this measure can be defined as follows [Kullback and Leibler 1951, Chow and Liu 1968]:

Definition 7.1.4 *Let A and B be two attributes and P a strictly positive probability measure. Then*

$$I_{\text{mut}}(A, B) = \sum_{a \in \text{dom}(A)} \sum_{b \in \text{dom}(B)} P(A = a, B = b) \log_2 \frac{P(A = a, B = b)}{P(A = a)\, P(B = b)},$$

is called the **mutual (Shannon) information** *or the* **(Shannon) cross entropy** *of A and B w.r.t. P.*

This measure is also known from decision tree induction [Quinlan 1986, Quinlan 1993], where it is usually called *(Shannon) information gain.* (This name indicates a close relation to the Hartley information gain.) Mutual information can be interpreted in several different ways, one of which is provided by the derivation from the Kullback–Leibler information divergence. Other interpretations are discussed in Section 7.2.4.

By recalling that the Kullback–Leibler information divergence is zero if and only if the two distributions coincide, we see that mutual information is zero if and only if the two attributes are independent, since their joint distribution is compared to an (assumed) independent distribution. In addition, the value of this measure is the larger, the more the two distributions differ, i.e. the more strongly dependent the attributes are. Therefore this measure can be used directly to test for (approximate) marginal independence.

This is demonstrated in Figure 7.6 for the three-dimensional example discussed above (cf. Figure 3.11 on page 75). On the left the three possible two-dimensional marginal distributions are shown, on the right the corresponding (assumed) independent distributions, to which they are compared by mutual information. Clearly, none of the three pairs of attributes are independent, although the dependence of A and C is rather weak.

In order to use mutual information to test for (approximate) conditional independence, we may proceed in a similar way as with Hartley information gain in the relational case: We compute this measure for each possible instantiation of the conditioning attributes and then sum these values weighted with the probability of the corresponding instantiation. That is, for one conditioning attribute, we may compute

$$\begin{aligned} &I_{\text{mut}}(A, B \mid C) \\ &= \sum_{c \in \text{dom}(C)} P(c) \sum_{a \in \text{dom}(A)} \sum_{b \in \text{dom}(B)} P(a, b \mid c) \log_2 \frac{P(a, b \mid c)}{P(a \mid c)\, P(b \mid c)}, \end{aligned}$$

where $P(c)$ is an abbreviation of $P(C = c)$ etc. With this measure it is easy to detect that, in the example distribution of Figure 3.11 on page 75, the attributes A and C are conditionally independent given the attribute B.

If we are not given a probability distribution (as we assumed up to now for the conditional independence graph approach), but a database of sample cases, it is clear that we have to estimate the conditioned joint distributions of the attributes, for which we want to test whether they are conditionally independent. Of course, this can lead to problems if the order of the tests is large, where the *order* of a conditional independence test is the number of conditioning attributes. If the number of conditioning attributes is large, there may be too few tuples having certain instantiation of these attributes to estimate reliably the conditioned joint distribution of the two test attributes. Actually this is a serious drawback of this method, although there are heuristic approaches to amend it (cf. Section 7.3), like, for instance, fixing an upper bound for the order of the tests to be carried out and assuming that all tests of higher order will fail if all tests with an order up to this bound failed.

Finally, let us consider the third approach to learn a graphical model, which is based on measuring only the strengths of (marginal) dependences. As in the relational case, we may use the same measure as for the (approximate) independence tests. For the simple example discussed above we may apply the Kruskal algorithm with mutual information providing the edge weights (cf. Figure 7.6). This leads to a construction of graph 5 of Figure 7.5, in agreement with the results of the other two learning methods.

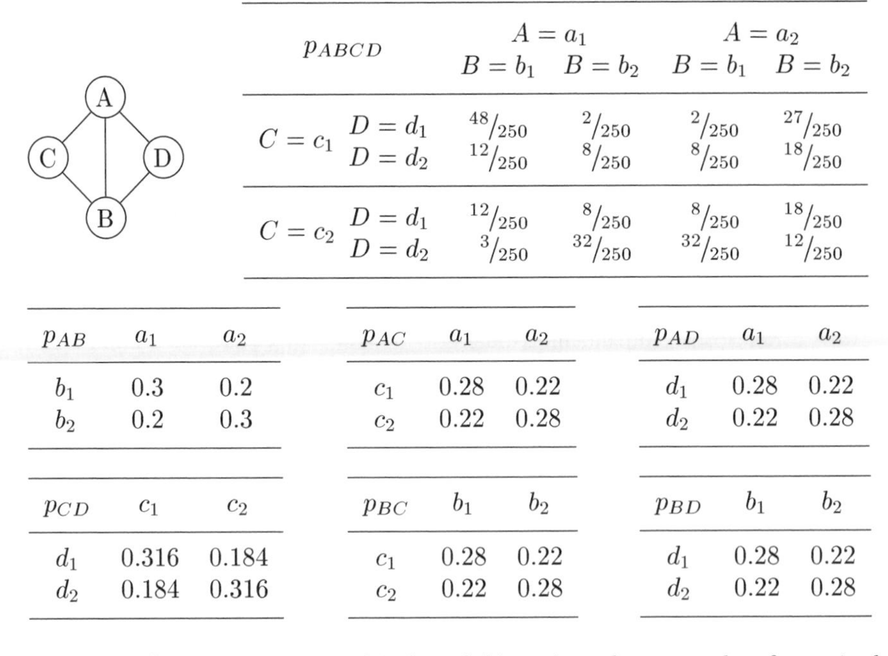

p_{ABCD}		$A = a_1$, $B = b_1$	$A = a_1$, $B = b_2$	$A = a_2$, $B = b_1$	$A = a_2$, $B = b_2$
$C = c_1$	$D = d_1$	$^{48}/_{250}$	$^{2}/_{250}$	$^{2}/_{250}$	$^{27}/_{250}$
	$D = d_2$	$^{12}/_{250}$	$^{8}/_{250}$	$^{8}/_{250}$	$^{18}/_{250}$
$C = c_2$	$D = d_1$	$^{12}/_{250}$	$^{8}/_{250}$	$^{8}/_{250}$	$^{18}/_{250}$
	$D = d_2$	$^{3}/_{250}$	$^{32}/_{250}$	$^{32}/_{250}$	$^{12}/_{250}$

p_{AB}	a_1	a_2
b_1	0.3	0.2
b_2	0.2	0.3

p_{AC}	a_1	a_2
c_1	0.28	0.22
c_2	0.22	0.28

p_{AD}	a_1	a_2
d_1	0.28	0.22
d_2	0.22	0.28

p_{CD}	c_1	c_2
d_1	0.316	0.184
d_2	0.184	0.316

p_{BC}	b_1	b_2
c_1	0.28	0.22
c_2	0.22	0.28

p_{BD}	b_1	b_2
d_1	0.28	0.22
d_2	0.22	0.28

Figure 7.7 Constructing a graphical model based on the strengths of marginal dependences can lead to suboptimal results.

Note that this method can be applied as well if we are given a database of sample cases, because we only have to estimate marginal distributions on (usually small) subspaces, which can be made reliable even with a moderate amount of data. Note also that in the probabilistic setting this method is even better justified than in the relational setting, since it can be shown that if there is a tree representing a decomposition of a given strictly positive probability distribution, then the Kruskal algorithm with either mutual information or the χ^2-measure (cf. Section 7.2.4) providing the edges weights will construct this tree. It can even be shown that with mutual information providing the edge weights, this approach finds the best tree-structured approximation of a given distribution w.r.t. the Kullback–Leibler information divergence [Chow and Liu 1968, Pearl 1988] (cf. Section 7.3.4).

If, however, the conditional independence graph is more complex than a tree, it can still happen that a suboptimal graph gets selected. This is demonstrated with the simple example shown in Figure 7.7. It is easy to check that the graph shown in the top left of this figure is a perfect map of the probability distribution shown in the top right: $C \perp\!\!\!\perp D \mid \{A, B\}$ is the only conditional independence that holds in the distribution. Suppose that we tried to find this conditional independence graph by the following algorithm, which seems to be a plausible approach to learning graphical models with a conditional independence graph that is more complex than a tree: First we construct a maximum weight spanning tree w.r.t. mutual information with the Kruskal algorithm. Then we enhance this skeleton graph with edges where we find that the conditional independences indicated by the graph do not hold in the distribution.[4] Unfortunately, with this approach the only edge that is missing in the perfect map, namely the edge $C - D$, is selected first, as can easily be seen from the marginal distributions.

7.1.3 Learning Possibilistic Networks

Learning possibilistic networks is even more closely related to learning relational networks than learning probabilistic networks. The reason is that the measures used in the relational case can be used directly to construct a corresponding measure for the possibilistic case, since a possibility distribution can be interpreted as a representation of a *set of relations*.

This interpretation is based on the notion of an α-cut—a notion that is transferred from the theory of fuzzy sets [Kruse *et al.* 1994].

Definition 7.1.5 *Let Π be a possibility measure on a sample space Ω. The* **α-cut** *of Π, written $[\Pi]_\alpha$, is the binary possibility measure*

$$[\Pi]_\alpha : 2^\Omega \to \{0,1\}, \qquad E \mapsto \begin{cases} 1, & \text{if } \Pi(E) \geq \alpha, \\ 0, & \text{otherwise.} \end{cases}$$

Of course, this definition can easily be adapted to possibility *distributions*, the α-cut of which is, obviously, a *relation*. As an example consider the three-dimensional possibility distribution discussed in Section 3.4 (cf. Figure 3.14 on page 84). For $0.04 < \alpha \leq 0.06$ the α-cut of this possibility distribution coincides with the relation studied in Section 3.2 (cf. Table 3.1 on page 56 and Figure 3.2 on page 57). If α is increased, tuples are removed from the α-cut (for example the tuple (a_2, b_3, c_3) together with four others as the value of α exceeds 0.06). If α is decreased, tuples are added (for example the tuple (a_1, b_3, c_2) together with three others as α falls below 0.04).

The notion of an α-cut is convenient, because it is obviously preserved by projection, i.e. by computing marginal possibility distributions. Whether we compute a marginal distribution and then determine an α-cut of the resulting marginal distribution, or whether we compute the α-cut first and then project the resulting relation does not matter: The result is always the same. The reason is, of course, that the projection operation used in the possibilistic case is the same as in the possibility-based formalization of the relational case (cf. Sections 3.2.5 and 3.4.1). Therefore we can treat the possibilistic case by

[4]This approach is inspired by an algorithm by [Rebane and Pearl 1987] for learning polytree-structured directed conditional independence graphs, which first constructs an undirected skeleton and then directs the edges (cf. Section 7.3.4).

drawing on the results for the relational case: We simply consider each α-cut in turn, which behaves exactly like a relation, and then we integrate over all values of α.

With this general paradigm in mind let us consider the first approach to learn a graphical model, i.e. the direct test for decomposability w.r.t. a given graph. In order to find an exact decomposition of a given multidimensional possibility distribution we have to find a graph w.r.t. which all α-cuts of the distribution are decomposable. Of course, this test may also be carried out by computing the possibility distribution that is represented by a given graph and by comparing it to the original distribution. If the two are identical, a suitable graph has been found.

However, the fact that we can test for possibilistic decomposability by testing for the relational decomposability of each α-cut provides an idea how to assess the quality of an approximate decomposition. In the relational case the quality of an approximate decomposition may be assessed by counting the number of additional tuples in the approximation (cf. Section 7.1.1). In the possibilistic case we may do the same for each α-cut and then we integrate over all values of α. This leads to the following measure for the "closeness" of an approximate decomposition to the original distribution:

$$\mathrm{diff}(\pi_1, \pi_2) = \int_0^1 \Big(\sum_{E \in \mathcal{E}} [\pi_2]_\alpha(E) - \sum_{E \in \mathcal{E}} [\pi_1]_\alpha(E) \Big) \, \mathrm{d}\alpha,$$

where π_1 is the original distribution, π_2 the approximation, and $\mathcal{E}$ their domain of definition. Obviously, this measure is zero if the two distributions coincide, and it is the larger, the more they differ.

It should be noted that the above measure presupposes that $\forall \alpha \in [0,1] : \forall E \in \mathcal{E} : [\pi_2]_\alpha(E) \geq [\pi_1]_\alpha(E)$, since otherwise the difference in the number of tuples would not have any significance (two relations can be disjoint and nevertheless have the same number of tuples). Therefore this measure cannot be used to compare arbitrary possibility distributions. However, for possibility distributions π_2 that are computed from approximate decompositions we know that $\forall E \in \mathcal{E} : \pi_2(E) \geq \pi_1(E)$ and this obviously implies $\forall \alpha \in [0,1] : \forall E \in \mathcal{E} : [\pi_2]_\alpha(E) \geq [\pi_1]_\alpha(E)$.

An alternative measure, which is very closely related to the above, can be derived from the notion of the *nonspecificity* of a possibility distribution [Klir and Mariano 1987], which is defined as follows:

Definition 7.1.6 *Let π be a (multidimensional) possibility distribution on a set $\mathcal{E}$ of events. Then*

$$\mathrm{nonspec}(\pi) = \int_0^{\sup_{E \in \mathcal{E}} \pi(E)} \log_2 \Big(\sum_{E \in \mathcal{E}} [\pi]_\alpha(E) \Big) \, \mathrm{d}\alpha$$

is called the **nonspecificity** *of the possibility distribution π.*

Recalling the paradigm that a possibility distribution can be seen as a set of relations—one for each value of α—this measure can easily be justified as a generalization of Hartley information (cf. Definition 7.1.1 on page 166) to the possibilistic setting [Higashi and Klir 1982, Klir and Folger 1988]. The nonspecificity of a possibility distribution π reflects the expected amount of information (measured in bits) that has to be added in order to identify the correct tuple within the relations $[\pi]_\alpha$ of alternatives, if we assume a uniform probability distribution on the set $[0, \sup_{E\in\mathcal{E}} \pi(E)]$ of possibility levels α [Gebhardt and Kruse 1996b].

Distributions π_2 that are computed from approximate decompositions obviously satisfy $\sup_{E\in\mathcal{E}} \pi_2(E) = \sup_{E\in\mathcal{E}} \pi_1(E)$, where π_1 is the original distribution. Consequently, we may construct from the nonspecificity measure the following measure for the "closeness" of an approximate decomposition to a given possibility distribution [Gebhardt and Kruse 1996b]:

Definition 7.1.7 *Let π_1 and π_2 be two possibility distributions on the same set $\mathcal{E}$ of events with $\forall E \in \mathcal{E} : \pi_2(E) \geq \pi_1(E)$. Then*

$$S_{\text{div}}(\pi_1, \pi_2) = \int_0^{\sup_{E\in\mathcal{E}} \pi_1(E)} \log_2\Big(\sum_{E\in\mathcal{E}} [\pi_2]_\alpha(E)\Big) - \log_2\Big(\sum_{E\in\mathcal{E}} [\pi_1]_\alpha(E)\Big)\, \mathrm{d}\alpha$$

is called the **specificity divergence** *of π_1 and π_2.*

As the measure discussed above, this measure is obviously zero if the two distributions coincide, and it is the larger, the more they differ. Note that this measure may also be written as

$$S_{\text{div}}(\pi_1, \pi_2) = \int_0^{\sup_{E\in\mathcal{E}} \pi_1(E)} \log_2 \frac{\sum_{E\in\mathcal{E}} [\pi_2]_\alpha(E)}{\sum_{E\in\mathcal{E}} [\pi_1]_\alpha(E)}\, \mathrm{d}\alpha$$

or simply as

$$S_{\text{div}}(\pi_1, \pi_2) = \text{nonspec}(\pi_2) - \text{nonspec}(\pi_1),$$

since the integral can be split into two parts, one part for the numerator and one for the denominator.

The name of this measure is, of course, chosen in analogy to the Kullback–Leibler information divergence, which is very similar. Note, however, that in contrast to the Kullback–Leibler information divergence, which can be used to compare arbitrary probability distributions (on the same set $\mathcal{E}$ of events), the specificity divergence presupposes $\forall E \in \mathcal{E} : \pi_2(E) \geq \pi_1(E)$. Note also that for a relation (which can be seen as a special, i.e. binary possibility distribution; cf. Chapter 3), this measure may also be called **Hartley information divergence**, because for the relational case nonspecificity obviously coincides with Hartley information.

As an illustration reconsider the possibility distribution shown in Figure 3.14 on page 84. The specificity divergence of the original distribution

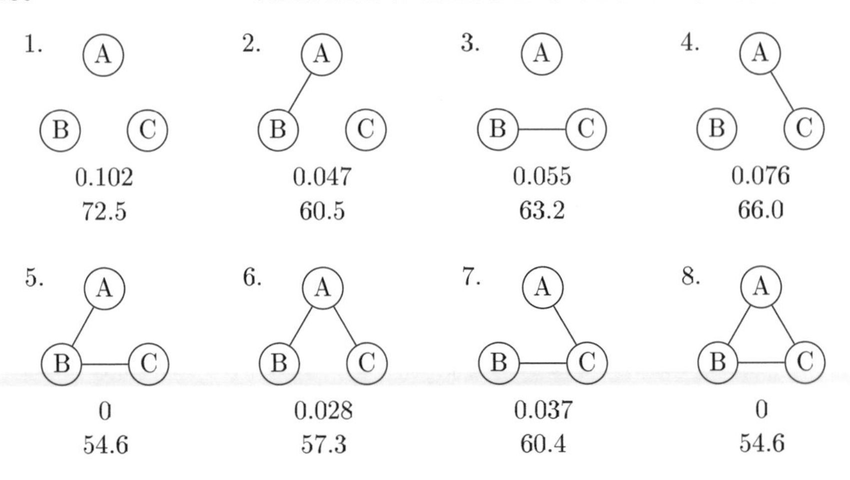

Figure 7.8 The specificity divergence of the original distribution and the possibility distributions represented by the eight possible candidate graphs (upper numbers) and the evaluation of the eight graphs on an example database that induces the possibility distribution (lower numbers).

and the possibility distributions represented by the eight possible graphs are shown in Figure 7.8 (upper numbers). Obviously, graph 5 would be selected, which is indeed the graph that represents a decomposition of the distribution (cf. Section 3.4.2). Note that, as in the relational and the probabilistic case, the complete graph 8 always receives or shares the best assessment, since it consists of only one clique and thus represents no real decomposition of the distribution.

Note also that in order to rank two graphs we need not compute the specificity divergences, but can confine to computing the nonspecificities of the distributions corresponding to the graphs, since the nonspecificity of the original distribution is fixed. This simplifies the assessment: Find the graph among those of acceptable complexity (w.r.t. maximal clique size) for which the corresponding distribution is most specific.

Up to now we have assumed that the multidimensional distribution, for which a decomposition is desired, is already given. However, as in the probabilistic setting, this is usually not the case in applications. Instead we are given a database of sample cases. Clearly, the situation parallels the probabilistic case: It is infeasible to estimate first the joint possibility distribution from the database, so that we can proceed as indicated above.

In order to derive a feasible indirect method we may reason as follows: An approximation of a given possibility distribution may be compared to this distribution by integrating the differences in the number of tuples for each α-cut over the possible values of α (see above). However, this formula can be transformed as follows:

$$\begin{aligned}
\operatorname{diff}(\pi_1, \pi_2) &= \int_0^1 \Big(\sum_{E \in \mathcal{E}} [\pi_2]_\alpha(E) - \sum_{E \in \mathcal{E}} [\pi_1]_\alpha(E) \Big) \,\mathrm{d}\alpha \\
&= \int_0^1 \sum_{E \in \mathcal{E}} [\pi_2]_\alpha(E) \,\mathrm{d}\alpha - \int_0^1 \sum_{E \in \mathcal{E}} [\pi_1]_\alpha(E) \,\mathrm{d}\alpha \\
&= \sum_{E \in \mathcal{E}} \int_0^1 [\pi_2]_\alpha(E) \,\mathrm{d}\alpha - \sum_{E \in \mathcal{E}} \int_0^1 [\pi_1]_\alpha(E) \,\mathrm{d}\alpha \\
&= \sum_{E \in \mathcal{E}} \pi_2(E) - \sum_{E \in \mathcal{E}} \pi_1(E).
\end{aligned}$$

Note, by the way, that this representation simplifies computing the value of this measure and that in order to rank two candidate graphs, we may discard the second sum, since it is identical for both graphs.

Unfortunately, the size of the joint domain (the size of the set $\mathcal{E}$ of events) still makes it impossible to compute the measure, even in this representation. However, we may consider restricting the set $\mathcal{E}$ from which this measure is computed, so that the computation becomes efficient. If we select a proper subset of events, the resulting ranking of different graphs may coincide with the ranking computed from the whole set $\mathcal{E}$ (although, of course, this cannot be guaranteed). A natural choice for such a subset is the set of events recorded in the database to learn from, because from these the distribution is induced and thus it is most important to approximate their degrees of possibility well. In addition, we may weight the degrees of possibility for these events by their frequency in order to capture their relative importance. That is, we may compute for a given database $D = (R, w_R)$

$$Q(G) = \sum_{t \in R} w_R(t) \cdot \pi_G(t)$$

to measure the quality of a candidate model G [Borgelt and Kruse 1997b]. Obviously, this measure is similar to the likelihood measure we used in the probabilistic case (cf. Section 7.1.2). Note, however, that this measure is to be minimized whereas the likelihood measure is to be maximized.

Of course, computing the value of this measure is simple only if all tuples in the database are precise, because only for a precise tuple a unique degree of possibility can be determined from the graphical model to evaluate. For an imprecise tuple some kind of approximation has to be used. We may, for instance, compute an aggregate, for example the average or the maximum, of the degrees of possibility of all precise tuples that are at least as specific as[5] an imprecise tuple [Borgelt and Kruse 1997b]. Since we are trying to minimize the value of the measure, it seems natural to choose pessimistically the

[5]The notion "at least as specific as" was introduced in Definition 5.1.6 on page 135.

A	B	C	freq.
a_1	?	c_2	10
a_1	b_1	?	20
a_1	b_2	?	10
a_1	b_3	?	10
a_2	b_1	?	20
a_2	b_3	?	10
a_3	b_1	?	10
a_3	b_2	?	10
a_4	b_2	?	10
a_4	b_3	?	10

Table 7.2 The imprecise tuples of the database used to compute the lower numbers in Figure 7.8 and their absolute frequency. Stars indicate missing values. All other tuples are precise. Their frequency is such that the degrees of possibility shown in Figure 3.14 on page 84 result.

maximum as the worst possible case. This choice has the additional advantage that it can be computed efficiently by simply propagating the evidence contained in an imprecise tuple in the given graphical model [Borgelt and Kruse 1997b], whereas other aggregates suffer from the fact that we have to compute explicitly the degree of possibility of the compatible precise tuples, the number of which can be very large.

As an example consider a database with 1000 tuples that induces the possibility distribution shown in Figure 3.14 on page 84. Such a database may consist of the imprecise tuples shown in Table 7.2 (question marks indicate missing values) having the absolute frequencies indicated in this table plus precise tuples of suitable absolute frequencies (which can easily be computed from the possibility distribution and the frequencies of the imprecise tuples). If we evaluate the eight possible graphs with the procedure outlined above, we arrive at the lower numbers shown in Figure 7.8 on page 180. Clearly, this method ranks the graphs in the same way as specificity divergence and thus the same graph, namely graph 5, is chosen as the search result.

Let us now turn to the second approach to learn a graphical model from data, namely finding an independence map by conditional independence tests. Formally, there is no difference to the relational case, since (conditional) possibilistic independence coincides with (conditional) relational independence, except that the latter is computed from a *binary* possibility measure. Thus it is not surprising that for the example of a possibility distribution shown in Figure 3.14 on page 84, graph 5 of Figure 7.8 is selected, since the only conditional independence that can be read from it, namely $A \perp\!\!\!\perp B \mid C$, holds in the distribution (as demonstrated in Section 3.4).

To find approximate decompositions we need, as in the relational and the probabilistic case, a measure for the strength of dependences. Drawing on the paradigm that a possibility distribution can be seen as a set of relations, we can construct such a measure [Gebhardt and Kruse 1996b], which may be

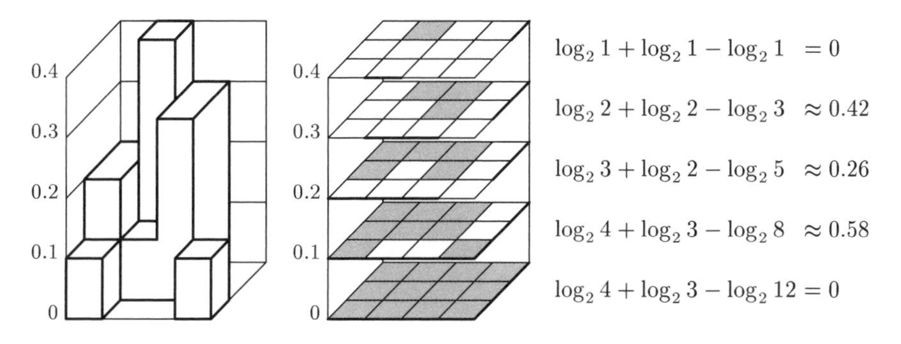

Figure 7.9 Illustration of the idea of specificity gain.

called *specificity gain* [Borgelt *et al.* 1996], from the Hartley information gain. The idea is illustrated in Figure 7.9 for a two-dimensional possibility distribution. For each α-cut of the possibility distribution the Hartley information gain is computed. These values are then aggregated by integrating over all values of α [Borgelt and Kruse 1997a].

Definition 7.1.8 *Let A and B be two attributes and Π a possibility measure.*

$$\begin{aligned} S_{\text{gain}}(A, B) \quad = \quad \int_0^{\sup \Pi} & \log_2 \Big(\sum_{a \in \text{dom}(A)} [\Pi]_\alpha(A = a) \Big) \\ & + \log_2 \Big(\sum_{b \in \text{dom}(B)} [\Pi]_\alpha(B = b) \Big) \\ & - \log_2 \Big(\sum_{a \in \text{dom}(A)} \sum_{b \in \text{dom}(B)} [\Pi]_\alpha(A = a, B = b) \Big) \, d\alpha \end{aligned}$$

is called the **specificity gain** *of A and B w.r.t. Π.*

In the example of Figure 7.9 the computation can be simplified due the fact that the α-cuts are the same for certain intervals of values of α (this is the case due to the finite number of tuples in the database to learn from). Hence we can compute the specificity gain as

$$\begin{aligned} S_{\text{gain}}(A, B) \quad = & \quad\ (0.1 - 0.0) \cdot (\log_2 4 + \log_2 3 - \log_2 8) \\ & + (0.2 - 0.1) \cdot (\log_2 3 + \log_2 2 - \log_2 5) \\ & + (0.3 - 0.2) \cdot (\log_2 2 + \log_2 2 - \log_2 3) \\ & + (0.4 - 0.3) \cdot (\log_2 1 + \log_2 1 - \log_2 1) \\ \approx & \quad 0.1 \cdot 0.58 + 0.1 \cdot 0.26 + 0.1 \cdot 0.42 + 0.1 \cdot 0 \quad = \quad 0.126. \end{aligned}$$

This simplification is useful to remember, because it can often be exploited in implementations of learning algorithms for possibilistic graphical models.

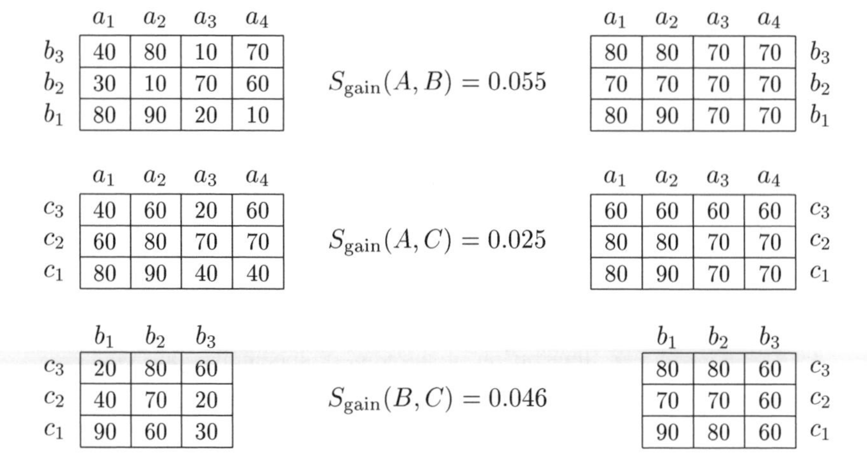

	a_1	a_2	a_3	a_4
b_3	40	80	10	70
b_2	30	10	70	60
b_1	80	90	20	10

$S_{\text{gain}}(A, B) = 0.055$

a_1	a_2	a_3	a_4	
80	80	70	70	b_3
70	70	70	70	b_2
80	90	70	70	b_1

	a_1	a_2	a_3	a_4
c_3	40	60	20	60
c_2	60	80	70	70
c_1	80	90	40	40

$S_{\text{gain}}(A, C) = 0.025$

a_1	a_2	a_3	a_4	
60	60	60	60	c_3
80	80	70	70	c_2
80	90	70	70	c_1

	b_1	b_2	b_3
c_3	20	80	60
c_2	40	70	20
c_1	90	60	30

$S_{\text{gain}}(B, C) = 0.046$

b_1	b_2	b_3	
80	80	60	c_3
70	70	60	c_2
90	80	60	c_1

Figure 7.10 Specificity gain in the simple example.

It should be noted that the specificity gain may also be derived from the specificity divergence in the same way as mutual (Shannon) information is derived from the Kullback–Leibler information divergence, namely by comparing two special distributions: the joint distribution and an assumed independent distribution. This is possible, because the α-cuts of the assumed independent distribution can be represented as

$$\forall a \in \text{dom}(A) : \forall b \in \text{dom}(B) :$$
$$[\pi_{AB}^{(\text{indep})}]_\alpha(A = a, B = b) \;=\; [\Pi]_\alpha(A = a) \cdot [\Pi]_\alpha(B = b),$$

(since $[\Pi]_\alpha(E)$ can assume only the values 0 and 1 the product coincides with the minimum) and thus specificity gain may be written as

$$S_{\text{gain}}(A, B) = \int_0^{\sup \Pi} \log_2 \frac{\left(\sum_{a \in \text{dom}(A)} [\Pi]_\alpha(A = a)\right)\left(\sum_{b \in \text{dom}(B)} [\Pi]_\alpha(B = b)\right)}{\sum_{a \in \text{dom}(A)} \sum_{b \in \text{dom}(B)} [\Pi]_\alpha(A = a, B = b)} \, d\alpha.$$

In addition, specificity gain may be written simply as

$$S_{\text{gain}}(A, B) = \text{nonspec}(\pi_A) + \text{nonspec}(\pi_B) - \text{nonspec}(\pi_{AB}),$$

since the integral may be split into three parts, which refer to the three marginal distributions π_A, π_B, and π_{AB}, respectively. This form is very convenient, since it can be exploited for an efficient implementation.

It is obvious that the specificity gain is zero if and only if the two attributes are independent, and that it is the larger, the more strongly dependent the two attributes are. Therefore this measure can be used directly to test for (approximate) marginal independence. This is demonstrated in Figure 7.10 for the simple example discussed above (cf. Figure 3.14 on page 84). On the left the three possible two-dimensional marginal distributions are shown, on the right the corresponding (assumed) independent distributions, to which they are compared by specificity gain. Clearly, none of the three pairs of attributes are independent. This result excludes the four graphs 1 to 4 of Figure 7.8.

In order to use specificity gain to test for (approximate) conditional independence, we may proceed in a similar way as with Hartley information gain in the relational case and mutual (Shannon) information in the probabilistic case: We compute this measure for each possible instantiation of the conditioning attributes and then sum the results weighted with the relative (marginal) degree of possibility of the corresponding instantiation. That is, for one conditioning attribute, we may compute

$$\begin{aligned} &S_{\text{gain}}(A, B \mid C) \\ &= \sum_{c \in \text{dom}(C)} \frac{\Pi(c)}{\sum_{c' \in \text{dom}(C)} \Pi(c')} \\ &\qquad \int_0^{\sup \Pi} \log_2 \frac{\left(\sum_{a \in \text{dom}(A)} [\Pi]_\alpha(a|c)\right)\left(\sum_{b \in \text{dom}(B)} [\Pi]_\alpha(b|c)\right)}{\sum_{a \in \text{dom}(A)} \sum_{b \in \text{dom}(B)} [\Pi]_\alpha(a, b|c)} \, d\alpha, \end{aligned}$$

where $\Pi(c)$ is an abbreviation of $\Pi(C = c)$ etc. Note that $\Pi(c)$ is normalized by dividing it by the sum of the degrees of possibility for all values of the attribute C. This is necessary, because (in contrast to the probabilistic case) this sum may exceed 1 and may differ for different (sets of) conditioning attributes. Hence, without normalization, it would not be possible to compare the value of this measure for different (sets of) conditions. With this measure it is easy to detect in the example distribution of Figure 3.14 on page 84 that the attributes A and C are conditionally independent given the attribute B.

It is clear that this approach suffers from the same drawbacks as the corresponding probabilistic approach: It can be computationally infeasible if the order of the conditional independence tests (i.e. the number of conditioning attributes) is large. However, it is clear as well that we can also amend it in the same way, for example, by fixing an upper bound for the order of the tests to be carried out and assuming that all tests of higher order will fail if all tests with an order up to this bound failed.

Finally, let us consider the third approach to learn a graphical model, namely to determine a suitable graph by measuring only the strengths of (marginal) dependences. As in the relational and the probabilistic case, we may use the same measure as for the (approximate) independence tests. For the simple example discussed above we may apply the Kruskal algorithm with specificity gain providing the edge weights (these edge weights are shown in Figure 7.10). This leads to a construction of graph 5 of Figure 7.8, in agreement with the results of the other two learning methods.

Note that, as in the probabilistic case, this method can be applied as well if we are given a database of sample cases, because we only have to determine marginal possibility distributions on (usually small) subspaces. (A method to compute these efficiently was discussed in Chapter 5.) Note also that this approach suffers from the same drawback as the corresponding relational approach, i.e. it may not find an exact decomposition although there is one: The relational example of Figure 7.4 on page 170 can be transferred directly to the possibilistic case, since, as already mentioned several times, a relation is a special possibility distribution.

7.1.4 Components of a Learning Algorithm

Up to now, there is no (computationally feasible) analytical method to construct optimal graphical models from a database of sample cases. In part this is due to the fact that there is no consensus about what constitutes an "optimal" model, because there is, as usual, a tradeoff between model accuracy and model simplicity. This tradeoff is hard to assess: How much accuracy must be gained in order to make a more complex model acceptable? However, the main reason is more technical, namely the number of candidate graphs, which is huge unless the number of attributes is *very* small and which makes it impossible to inspect all possible graphs (cf. Section 7.3.1). Although there are methods by which large sets of graphs can be excluded with single tests (cf. the conditional independence tests mentioned in the preceding sections), the number of tests is often still too large to be carried out exhaustively or the tests require too much data to be reliable.

Since there is no (computationally feasible) analytical method, all algorithms for constructing a graphical model perform some kind of (heuristic) search, evaluating networks or parts of networks as they go along and finally stopping with the best network found. Therefore an algorithm for learning a graphical model from data usually consists of

1. an *evaluation measure* (to assess the quality of a given network) and
2. a *search method* (to traverse the space of possible networks).

It should be noted, though, that restrictions of the search space introduced by an algorithm and special properties of the evaluation measure used sometimes disguise the fact that a search through the space of possible network structures is carried out. For example, by conditional independence tests all graphs missing certain edges can be excluded without inspecting these graphs explicitly. Greedy approaches try to find good edges or subnetworks and combine them in order to construct an overall model and thus may not appear to be searching. Nevertheless the above characterization is apt, since an algorithm that does not explicitly search the space of possible networks usually carries out a (heuristic) search on a different level, guided by an evaluation measure. For example, some greedy approaches search for the best set of parents of an attribute by measuring the strength of dependence on candidate parent attributes; conditional independence test approaches search the space of all possible conditional independence statements also measuring the strengths of (conditional) dependences.

Although the two components of a learning algorithm for graphical models usually cooperate closely, since, for instance, in a greedy algorithm the search is guided by the evaluation measure, they can be treated independently. The reason is that most search methods and evaluation measures can be combined freely. Therefore we focus on evaluation measures in Section 7.2, whereas search methods are discussed in Section 7.3.

7.2 Evaluation Measures

An *evaluation measure* serves to assess the quality of a given candidate graphical model w.r.t. a given database of sample cases (or w.r.t. a given distribution), so that it can be determined which of a set of candidate graphical models best fits the given data. In this section we are going to discuss systematically a large variety of evaluation measures for relational, probabilistic, and possibilistic graphical models (although restricted to attributes with a finite set of values) and the ideas underlying them.

7.2.1 General Considerations

Although the evaluation measures discussed in the following are based on a wide range of ideas and thus are bound to differ substantially w.r.t. the computations that have to be carried out, several of them share certain general characteristics, especially w.r.t. how they are applied in certain situations. Therefore it is useful to start this section with some general considerations, so that we need not repeat them for each measure to which they apply. Most of these considerations are based on the following distinction:

Apart from the obvious division into relational, probabilistic, and possibilistic measures, evaluation measures may also be classified w.r.t. whether they are *holistic* or *global*, i.e. can be computed only for a graphical model as a whole, or whether they are *decomposable* or *local*, i.e. can be computed by aggregating local assessments of subnetworks or even single edges. Drawing on the examples of Section 7.1, the number of additional tuples is a global evaluation measure for a relational graphical model, since these tuples cannot be determined from a single subspace relation, but only from the combination of all subspace relations of the graphical model. In contrast to this, the mutual (Hartley or Shannon) information is a local evaluation measure, since it is computed for single edges.

It is clear that local evaluation measures are desirable, not only because they are usually much simpler to compute, but also because their decomposability can be exploited, for example, by greedy search methods (cf. Section 7.3.4). In addition, local evaluation measures are advantageous in the probabilistic case if the database to learn from has missing values and we do not want to employ costly methods like expectation maximization to handle them (cf. Section 5.2). With a global evaluation measure all we can do in such a case is to discard *all* incomplete tuples of the database, thus throwing away a lot of valuable information. With a local evaluation measure, we only have to discard tuples selectively w.r.t. the component of the graphical model to be evaluated. That is, for each component all tuples of the database that do not miss a value *for the attributes underlying the component* can be taken into account [Borgelt and Kruse 1997b].

Global evaluation measures need no further consideration, since the application of each of them is, in a way, unique. Local evaluation measures, however, share that they all must be aggregated over the components they are computed on. In addition, all of them are either derived from a measure for the strength of the dependence of two attributes or at least can be interpreted as such a measure, partly because several of them were originally devised for feature selection or for decision tree induction (as can be seen from the references given for these measures).

It is clear that a measure for the strength of the dependence of two attributes is sufficient if we only have to evaluate single edges of a conditional independence graph. Note, however, that we have to take care of the fact that certainly any such measure can be used to evaluate a directed edge, but not all are suited to evaluate an undirected edge. The reason is that for evaluating an undirected edge the measure should be required to be *symmetric*, i.e. it should remain unchanged if the two attributes are exchanged, simply because an undirected edge does not distinguish between the two attributes it connects and thus there is no way to assign the attributes based on the properties of the edge. This is no real constraint, though, because any non-symmetric evaluation measure can be turned into a symmetric one by simply averaging its value for the two possible cases.

Unfortunately, evaluating single edges is sufficient only if the graphical model to be assessed is based on a (directed or undirected) tree, because only in this case the components of the decomposition are associated with single edges. If the graphical model is based on a polytree or on a multiply connected graph, the components of the decomposition are the conditional distributions of a chain rule decomposition or the functions on the maximal cliques (usually represented by marginal distributions, cf. Section 4.1.6). In such cases evaluating single edges is not appropriate.

To see this, recall from Section 7.1 the intuition underlying the learning approach based on measuring the strengths of marginal dependences, especially, reconsider this approach in the relational example. The rationale was to select those *subspaces*, for which the attributes scaffolding it are strongly dependent. However, it is clear that three or more attributes that are strongly dependent need not exhibit strong pairwise dependence. As an example consider the extreme case of three binary attributes, one of which is computed as the exclusive or of the other two. In this case all pairs of attributes are independent, but all three are, of course, strongly dependent, and thus measuring only the strengths of mutual dependences does not assess the situation correctly. Note that in this case it does not matter whether we consider a maximal clique with these attributes or one of them conditioned on the other two: In any case we have to take all three attributes into account.

Consequently, we have to find ways of extending a measure for the strength of the dependence of two attributes to a measure for the strength of the dependence of one attribute on several others (i.e. its parents in a directed acyclic graph) and to a measure for the strength of the joint dependence of several attributes (i.e. the attributes of a maximal clique of an undirected graph). The former is always possible: We only have to combine all conditioning attributes into one pseudo-attribute, the values of which are all distinct instantiations of the conditioning attributes. Thus we reduce this case to the two-attribute case. The latter, however, is usually much harder to achieve, because the resulting measure must exhibit a high symmetry, i.e. it must be invariant under any permutation of the attributes, since in a maximal clique no attribute can be distinguished from any other.

As an example of such extensions consider mutual (Shannon) information (cf. Definition 7.1.4 on page 174) for n attributes $A_1, \ldots, A_n$. Combining the attributes $A_2, \ldots, A_n$ into one pseudo-attribute yields the formula

$$\begin{aligned} I_{\text{mut}}^{(1)}(A_1, \ldots, A_n) &= \sum_{a_1 \in \text{dom}(A_1)} \cdots \sum_{a_n \in \text{dom}(A_n)} P(A_1 = a_1, \ldots, A_n = a_n) \\ &\qquad \log_2 \frac{P(A_1 = a_1, \ldots, A_n = a_n)}{P(A_1 = a_1) \cdot P(A_2 = a_2, \ldots, A_n = a_n)}. \end{aligned}$$

On the other hand, treating all attributes equally yields

$$\begin{aligned} I_{\text{mut}}^{(2)}(A_1, \ldots, A_n) &= \sum_{a_1 \in \text{dom}(A_1)} \cdots \sum_{a_n \in \text{dom}(A_n)} P(A_1 = a_1, \ldots, A_n = a_n) \\ &\qquad \log_2 \frac{P(A_1 = a_1, \ldots, A_n = a_n)}{\prod_{k=1}^{n} P(A_k = a_k)}. \end{aligned}$$

Obviously, the former is best suited for evaluating an attribute (here A_1) conditioned on its parent attributes (here $A_2, \ldots, A_n$), whereas the latter is better suited for evaluating a maximal clique of attributes.

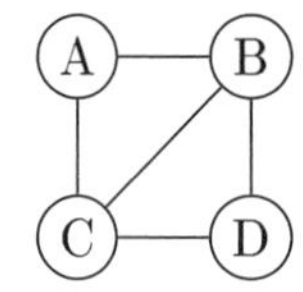

Figure 7.11 A simple undirected graph with overlapping maximal cliques.

Having seen this example, it should not be surprising that several evaluation measures can be extended in the same manner. Since the extensions are usually very easy to find, we do not consider them explicitly. In the following we state all evaluation measures w.r.t. two attributes A and C, where C is the child attribute if we have a directed edge and A is the parent.

After an assessments of all components of a graphical model has been computed, they have to be aggregated. Usually this is done by simply summing them. Such a simple aggregation is, obviously, satisfactory if the graph is a directed acyclic graph, because in this case the components of the decomposition are clearly "separated" from each other by the conditioning and thus there can be no interaction between the evaluations of these components. For undirected graphs, however, the situation is different. If two cliques overlap on more than one attribute, then the dependence of the attributes in the overlap is, in a way, "counted twice", since it has an influence on the assessment of both cliques. As a consequence the quality of the graph may be overrated.

As an example consider the simple undirected graph shown in Figure 7.11. The maximal cliques are induced by the node sets $\{A, B, C\}$ and $\{B, C, D\}$, which overlap on $\{B, C\}$. Thus the mutual dependence of B and C is, "counted twice" and hence the graph may be overrated compared to one in which the cliques overlap only on single attributes. In order to cope with this problem one may consider deducing the assessments of overlaps from the summed assessments of the maximal cliques. (Note that this may require fairly complicated computations in case the overlap of several pairs of cliques has a nonempty intersection, since we may have to re-add the assessment of the intersection of overlaps etc.) Of course, this is only a heuristic approach, since it may not be justifiable for all evaluation measures. (It can be justified, though, for mutual (Shannon) information; see below.)

An alternative to cope with the evaluation problem for undirected graphs is to direct the edges and then to apply the "directed version" of the chosen evaluation measure. Unfortunately, not all undirected graphs can be turned into an equivalent directed acyclic graph (cf. Section 4.1.4) and thus this approach may not be applicable. That it is applicable can be guaranteed, though, if the undirected graphs has hypertree structure,[6] a circumstance that may be seen as another argument in favor of such graphs. (A simple algorithm to construct the corresponding directed acyclic graph is given in Section 7.3.2.) However, even in this case the resulting graph need not be

[6]The notion of hypertree structure was introduced in Definition 4.1.23 on page 110.

unique and thus the assessment may depend on the edge directions chosen.

Nevertheless, the idea to turn an undirected graph into a directed acyclic one is useful, since it can often be used to justify the approach mentioned above, namely to deduce the assessments of the overlaps from the summed assessments of the cliques. As an example reconsider the simple undirected graph of Figure 7.11. If we apply the deduction method to a mutual (Shannon) information evaluation of this graph, we get

$$\begin{aligned}
Q_1 &= \sum_{a\in\mathrm{dom}(A)} \sum_{b\in\mathrm{dom}(B)} \sum_{c\in\mathrm{dom}(c)} P(a,b,c) \log_2 \frac{P(a,b,c)}{P(a)P(b)P(c)} \\
&+ \sum_{b\in\mathrm{dom}(B)} \sum_{c\in\mathrm{dom}(C)} \sum_{d\in\mathrm{dom}(D)} P(b,c,d) \log_2 \frac{P(b,c,d)}{P(b)P(c)P(d)} \\
&- \sum_{b\in\mathrm{dom}(B)} \sum_{c\in\mathrm{dom}(C)} P(b,c) \log_2 \frac{P(b,c)}{P(b)P(c)},
\end{aligned}$$

where $P(a)$ is an abbreviation of $P(A = a)$ etc. On the other hand, if we turn the graph into a directed acyclic graph by directing the edges (A, B) and (A, C) towards A, the edges (B, D) and (C, D) towards D, and the edge (B, C) towards C (although the latter direction does not matter), we get

$$\begin{aligned}
Q_2 &= \sum_{b\in\mathrm{dom}(B)} \sum_{c\in\mathrm{dom}(C)} P(b,c) \log_2 \frac{P(b,c)}{P(b)P(c)} \\
&+ \sum_{a\in\mathrm{dom}(A)} \sum_{b\in\mathrm{dom}(B)} \sum_{c\in\mathrm{dom}(c)} P(a,b,c) \log_2 \frac{P(a,b,c)}{P(a)P(b,c)} \\
&+ \sum_{b\in\mathrm{dom}(B)} \sum_{c\in\mathrm{dom}(C)} \sum_{d\in\mathrm{dom}(D)} P(b,c,d) \log_2 \frac{P(b,c,d)}{P(d)P(b,c)}.
\end{aligned}$$

By exploiting $\log_2 \frac{x}{y} = \log_2 x - \log_2 y$ it is easy to verify that these two expressions are equivalent. Hence deducing the assessment of the overlap yields a consistent and plausible result in this case. Other measures may be treated in a similar way, for example, *specificity gain* (cf. Section 7.2.5).

7.2.2 Notation and Presuppositions

Since a graphical model is usually evaluated w.r.t. a database of sample cases, we assume that we are given a database $D = (R, w_R)$ as defined in Definition 5.1.5 on page 135 and that the values needed to compute the chosen evaluation measure are determined from this database.

As already indicated above, we will state all local evaluation measures w.r.t. two attributes C and A, with the attribute C taking the place of the child, if a directed edge is to be evaluated, and attribute A taking the place of the parent. The extension to more than two attributes is usually straightforward (see above) and thus omitted. Carrying on the restriction to attributes with a finite domain, we assume that the domain of attribute A has n_A values, i.e. $\text{dom}(A) = \{a_1, \ldots, a_{n_A}\}$, and that the domain of attribute C has n_C values, i.e. $\text{dom}(C) = \{c_1, \ldots, c_{n_C}\}$.

To simplify the notation, we use the following abbreviations to state local evaluation measures for relational graphical models:

$r_{i.}$ Indicator whether there is a tuple having the value c_i for attribute C, i.e. $r_{i.} = 1$ iff $\exists t \in R : t|_C = c_i$, and $r_{i.} = 0$ otherwise.

$r_{.j}$ Indicator whether there is a tuple having the value a_j for attribute A, i.e. $r_{.j} = 1$ iff $\exists t \in R : t|_A = a_j$, and $r_{.j} = 0$ otherwise.

r_{ij} Indicator whether there is a tuple having the value c_i for attribute C and the value a_j for attribute A, i.e. $r_{ij} = 1$ iff $\exists t \in R : t|_C = c_i \wedge t|_A = a_j$, and $r_{ij} = 0$ otherwise.

From the above lists it is already clear that we always use the index i with values of attribute C and index j with values of attribute A, so that the index is sufficient to identify the attribute referred to.

For stating some of the evaluation measures it is necessary to refer to the number of sample cases in the database having certain properties. For these we use the following abbreviations:

$N_{..}$ The total number of sample cases, i.e. $N_{..} = \sum_{t \in R} w_R(t)$.

$N_{i.}$ Absolute frequency of the attribute value c_i, i.e. $N_{i.} = \sum_{t \in \{s \in R | s|_C = c_i\}} w_R(t)$.

$N_{.j}$ Absolute frequency of the attribute value a_j, i.e. $N_{.j} = \sum_{t \in \{s \in R | s|_A = a_j\}} w_R(t)$.

N_{ij} Absolute frequency of the combination of the values c_i and a_j, i.e. $N_{ij} = \sum_{t \in \{s \in R | s|_C = c_i \wedge s|_A = a_j\}} w_R(t)$.

Obviously, it is $N_{i.} = \sum_{j=1}^{n_A} N_{ij}$ and $N_{.j} = \sum_{i=1}^{n_C} N_{ij}$.

Some evaluation measures for probabilistic graphical models are defined directly in probabilities. Since we assume that we are given a database of sample cases, these probabilities have to be estimated from the database. In general, we assume that maximum likelihood estimation is used for this, although it is clear that other estimation methods, for example, Laplace-corrected maximum likelihood estimation, may also be employed (cf. Section 5.2). For these probabilities we use the following abbreviations:

$p_{i.}$ (Estimated) probability of the attribute value c_i, i.e. $p_{i.} = \frac{N_{i.}}{N_{..}}$.

$p_{.j}$ (Estimated) probability of the attribute value a_j, i.e. $p_{.j} = \frac{N_{.j}}{N_{..}}$.

p_{ij} (Estimated) probability of the combination of the attribute values c_i and a_j, i.e. $p_{ij} = \frac{N_{ij}}{N_{..}}$.

$p_{i|j}$ (Estimated) conditional probability of the attribute value c_i given that attribute A has the value a_j, i.e. $p_{i|j} = \frac{p_{ij}}{p_{.j}} = \frac{N_{ij}}{N_{.j}}$.

$p_{j|i}$ (Estimated) conditional probability of the attribute value a_j given that attribute C has the value c_i, i.e. $p_{j|i} = \frac{p_{ij}}{p_{i.}} = \frac{N_{ij}}{N_{i.}}$.

The evaluation measures for possibilistic graphical models are usually defined in degrees of possibility. For these we use the abbreviations:

$\pi_{i.}$ (Database-induced) degree of possibility of the attribute value c_i.

$\pi_{.j}$ (Database-induced) degree of possibility of the attribute value a_j.

π_{ij} (Database-induced) degree of possibility of the combination of the attribute values c_i and a_j.

All of these values are assumed to be computed as described in Section 5.4, i.e. by determining the maxima of the elementary degrees of possibility over all values of all other attributes.

7.2.3 Relational Evaluation Measures

There are only a few relational evaluation measures, mainly because the information provided by a relation is so scarce, but also because relational decomposition is usually studied by completely different means in database theory, namely by exploiting known functional dependences between attributes [Maier 1983, Date 1986, Ullman 1988]. The few there are, however, are important, because from each of them at least one evaluation measure for the possibilistic case can be derived, for example, by drawing on the α-cut view[7] of a possibility distribution (cf. Section 7.1.3).

Hartley Information Gain

Hartley information gain was already defined in Definition 7.1.2 on page 167. Restated with the abbreviations introduced in the preceding section it reads:

$$\begin{aligned} I_{\text{gain}}^{(\text{Hartley})}(C, A) &= \log_2\left(\sum_{i=1}^{n_C} r_{i.}\right) + \log_2\left(\sum_{j=1}^{n_A} r_{.j}\right) - \log_2\left(\sum_{i=1}^{n_C}\sum_{j=1}^{n_A} r_{ij}\right) \\ &= \log_2 \frac{\left(\sum_{i=1}^{n_C} r_{i.}\right)\left(\sum_{j=1}^{n_A} r_{.j}\right)}{\left(\sum_{i=1}^{n_C}\sum_{j=1}^{n_A} r_{ij}\right)}. \end{aligned}$$

The idea underlying this measure was discussed in detail in Section 7.1.1 (cf. Figure 7.2 on page 166): If the two attributes are relationally independent, then the numerator and the denominator of the fraction are equal and thus the measure is 0. The more strongly dependent the two attributes are, the

[7]The notion of an α-cut was introduced in Definition 7.1.5 on page 177.

smaller the denominator and thus the larger the measure. Note that Hartley information gain may also be written as

$$I_{\text{gain}}^{(\text{Hartley})}(C,A) = H^{(\text{Hartley})}(C) + H^{(\text{Hartley})}(A) - H^{(\text{Hartley})}(C,A)$$

if we define the *Hartley entropy* of an attribute C as

$$H^{(\text{Hartley})}(C) = \log_2\left(\sum_{i=1}^{n_C} r_{i.}\right)$$

(cf. Definition 7.1.1 on page 166) and provide an analogous definition for the entropy of the combination of two attributes C and A. This indicates a direct connection to Shannon information gain, which is studied in Section 7.2.4. In that section some variants of Shannon information gain will be considered, which were originally devised to overcome its bias towards many-valued attributes. (This bias was discovered when Shannon information gain was used for decision tree induction [Quinlan 1993].) If Hartley information gain is written as above, analogous variants can be constructed. In this way we obtain the *Hartley information gain ratio*

$$I_{\text{gr}}^{(\text{Hartley})}(C,A) = \frac{I_{\text{gain}}^{(\text{Hartley})}(C,A)}{H^{(\text{Hartley})}(A)}$$

and two versions of a *symmetric Hartley information gain ratio*, namely

$$I_{\text{sgr1}}^{(\text{Hartley})}(C,A) = \frac{I_{\text{gain}}^{(\text{Hartley})}(C,A)}{H^{(\text{Hartley})}(C,A)} \quad \text{and}$$

$$I_{\text{sgr2}}^{(\text{Hartley})}(C,A) = \frac{I_{\text{gain}}^{(\text{Hartley})}(C,A)}{H^{(\text{Hartley})}(C) + H^{(\text{Hartley})}(A)}.$$

We do not provide an explicit justification of these measures here, because it can be derived from the justification of the more general measures based on Shannon information gain by noting that Hartley entropy is only a special case of Shannon entropy (cf. Section 7.2.4).

Hartley information gain, as it was described above, is computed from the joint relation of the values of the attributes C and A. However, we may also consider computing an analogous measure from the conditional relations of the values of attribute C given the values of attribute A, especially if we have to evaluate a directed edge $A \rightarrow C$. That is, we may compute

$$\begin{aligned} I_{\text{cgain}}^{(\text{Hartley})}(C,A) &= \frac{1}{\sum_{j=1}^{n_A} r_{.j}} \sum_{j=1}^{n_A} r_{.j}\left(\log_2\left(\sum_{i=1}^{n_C} r_{i.}\right) - \log_2\left(\sum_{i=1}^{n_C} r_{ij}\right)\right) \\ &= \frac{1}{\sum_{j=1}^{n_A} r_{.j}} \sum_{j=1}^{n_A} r_{.j} \log_2 \frac{\sum_{i=1}^{n_C} r_{i.}}{\sum_{i=1}^{n_C} r_{ij}}. \end{aligned}$$

This measure may be called *conditional Hartley information gain*, since it is computed from *conditional* relations. Note that in order to emphasize in the notation that conditional relations are considered we may write $r_{i|j}$ instead of r_{ij} by exploiting Definition 3.2.9 on page 70.

Reconsidering the illustration of ordinary Hartley information gain in Figure 7.2 on page 166 (with attribute C replacing attribute B) we see that this measure evaluates the relation "column by column": For each conditional relation the number of tuples is compared to the number of tuples in the marginal relation (this comparison is done in the argument of the logarithm). If these numbers are the same for all conditional relations, the two attributes are, obviously, independent. The more they differ, the more strongly dependent the two attributes are. The results of these comparisons are aggregated over all conditional relations containing at least one tuple (that only non-empty relations are considered is achieved with the factor $r_{.j}$). This aggregate is normalized by dividing it by $\sum_{j=1}^{n_A} r_{.j}$ to make it independent of the number of possible values of the attribute A.

Note that, in contrast to the ordinary Hartley information gain, this measure is, in general, not symmetric. Hence we cannot obtain symmetric gain ratios as for ordinary Hartley information gain, but have to rely on

$$I_{\text{scgr}}^{(\text{Hartley})}(C, A) = \frac{I_{\text{cgain}}^{(\text{Hartley})}(C, A) + I_{\text{cgain}}^{(\text{Hartley})}(A, C)}{H^{(\text{Hartley})}(C) + H^{(\text{Hartley})}(A)}.$$

Number of Additional Tuples

Evaluating a given relational graphical model by simply counting the additional tuples in the relation represented by the model compared to the original relation was already discussed extensively in Section 7.1.1. Therefore we only mention it here for reasons of completeness. Note that this measure is a global evaluation measure—in contrast to the Hartley information gain and its variants, which are all local evaluation measures.

7.2.4 Probabilistic Evaluation Measures

There is an abundance of probabilistic evaluation measures, although only few of them have been used for learning probabilistic graphical models from data yet. One reason for this may be that many probabilistic evaluation measures were originally developed either as independence tests in statistics or for the purposes of feature selection and decision tree induction in machine learning [Borgelt and Kruse 1998b] and that the connection to learning graphical models was simply not recognized. However, since all that is required of an evaluation measure in order to make it usable for learning graphical models is that it yields an assessment of the strength of the dependence of two attributes, there is no reason why one should not consider using them.

Shannon Information Gain

Shannon information gain was already defined in Definition 7.1.4 on page 174, although we used the synonymous names *mutual Shannon information* and *Shannon cross entropy* in that definition. Restated with the abbreviations introduced in Section 7.2.2 it reads:

$$I_{\text{gain}}^{(\text{Shannon})}(C, A) = \sum_{i=1}^{n_C} \sum_{j=1}^{n_A} p_{ij} \log_2 \frac{p_{ij}}{p_{i.}p_{.j}}.$$

In Section 7.1.2 Shannon information gain was interpreted as a measure of the difference between the joint probability distribution (represented by p_{ij}) and the distribution that can be computed from the marginal distributions under the assumption that C and A are independent (represented by $p_{.i}p_{.j}$). Clearly, if C and A are actually independent, this measure is zero. It can be shown (see below) that this is indeed the only case in which it is zero. For all other joint distributions this measure is greater than zero and it is the larger, the more strongly dependent the two attributes are.

A different interpretation of this measure, which is preferred in connection to decision tree induction [Quinlan 1993], is based on the notion of the *Shannon entropy* of a set of alternatives [Shannon 1948].

Definition 7.2.1 *Let $S = \{s_1, \ldots, s_n\}$ be a finite set of alternatives having positive probabilities $P(s_i)$, $i = 1, \ldots, n$, satisfying $\sum_{i=1}^{n} P(s_i) = 1$. Then*

$$H^{(\text{Shannon})}(S) = -\sum_{i=1}^{n} P(s_i) \log_2 P(s_i).$$

is called the **Shannon entropy** *or* **Shannon information** *of S.*

Intuitively, Shannon entropy can be interpreted as the expected number of yes/no-questions that have to be asked in order to determine the obtaining alternative. As an example consider the five alternatives $s_1, \ldots, s_5$ with probabilities as shown in Figure 7.12, so that the Shannon entropy of this set of alternatives is approximately 2.15. Suppose there is an oracle, which knows the obtaining alternative, but which will respond only if the question asked can be answered with "yes" or "no". A better question scheme than asking for one alternative after the other is easily found (cf. Section 7.1.1): We divide the alternatives into two subsets of about equal size. Then we choose arbitrarily one of the two sets and ask whether the obtaining alternative is contained in it. Independent of the answer the alternatives in one of the subsets can be ruled out. The other subset is divided again and the interrogation proceeds in this way until only one alternative remains. Obviously, the number of questions needed with this scheme is bounded by $\lceil \log_2 n \rceil$, where n is the number of alternatives. That is, in our example we can determine the obtaining alternative with at most 3 questions. A more detailed analysis shows that

$P(s_1) = 0.40,\ P(s_2) = 0.19,\ P(s_3) = 0.16,\ P(s_4) = 0.15,\ P(s_5) = 0.10$

Shannon Entropy: 2.15 bits/symbol

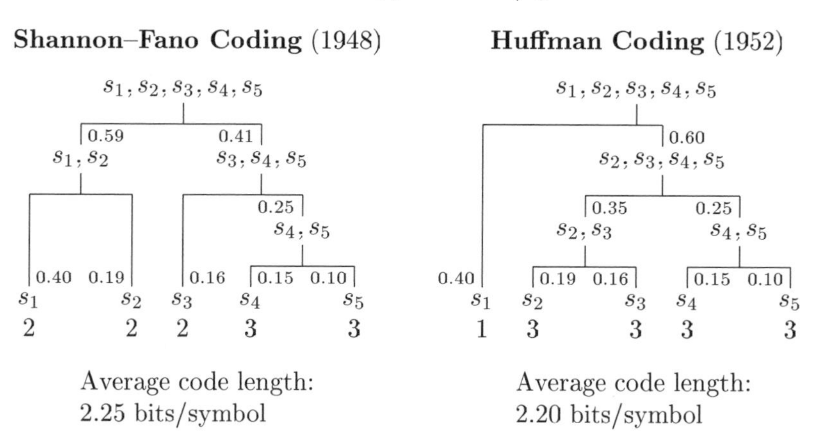

Figure 7.12 Question/coding schemes based on a probability distribution.

the expected number of questions is at most 2.59, even if we choose the most disadvantageous divisions of the set of alternatives: If we divide it into the sets $\{s_1, s_2, s_3\}$ and $\{s_4, s_5\}$ and the first of these into $\{s_1, s_2\}$ and $\{s_3\}$ if the oracle's answer indicates that it contains the obtaining alternative, we need $3P(s_1) + 3P(s_2) + 2P(s_3) + 2P(s_4) + 2P(s_5) = 2.59$ questions on average.

It is clear that this question scheme can be improved by exploiting the known probabilities of the alternatives. The basic idea is that we should strive for a question scheme, with which we have to ask only few questions in order to identify a highly probable alternative as the obtaining one, while we accept more questions for identifying improbable alternatives. It is plausible that in this way the average number of questions can be reduced.

Of course, the problem is how to find a good scheme with this property. A simple, though not optimal, method has been suggested by [Shannon 1948] and Fano: Instead of dividing the set of alternatives into two subsets of about *equal size*, we sort the alternatives w.r.t. their probability and then divide them into subsets of about *equal probability* respecting the order of the probabilities. For the simple example considered above the resulting question scheme is shown on the left in Figure 7.12. With this scheme we need $2P(s_1) + 2P(s_2) + 2P(s_3) + 3P(s_4) + 3P(s_5) = 2.25$ questions on average. However, this is not the best possible method as can already be guessed from the fact that in the left branch of the division tree the two subsets differ considerably w.r.t. their probabilities.

A better method has been found by [Huffman 1952]: Instead of starting with the complete set of alternatives and recursively dividing it, thus constructing the tree top-down, one element sets are taken as the starting point and the tree is constructed bottom-up. In each step the two sets having the smallest probability are merged, until all alternatives are contained in a single set. For the simple example the resulting question scheme is shown on the right in Figure 7.12. We actually obtained a better question scheme, since with it we only need $1P(s_1) + 3P(s_2) + 3P(s_3) + 3P(s_4) + 3P(s_5) = 2.20$ questions on average (compared to 2.25 for the Shannon–Fano scheme).

[Huffman 1952] has shown that his method is optimal if we have to determine the obtaining alternative in a single instance. If, however, we are concerned with a sequence of (independent) situations, for each of which the set of alternatives and their respective probabilities are the same, even this sophisticated method can be improved upon. The idea is to process this sequence not instance by instance, applying Huffman's scheme in each case, but to combine two or more consecutive instances and to ask directly for the combination of alternatives obtaining in such a subsequence. Although the question tree is enlarged by this, the expected number of questions per identification is reduced, since with each interrogation the obtaining alternative is determined not only for one, but for several situations, and it is reduced the more, the more situations are considered in combination. However, the expected number of questions cannot be made arbitrarily small. As [Shannon 1948] showed for the general case, there is an ultimate lower bound for the expected number of questions. This lower bound is the Shannon entropy, which is 2.15 questions for the simple example.

W.r.t. the following sections, especially the section on the measures based on the minimum description length principle, it is useful to note that a question scheme can also be interpreted as a coding rule. In this case the alternatives considered are symbols that may appear in a message, for instance, the letters of the alphabet. Each of the symbols has a probability of occurrence. For example, the letter "e" is much more frequent than the letter "q". A message is coded by mapping the symbols it consists of to sequences of zeros and ones, which are then transmitted. Such a mapping can easily be obtained by simply assigning a 0 to the answer "no" and a 1 to the answer "yes". With this assignment each symbol can be coded by the sequence of answers that would lead to its identification as the obtaining alternative. (It is clear that a coded message formed by concatenating the codes for individual symbols that were obtained in this way can easily be decoded, even if the length of the code is not the same for all symbols.) The Shannon entropy is a lower bound for the average number of bits needed to encode a symbol, provided the symbols are independent of each other.[8]

Note also that *Hartley entropy* or *Hartley information*, which was studied in Section 7.2.3, is only a special case of Shannon entropy. It results if all

[8]Note that this is not the case for normal texts, in which, for instance, the letter "u" is not very likely to be followed by the overall most frequent letter "e". If there is a dependence on the context, the question/coding scheme can be improved upon.

probabilities are equal, i.e. if $\forall i \in \{1, \ldots, n\} : P(s_i) = \frac{1}{n}$, namely

$$H^{(\text{Shannon})}(S_u) = -\sum_{i=1}^{n} \frac{1}{n} \log_2 \frac{1}{n} = -\log_2 \frac{1}{n} = \log_2 n = H^{(\text{Hartley})}(S_u),$$

where the index u is meant to indicate that the probability distribution on S is uniform. Hence most things considered w.r.t. Shannon information gain can be transferred more or less directly to Hartley information gain.

From the definition of the Shannon entropy it is immediately clear that we can write the Shannon information gain as (in the following we drop the upper index "(Shannon)", since no confusion is to be expected)

$$I_{\text{gain}}(C, A) = H(C) + H(A) - H(C, A),$$

if we define the Shannon entropy of an attribute C as

$$H(C) = -\sum_{i=1}^{n_C} P(C = c_i) \log_2 P(C = c_i)$$

and provide an analogous definition for the Shannon entropy of the combination of two attributes C and A. (Note that we can interpret the events underlying the conditions $C = c_i$ as the alternatives referred to in the definition.) If it is written in this way, we see that Shannon information gain measures how many questions can be saved by asking directly for the value combination of C and A instead of asking for each of the values independently. This is directly analogous to the intuitive interpretation of Hartley information gain that was discussed w.r.t. Figure 7.2 on page 166.

Another way to understand Shannon information gain is to write it as

$$I_{\text{gain}}(C, A) = H(C) - H(C \mid A), \qquad \text{where}$$

$$H(C \mid A) = -\sum_{j=1}^{n_A} p_{.j} \sum_{i=1}^{n_C} \frac{p_{ij}}{p_{.j}} \log_2 \frac{p_{ij}}{p_{.j}} = -\sum_{j=1}^{n_A} p_{.j} \sum_{i=1}^{n_C} p_{i|j} \log_2 p_{i|j}$$

is the *expected (Shannon) entropy of C given A*. That is, $H(C \mid A)$ is the expected value of the average number of questions we have to ask in order to determine the value of attribute C if the value of attribute A becomes known. Subtracting this number from the number of questions we need without knowing the value of A (or disregarding it), we get the expected reduction of the average number of questions we have to ask. Using the interpretation of the Shannon entropy as an average code length per symbol, we may also say that Shannon information gain measures the expected reduction of the message length if an "unconditional" coding of the values of attribute C is replaced by a "conditional" coding that takes into account the value of attribute A.

It should be noted that this way of writing Shannon information gain is directly analogous to conditional Hartley information gain. However, in contrast to conditional Hartley information gain, which differs from its ordinary version, conditional Shannon information gain is identical to its ordinary version, as a simple calculation shows.

Although Shannon information gain is a well-founded measure for the strength of the dependence of two attributes, it has an unpleasant property: When it was used for decision tree induction, it was discovered that it is biased towards many-valued attributes [Quinlan 1993]. That is, it is likely that $I_{\text{gain}}(C, A_1) \leq I_{\text{gain}}(C, A_2)$ if the attribute A_2 has more possible values than A_1 and the probabilities are estimated from a database of sample cases. In other words: W.r.t. a given database two attributes can appear to be more strongly dependent than two others, simply because the former have more possible values. The reasons for this effect are twofold: The first is that Shannon information gain can only increase if the number of values of an attribute is increased, for example by splitting them. Formally this can be studied by comparing an attribute A to the combination of A and another attribute B.

Lemma 7.2.2 *Let A, B, and C be three attributes with finite domains and let their joint probability distribution be strictly positive, i.e.* $\forall a \in \text{dom}(A) : \forall b \in \text{dom}(B) : \forall c \in \text{dom}(C) : P(A = a, B = b, C = c) > 0$. *Then*

$$I_{\text{gain}}(C, AB) \;\geq\; I_{\text{gain}}(C, B),$$

with equality obtaining only if the attributes C and A are conditionally independent given B.

Proof. The proof, which is a mainly technical task, can be found in Section A.10 in the appendix. We provide a full proof (derived from a proof in [Press *et al.* 1992] that Shannon information gain is always non-negative), because it is rarely spelled out clearly.

Note that with the above lemma it is easily established that Shannon information gain is always nonnegative and zero only for independent attributes: Assume that attribute B has only one value. In this case it is $I_{\text{gain}}(C, B) = 0$, since the joint distribution on the values of the two attributes clearly coincides with the distribution on the values of C. In addition, the combination of the attributes A and B is obviously indistinguishable from A alone and thus we get $I_{\text{gain}}(C, AB) = I_{\text{gain}}(C, A)$. Consequently, we have as a corollary:

Corollary 7.2.3 *Let C and A be two attributes with finite domains and let their joint probability distribution be strictly positive, i.e.* $\forall c \in \text{dom}(C) : \forall a \in \text{dom}(A) : P(C = c, A = a) > 0$. *Then*

$$I_{\text{gain}}(C, A) \;\geq\; 0,$$

with equality obtaining only if C and A are (marginally) independent.

The second reason for the bias of Shannon information gain towards many-valued attributes is the quantization of the probabilities that is caused by the fact that they are estimated from a database. Since the database contains only a finite number of sample cases, the probabilities are restricted to a finite set of rational numbers. However, under these circumstances it is clear that Shannon information gain can increase even if an attribute A and a "split attribute" B are actually independent, simply because the probabilities needed to represent the independence may not be in this set of rational numbers. In such a case an attribute that is equivalent to the combination of A and B is judged to be more strongly dependent on an attribute C than the attribute A alone, merely because this attribute has more possible values.

In order to compensate the unpleasant bias towards many-valued attributes, several normalized variants of Shannon information gain have been suggested. The most widely known of these suggestions is the *(Shannon) information gain ratio* [Quinlan 1993], which is defined as

$$I_{\text{gr}}(C, A) = \frac{I_{\text{gain}}(C, A)}{H(A)} = \frac{H(C) + H(A) - H(CA)}{H(A)}.$$

The idea underlying this normalization is that an attribute with a larger number of values not only yields a higher information gain, but also has a higher entropy. Hence it is hoped that by dividing the information gain by the attribute entropy the two effects cancel each other.

Note that the so-called *uncertainty coefficient* [Press *et al.* 1992] is equivalent to the (Shannon) information gain, although it is defined as

$$U(C \mid A) = \frac{H(C) - H(C \mid A)}{H(C)},$$

since it is obvious that $H(C) + H(C \mid A) = H(A) + H(A \mid C)$ and thus $U(C \mid A) = I_{\text{gr}}(A, C)$. Hence the only difference is the "direction" of the measure (recall that for evaluating directed graphs we associate C with the child attribute of a directed edge and A with its parent).

However, the advantage of this way of writing the measure is that we can see that $I_{\text{gr}}(A, C)$ (as well as $I_{\text{gr}}(C, A)$) must be less than or equal to 1: By assuming that one of the attributes has only one value, we can easily derive from Corollary 7.2.3 that a (conditional) entropy cannot be negative. Consequently, the numerator of the uncertainty coefficient cannot be greater than $H(C)$ and thus the fraction cannot be greater than 1. Since from Corollary 7.2.3 we also known that $I_{\text{gain}}(C, A) \geq 0$, we have $0 \leq I_{\text{gr}}(C, A) \leq 1$.

The (Shannon) information gain ratio is not symmetric, i.e. in general it is $I_{\text{gr}}(C, A) \neq I_{\text{gr}}(A, C)$. Hence it is not suited to evaluate undirected edges. The *symmetric information gain ratio* [Lopez de Mantaras 1991]

$$I_{\text{sgr1}}(C, A) = \frac{I_{\text{gain}}(C, A)}{H(CA)} = \frac{H(C) + H(A) - H(CA)}{H(CA)}$$

avoids this drawback by dividing the (Shannon) information gain by the entropy of the combination of the attributes A and C. With a similar argument as above we obtain $0 \leq I_{\text{sgr1}}(C, A) \leq 1$.

Another symmetric normalization can be achieved by dividing by the sum of the individual entropies [Borgelt and Kruse 1998b], i.e.

$$I_{\text{sgr2}}(C, A) = \frac{I_{\text{gain}}(C, A)}{H(A) + H(C)} = \frac{H(C) + H(A) - H(CA)}{H(A) + H(C)}.$$

Of course, both of these normalizations compensate the bias of (Shannon) information gain towards many-valued attributes.

Note that the second symmetric information gain ratio is almost identical to the *symmetric uncertainty coefficient*, which is defined as a kind of weighted average of the uncertainty coefficients $U(C \mid A)$ and $U(A \mid C)$:

$$U(C, A) = \frac{H(C)U(C \mid A) + H(A)U(A \mid C)}{H(C) + H(A)} = 2\frac{H(C) + H(A) - H(A, C)}{H(C) + H(A)}.$$

From the left part of this formula it is clear that $0 \leq U(C, A) \leq 1$, since, as shown above, $0 \leq U(C \mid A) \leq 1$. Hence we have $0 \leq I_{\text{sgr2}}(C, A) \leq \frac{1}{2}$.

Quadratic Information Gain

As described in the preceding section, Shannon information gain can be based on Shannon entropy. However, Shannon entropy is not the only known type of entropy, since the concept of an entropy measure has been extended, for example by [Daróczy 1970]. His *generalized entropy* is defined as follows:

$$\begin{aligned} H_\beta^{(\text{general})}(S) &= \frac{2^{\beta-1}}{2^{\beta-1} - 1} \sum_{i=1}^{n} P(s_i) \left(1 - P(s_i)^{\beta-1}\right) \\ &= \frac{2^{\beta-1}}{2^{\beta-1} - 1} \left(1 - \sum_{i=1}^{n} P(s_i)^\beta\right). \end{aligned}$$

From this generalized entropy the Shannon entropy can be derived as

$$H^{(\text{Shannon})}(S) = \lim_{\beta \to 1} H_\beta^{(\text{general})}(S) = -\sum_{i=1}^{n} P(s_i) \log_2 P(s_i).$$

Another often used specialized version is the *quadratic entropy*

$$H^2(S) = H_{\beta=2}^{(\text{general})} = 2\sum_{i=1}^{n} P(s_i)(1 - P(s_i)) = 2 - 2\sum_{i=1}^{n} P(s_i)^2.$$

An intuitive interpretation of the quadratic entropy is the following: In order to determine the obtaining alternative we do not ask an imagined oracle as with Shannon entropy, but we simply guess. In doing so, we respect the known probabilities of the alternatives, i.e. we choose each alternative with its respective probability. Of course, we cannot be sure to guess the correct alternative. But we can determine how often our guess will probably be wrong. If our guess is the alternative s_i, the guess will be wrong with probability $1-P(s_i)$, because this is the probability that an alternative other than s_i is the obtaining one. Since we choose each alternative s_i with its probability $P(s_i)$, the probability that our guess is wrong is

$$P(\text{wrong guess}) = \sum_{i=1}^{n} P(s_i)(1 - P(s_i)).$$

The only difference of the quadratic entropy to this formula is the additional factor 2. Therefore, since the above expression is a probability, we can infer that it is $0 \leq H^2(S) \leq 2$. (Actually it must be strictly less than 2, since it is impossible that the probability of a correct guess vanishes.)

It is clear that the quadratic entropy can be used in direct analogy to the Shannon entropy to derive the *quadratic information gain*

$$I^2_{\text{gain}}(C, A) = H^2(C) + H^2(A) - H^2(CA).$$

For this measure a similar lemma holds as for Shannon information gain.

Lemma 7.2.4 *Let A, B, and C be attributes with finite domains. Then*

$$I^2_{\text{gain}}(C, AB) \;\geq\; I^2_{\text{gain}}(C, B).$$

Proof. The proof, which is a mainly technical task, can be found in Section A.11 in the appendix.

Note that the lemma is not restricted to strictly positive distributions and that we do not have the assertion that equality only holds if the attributes C and A are conditionally independent given B. Indeed, as the proof of this lemma shows, the two measures cannot be equal unless at least one of the attributes A and C has only one value with a non-vanishing probability.

In analogy to Shannon information gain we have the following corollary if we consider an attribute B with only one value. In this case it is $H^2(B) = 0$, $H^2(AB) = H^2(A)$, $H^2(CB) = H^2(C)$, and $H^2(CAB) = H^2(CA)$.

Corollary 7.2.5 *Let C and A be attributes with finite domains. Then it is*

$$I^2_{\text{gain}}(C, A) \;\geq\; 0.$$

From Lemma 7.2.4 it is to be expected that quadratic information gain is even more strongly biased towards many-valued attributes than Shannon information gain. In order to counteract this bias normalized versions may be considered. However, since they are defined in exact analogy to the normalized variants of Shannon information gain, we do not state them explicitly.

Gini Index

The conditional version of Shannon information gain is identical to the ordinary version, but with Hartley information gain we already saw a measure for which these two versions differed. For quadratic information gain they differ, too, and this provides us with another measure. This measure is well known from decision tree induction, where it is usually called the *Gini index* [Breiman *et al.* 1984, Wehenkel 1996]. It can be defined as

$$\begin{aligned}\text{Gini}(C, A) = \frac{1}{2}\left(H^2(C) - H^2(C|A)\right) &= 1 - \sum_{i=1}^{n_C} p_{i.}^2 - \sum_{j=1}^{n_A} p_{.j}\left(1 - \sum_{i=1}^{n_C} p_{i|j}^2\right) \\ &= \sum_{j=1}^{n_A} p_{.j} \sum_{i=1}^{n_C} p_{i|j}^2 - \sum_{i=1}^{n_C} p_{i.}^2.\end{aligned}$$

(Note that the factor $\frac{1}{2}$ only removes the factor 2 of the quadratic entropy.)

Drawing on the interpretation given for the quadratic entropy in the preceding section, we may say that the Gini index measures the expected reduction of the probability of a wrong guess.

In analogy to Shannon information gain and quadratic information gain the Gini index may be normalized in order to remove a bias towards many-valued attributes. We confine ourselves here to the symmetric version suggested by [Zhou and Dillon 1991].

$$\begin{aligned}\text{Gini}_{\text{sym}}(C, A) &= \frac{H^2(C) - H^2(C|A) + H^2(A) - H^2(A|C)}{H^2(C) + H^2(A)} \\ &= \frac{\sum_{i=1}^{n_C} p_{i.} \sum_{j=1}^{n_A} p_{j|i}^2 + \sum_{j=1}^{n_A} p_{.j} \sum_{i=1}^{n_C} p_{i|j}^2 - \sum_{i=1}^{n_C} p_{i.}^2 - \sum_{j=1}^{n_A} p_{.j}^2}{2 - \sum_{i=1}^{n_C} p_{i.}^2 - \sum_{j=1}^{n_A} p_{.j}^2}\end{aligned}$$

Note that the numerator is the sum of the two possible "directions" of the Gini index, since in general it is $\text{Gini}(C, A) \neq \text{Gini}(A, C)$.

Another way to reduce or even eliminate a bias towards many-valued attributes is the *modified Gini index*, which was suggested by [Kononenko 1994, Kononenko 1995] and which is closely related to the relief measure, which is discussed in the next section. It is defined as

$$\text{Gini}_{\text{mod}}(C, A) = \sum_{j=1}^{n_A} \frac{p_{.j}^2}{\sum_{j=1}^{n_A} p_{.j}^2} \sum_{i=1}^{n_C} p_{i|j}^2 - \sum_{i=1}^{n_C} p_{i.}^2.$$

The only difference from the ordinary Gini index is that it uses

$$\frac{p_{.j}^2}{\sum_{j=1}^{n_A} p_{.j}^2} \qquad \text{instead of} \qquad \frac{p_{.j}}{\sum_{j=1}^{n_A} p_{.j}} = p_{.j}.$$

By squaring the probabilities of the values of the attribute A, more probable values have a higher influence on the value of the measure. This also reduces the bias towards many-valued attributes, as can be made plausible by considering a uniform distribution on the values of the attribute A.

Relief Measure

The *relief measure* [Kira and Rendell 1992, Kononenko 1994, Kononenko 1995], which is closely related to the Gini index, has been devised mainly for feature selection and decision tree induction and thus it is strongly aimed at classification tasks. The idea underlying it is to assess a (descriptive) attribute A based on how well suited it is to predict the values of the (class) attribute C. Obviously, a good prediction can be achieved if the values of A correspond well with single values of C. At best, each value of A occurs always in conjunction with the same value of C.

The strength of such a correspondence can be measured by considering the probability that two sample cases having different values for the attribute C also have different values for the attribute A (this probability should be as large as possible) and the probability that two sample cases having the same value for the attribute C have different values for the attribute A (this probability should be as small as possible). Hence, the quality of an attribute A can be assessed by computing

$$\text{Relief}(C, A) = P(t_1|_A \neq t_2|_A \mid t_1|_C \neq t_2|_C) - P(t_1|_A \neq t_2|_A \mid t_1|_C = t_2|_C),$$

where t_1 and t_2 are two tuples that represent two sample cases. This difference can easily be transformed into

$$\begin{aligned}
\text{Relief}(C, A) &= \quad (1 - P(t_1|_A = t_2|_A \mid t_1|_C \neq t_2|_C)) \\
&\quad -(1 - P(t_1|_A = t_2|_A \mid t_1|_C = t_2|_C)) \\
&= \frac{P(t_1|_A = t_2|_A \wedge t_1|_C = t_2|_C)}{P(t_1|_C = t_2|_C)} \\
&\quad - \frac{P(t_1|_A = t_2|_A) - P(t_1|_A = t_2|_A \wedge t_1|_C = t_2|_C)}{1 - P(t_1|_C = t_2|_C)}.
\end{aligned}$$

Next we exploit the obvious relations

$$P(t_1|_A = t_2|_A) = \sum_{j=1}^{n_A} p_{.j}^2,$$

$$P(t_1|_C = t_2|_C) = \sum_{i=1}^{n_C} p_{i.}^2, \qquad \text{and}$$

$$P(t_1|_A = t_2|_A \wedge t_1|_C = t_2|_C) = \sum_{i=1}^{n_C} \sum_{j=1}^{n_A} p_{ij}^2.$$

These hold, since the probability that two sample cases have the same value a_j is clearly $p_{.j}^2$. Hence the probability that they have the same value, whichever it may be, is $\sum_{j=1}^{n_A} p_{.j}^2$. The two other cases are analogous. Thus we arrive at

$$\begin{aligned}
\mathrm{Relief}(C, A) &= \frac{\sum_{i=1}^{n_C}\sum_{j=1}^{n_A} p_{ij}^2}{\sum_{i=1}^{n_C} p_{i.}^2} - \frac{\sum_{j=1}^{n_A} p_{.j}^2 - \sum_{i=1}^{n_C}\sum_{j=1}^{n_A} p_{ij}^2}{1 - \sum_{i=1}^{n_C} p_{i.}^2} \\
&= \frac{\sum_{i=1}^{n_C}\sum_{j=1}^{n_A} p_{ij}^2 - \sum_{j=1}^{n_A} p_{.j}^2 \sum_{i=1}^{n_C} p_{i.}^2}{\left(\sum_{i=1}^{n_C} p_{i.}^2\right)\left(1 - \sum_{i=1}^{n_C} p_{i.}^2\right)}.
\end{aligned}$$

Using the equation

$$\sum_{i=1}^{n_C}\sum_{j=1}^{n_A} p_{ij}^2 = \sum_{j=1}^{n_A} p_{.j}^2 \sum_{i=1}^{n_C} p_{i|j}^2$$

we may also derive

$$\begin{aligned}
\mathrm{Relief}(C, A) &= \frac{\sum_{j=1}^{n_A} p_{.j}^2 \left(\sum_{j=1}^{n_A} \frac{p_{.j}^2}{\sum_{j=1}^{n_A} p_{.j}^2} \sum_{i=1}^{n_C} p_{i|j}^2 - \sum_{i=1}^{n_C} p_{i.}^2\right)}{\left(\sum_{i=1}^{n_C} p_{i.}^2\right)\left(1 - \sum_{i=1}^{n_C} p_{i.}^2\right)} \\
&= \frac{\left(\sum_{j=1}^{n_A} p_{.j}^2\right) \mathrm{Gini}_{\mathrm{mod}}(C, A)}{\left(\sum_{i=1}^{n_C} p_{i.}^2\right)\left(1 - \sum_{i=1}^{n_C} p_{i.}^2\right)},
\end{aligned}$$

by which the close relation to the Gini index is revealed.

Weight of Evidence

The *weight of evidence* [Michie 1989] was originally defined for binary attributes C, i.e. attributes with $\mathrm{dom}(C) = \{c_1, c_2\}$, as

$$w_{\mathrm{Evid}}(c_i, A) = \sum_{j=1}^{n_A} p_{.j} \left| \log_2 \frac{p_{i|j}/(1 - p_{i|j})}{p_{i.}/(1 - p_{i.})} \right|, \qquad i = 1, 2.$$

(Note that it is easy to verify that always $w_{\mathrm{Evid}}(c_1, A) = w_{\mathrm{Evid}}(c_2, A)$.)

The idea underlying this measure is to compare the odds of a bet on a value c_i, i.e. the quotient $\text{odds}(p) = p/(1-p)$, if we know the value of the attribute A to the odds of a bet if we do not know it. Obviously, the intention is again to assess how well suited a (descriptive) attribute A is to predict the value of a (class) attribute C, by which it is revealed that this measure, too, was devised for classification purposes. It is clear that an attribute A is judged to be the better, the greater the value of this measure.

The weight of evidence can easily be extended to attributes C with more than two values by defining it as [Kononenko 1995]

$$\begin{aligned} w_{\text{Evid}}(C, A) &= \sum_{i=1}^{n_C} p_{i.} w_{\text{Evid}}(c_i, A) \\ &= \sum_{i=1}^{n_C} p_{i.} \sum_{j=1}^{n_A} p_{.j} \left| \log_2 \frac{p_{i|j}/(1-p_{i|j})}{p_{i.}/(1-p_{i.})} \right|, \end{aligned}$$

i.e. by computing the weighted average over all values c_i.

Relevance

As the relief measure and the weight of evidence the *relevance* [Baim 1988] is also a measure devised for classification purposes. It is defined as

$$\begin{aligned} R(C, A) &= 1 - \frac{1}{n_C - 1} \sum_{j=1}^{n_A} \sum_{i=1, i \neq i_{\max}(j)}^{n_C} \frac{p_{ij}}{p_{i.}} \\ &= 1 - \frac{1}{n_C - 1} \sum_{j=1}^{n_A} p_{.j} \left(\sum_{i=1}^{n_C} \frac{p_{i|j}}{p_{i.}} - \max_i \frac{p_{i|j}}{p_{i.}} \right), \end{aligned}$$

where $c_{i_{\max}(j)}$ is the most probable value of the attribute C given that the attribute A has the value a_j. This measure is based on the same idea as the relief measure, namely that for a reliable identification of the value of a (class) attribute C it is best if each value of a (descriptive) attribute A uniquely indicates a value of C. Consequently, the conditional probability $p_{i|j}$ of the most probable value c_i of C given that the attribute A has the value a_j should be as large as possible. However, in order to avoid giving false merits to an attribute A for a value c_i with a high prior probability, these probabilities are considered relative to the prior probability $p_{i.}$.

χ^2 Measure

In Section 7.1.2 we interpreted Shannon information gain as a measure for the difference of the actual joint distribution and an assumed independent distribution of two attributes C and A. The χ^2 *measure*, which is well known in statistics, does the same, but instead of the pointwise quotient (as Shannon information gain does) it computes the pointwise squared difference of the two distributions. The χ^2 measure is usually defined as

$$\begin{aligned}
\chi^2(C, A) &= \sum_{i=1}^{n_C} \sum_{j=1}^{n_A} \frac{(E_{ij} - N_{ij})^2}{E_{ij}} \qquad \text{where } E_{ij} = \frac{N_{i.}\, N_{.j}}{N_{..}} \\
&= \sum_{i=1}^{n_C} \sum_{j=1}^{n_A} \frac{N_{..}^2 \left(\frac{N_{i.}}{N_{..}} \frac{N_{.j}}{N_{..}} - \frac{N_{ij}}{N_{..}} \right)^2}{N_{..} \frac{N_{i.}}{N_{..}} \frac{N_{.j}}{N_{..}}} \\
&= N_{..} \sum_{i=1}^{n_C} \sum_{j=1}^{n_A} \frac{(p_{i.}\, p_{.j} - p_{ij})^2}{p_{i.}\, p_{.j}}.
\end{aligned}$$

With the above transformation it is obvious that the numerator of the fraction is the squared difference of the actual joint distribution and the assumed independent distribution. The denominator serves to weight these pointwise differences. In order to render this measure independent of the number of sample cases, the factor $N_{..}$ is often discarded.

For the χ^2 measure we have a direct analog of Lemma 7.2.2:

Lemma 7.2.6 *Let A, B, and C be three attributes with finite domains and let their joint probability distribution be strictly positive, i.e.* $\forall a \in \mathrm{dom}(A) : \forall b \in \mathrm{dom}(B) : \forall c \in \mathrm{dom}(C) : P(A = a, B = b, C = c) > 0$. *Then*

$$\chi^2(C, AB) \geq \chi^2(C, B),$$

with equality obtaining only if the attributes C and A are conditionally independent given B.

Proof. The proof, which is mainly technical, can be found in Section A.12 in the appendix.

Note that we need no corollary in this case, because from the definition of the χ^2 measure it is already obvious that $\chi^2(C, A) \geq 0$.

The above lemma indicates that the χ^2 measure, like Shannon information gain and quadratic information gain, is biased towards many-valued attributes. However, in this case there is no simple way to eliminate this bias, because there is no obvious normalization.

A closely related measure, which differs from the χ^2 measure only in the way in which it weights the squared differences, is

$$d^2_{\text{weighted}}(C, A) = \sum_{i=1}^{n_C} \sum_{j=1}^{n_A} p_{ij}\, (p_{i.}\, p_{.j} - p_{ij})^2.$$

Although this measure is fairly obvious and, in a way, more natural than the χ^2 measure, it seems to have been neglected in the literature.

Bayesian–Dirichlet Metric

The *Bayesian–Dirichlet metric* is the result of a Bayesian approach to learning Bayesian networks from data, i.e. an approach that is based on Bayes' rule. It was first derived in the special form of the *K2 metric* by [Cooper and Herskovits 1992], which was later generalized by [Heckerman *et al.* 1995].

The derivation of this measure starts with a global consideration of the probability of a directed acyclic graph given a database of sample cases. Thus the explanation of this measure is more complex than that of the measures discussed above and involves an extension of the notation introduced in Section 7.2.2. However, we try to be as consistent as possible with our usual notation in order to avoid confusion.

The idea of the K2 metric is the following [Cooper and Herskovits 1992]: We are given a database D of sample cases over a set of attributes, each having a finite domain. It is assumed (1) that the process that generated the database can be accurately modeled by a Bayesian network, (2) that given a Bayesian network model cases occur independently, and (3) that all cases are complete, i.e. there are no missing or imprecise values. With these assumptions we can compute from the directed acyclic graph $\vec{G}$ and the set of conditional probabilities Θ underlying a given Bayesian network the probability of the database D. That is, we can compute $P(D \mid \vec{G}, \Theta)$, i.e. the *likelihood* of the database given the model (cf. Section 7.1.2). From this probability we can determine the probability of the Bayesian network given the database via Bayes' rule:[9]

$$P(\vec{G}, \Theta \mid D) = \frac{P(D \mid \vec{G}, \Theta) \cdot f(\vec{G}, \Theta)}{P(D)}.$$

However, this is not exactly what we want. Since we are concerned only with learning the *structure* of a Bayesian network, we should eliminate the conditional probabilities Θ. This is done by simply integrating the above formula over all possible choices of Θ. Thus we get

$$P(\vec{G} \mid D) = \frac{1}{P(D)} \int_{\Theta} P(D \mid \vec{G}, \Theta) f(\vec{G}, \Theta) \, \mathrm{d}\Theta.$$

Of course, to evaluate this formula, we need to know the prior probabilities $f(\vec{G}, \Theta)$ of the Bayesian network and $P(D)$ of the database. Fortunately, though, the prior probability of the database can be dispensed with, because we only need to be able to *compare* graph structures. For this the joint probability of the graph and the database is sufficient, since obviously

$$\frac{P(\vec{G}_1 \mid D)}{P(\vec{G}_2 \mid D)} = \frac{P(\vec{G}_1, D)}{P(\vec{G}_2, D)}.$$

[9]Note that we need a probability density function f for the prior probability of the Bayesian network, since the space of the conditional probabilities Θ is continuous.

Often this quotient, which is usually called a *Bayes factor*, is used explicitly to evaluate different graphs $\vec{G}_1$ w.r.t. a given *reference structure* $\vec{G}_2$. A commonly used reference structure is a graph without any edges. In the following, however, we confine ourselves to $P(\vec{G}, D)$. Starting from the formula derived above and applying the product rule of probability theory to the density $f(\vec{G}, \Theta)$, we arrive at

$$P(\vec{G}, D) = \int_{\Theta} P(D \mid \vec{G}, \Theta) f(\Theta \mid \vec{G}) P(\vec{G}) \, \mathrm{d}\Theta$$

as an assessment of the quality of a network structure $\vec{G}$ given a database D of sample cases. $f(\Theta \mid \vec{G})$ is a density function on the space of possible conditional probabilities and $P(\vec{G})$ is the prior probability of the graph $\vec{G}$.

In order to be able to evaluate this formula, it is assumed that all possible graphs $\vec{G}$ are equally likely and that the density functions $f(\Theta \mid \vec{G})$ are marginally independent for all pairs of attributes and for all pairs of instantiations of the parents of an attribute. This enables us to write the density function $f(\Theta \mid \vec{G})$ as a product of density functions with one factor for each attribute and each instantiation of its parents. This yields

$$P(\vec{G}, D) = \gamma \prod_{k=1}^{r} \prod_{j=1}^{m_k} \int_{\theta_{ijk}} \cdots \int \left(\prod_{i=1}^{n_k} \theta_{ijk}^{N_{ijk}} \right) f(\theta_{1jk}, \ldots, \theta_{n_k jk}) \, \mathrm{d}\theta_{1jk} \ldots \mathrm{d}\theta_{n_k jk},$$

where γ is a constant that represents the identical prior probability of each graph, r is the number of attributes used to describe the domain under consideration, m_k is the number of distinct instantiations of the parents of the attribute A_k in the graph $\vec{G}$, and n_k is the number of values of attribute A_k. θ_{ijk} is the probability that attribute A_k assumes the ith value of its domain, given that its parents are instantiated with the jth combination of values, and N_{ijk} is the number of cases in the database, in which the attribute A_k is instantiated with its ith value and its parents are instantiated with the jth value combination. Note that the notation N_{ijk} is consistent with the notation introduced in Section 7.2.2, because the additional index k only distinguishes the N_{ij} defined in that section for different child attributes. In the following this index is dropped, since we confine ourselves to single factors of the outermost product of the above expression, i.e. we consider only the contribution of the assessment of a single attribute and its parents in the graph to the overall quality of the graph $\vec{G}$. In addition, we confine ourselves to a child attribute C having only one parent A, since the (re)extension to more than one parent is obvious (cf. Section 7.2.1). Thus the following considerations are in line with the paradigm of the preceding sections.

In order to actually compute the factors of the above product, we need still another assumption, namely an assumption about the probability density function $f(\theta_{1j}, \ldots, \theta_{n_C j})$ on the space of the conditional probabilities. [Cooper and Herskovits 1992] assumed a uniform distribution, i.e. they assumed that $f(\theta_{1j}, \ldots, \theta_{n_C j}) = \gamma_j$, where γ_j is a constant, so that

$$\int \cdots \int_{\theta_{ij}} \gamma_j \, \mathrm{d}\theta_{1j} \ldots \mathrm{d}\theta_{n_C j} = 1.$$

This leads to $\gamma_j = (n_C - 1)!$. Intuitively, this formula can be made plausible as follows: Since the θ_{ij} are conditional probabilities, their sum must be one, i.e. $\sum_{i=1}^{n_C} \theta_{ij} = 1$. Hence we only have $n_C - 1$ free parameters. Suppose first that $n_C = 2$, so that there is only one free parameter. This parameter can be chosen freely from the interval $[0, 1]$ and thus the uniform density function on the parameter space is $f(x) \equiv 1$. Suppose next that $n_C = 3$. In this case we have two free parameters, the choices of which can be visualized as points in a plane. Since their sum must not exceed one, the points we may choose are restricted to the triangle that is formed by the coordinate axes and the line $x + y = 1$. This triangle clearly has an area of $\frac{1}{2}$ and thus the uniform distribution on the possible parameters must be $f(x, y) \equiv 2$. For $n_C = 4$ and thus three parameters the parameter space is the pyramid formed by the coordinate planes and the plane $x + y + z = 1$, which has the volume $\frac{1}{6}$ and thus the uniform density function is $f(x, y, z) \equiv 6$. In general, the hyper-pyramid in $n_C - 1$ dimensions that defines the parameter space has a hyper-volume of $\frac{1}{(n_C - 1)!}$, and thus the density function must have the value $(n_C - 1)!$. A formal justification of $\gamma_j = (n_C - 1)!$ is obtained with Dirichlet's integral [Dirichlet 1839]

$$\int \cdots \int_{\theta_{ij}} \prod_{i=1}^{n_C} \theta_{ij}^{N_{ij}} \, \mathrm{d}\theta_{1j} \ldots \mathrm{d}\theta_{n_C j} = \frac{\prod_{i=1}^{n_C} \Gamma(N_{ij} + 1)}{\Gamma(N_{.j} + n_C)},$$

where Γ is the well-known generalized factorial

$$\Gamma(x) = \int_0^\infty e^{-t} t^{x-1} \mathrm{d}t, \qquad \forall n \in \mathbb{N} : \Gamma(n+1) = n!,$$

by simply choosing $N_{ij} = 0$. This integral also provides us with means to evaluate the resulting formula. We arrive at [Cooper and Herskovits 1992]

$$\begin{aligned} K_2(C, A) &= \prod_{j=1}^{n_A} \int \cdots \int_{\theta_{ij}} \left(\prod_{i=1}^{n_C} \theta_{ij}^{N_{ij}} \right) (n_C - 1)! \, \mathrm{d}\theta_{1j} \ldots \mathrm{d}\theta_{n_C j} \\ &= \prod_{j=1}^{n_A} \frac{(n_C - 1)!}{(N_{.j} + n_C - 1)!} \prod_{i=1}^{n_C} N_{ij}!. \end{aligned}$$

This measure is known as the *K2 metric*, since it is used in the K2 algorithm [Cooper and Herskovits 1992]. Clearly, the greater the value of this evaluation measure (that is, its product over all variables), the better the corresponding graph $\vec{G}$. To simplify the computation often the logarithm of the above function is used:

$$\log_2(K_2(C, A)) = \sum_{j=1}^{n_A} \log_2 \frac{(n_C - 1)!}{(N_{.j} + n_C - 1)!} + \sum_{j=1}^{n_A} \sum_{i=1}^{n_C} \log_2 N_{ij}!.$$

In addition, since the value of this measure depends on the number of sample cases in the database, one may consider dividing this logarithm by the total number $N_{..}$ of sample cases [Borgelt *et al.* 1996].

As already mentioned at the beginning of this section the K2 metric was generalized to the *Bayesian–Dirichlet metric* by [Heckerman *et al.* 1995]. The idea underlying this generalization is very simple. Instead of assuming a uniform distribution on the space of the conditional probabilities, an explicit prior density function is used. For simplicity, this prior density function is assumed to be representable as a Dirichlet distribution, i.e. as

$$f(\theta_{1j}, \ldots, \theta_{n_C j}) = \prod_{i=1}^{n_C} \theta_{ij}^{N'_{ij} - 1}$$

for appropriate values N'_{ij}. It should be noted, though, that it is possible to justify this somewhat arbitrary choice by plausible additional assumptions [Heckerman *et al.* 1995]. Inserting the above product instead of the term $(n_C - 1)!$ into the formulae derived above, we get

$$\text{BD}(C, A) = \prod_{j=1}^{n_A} \frac{\Gamma(N'_{.j})}{\Gamma(N_{.j} + N'_{.j})} \prod_{i=1}^{n_C} \frac{\Gamma(N_{ij} + N'_{ij})}{\Gamma(N'_{ij})},$$

where $N'_{.j} = \sum_{i=1}^{n_C} N'_{ij}$. Note that it is necessary to use the Γ-function, since the N'_{ij} need not be integer numbers. Note also that the K2 metric is a special case of this measure, which results for $\forall i, j : N'_{ij} = 1$.

Intuitively, the N'_{ij} can be seen as derived from a database (other than the one to learn from) representing prior knowledge about the domain of consideration. Of course, such an interpretation is consistent only, if certain conditions hold. For example, $N'_{..} = \sum_{i=1}^{n_C} \sum_{j=1}^{n_A} N'_{ij}$ (or the extended version for more than one parent) must be the same for all attributes C, since, obviously, the size of the imagined database must be the same for all attributes. In addition, the frequency of an attribute value must not depend on whether the attribute is considered as a child or as a parent etc.

This consideration brings us directly to the notion of *likelihood equivalence*. An evaluation measure is called *likelihood equivalent* if it assigns the same value to all Markov equivalent graphs, where two graphs are called *Markov equivalent* if they represent the same set of conditional independence statements. As a very simple example of Markov equivalent graphs consider the three graphs $A \to B \to C$, $A \leftarrow B \to C$, and $A \leftarrow B \leftarrow C$. All three represent only the conditional independence statement $A \perp\!\!\!\perp C \mid B$.[10] Since a database of sample cases does not provide us with any information by which we could distinguish between two Markov equivalent graphs, it is desirable that an evaluation measure is likelihood equivalent.

The Bayesian–Dirichlet metric is likelihood equivalent if the N'_{ij} can be interpreted consistently as derived from a database representing prior information. This can be made plausible as follows: Consider two Markov equivalent graphs. For each choice of the probability parameters for one of them there is a corresponding choice for the other, since any probability distribution representable by one graph must be representable by the other. Next we exploit the fact that the Bayesian–Dirichlet metric is basically the computation of the likelihood of a database of sample cases (see above). It is clear that the likelihood of a database must be the same for two Bayesian networks that represent the same probability distribution, simply because the probability of the cases is read from this distribution. Since there is a 1-to-1 relation of the possible choices of probability parameters, it is also clear the integration over all possible choices cannot lead to a difference in the assessment. Finally, since the prior probabilities of all graphs are assumed to be the same, the two graphs must receive the same score.

The K2 metric is *not* likelihood equivalent, as can be seen from the fact that the values $\forall i, j : N'_{ij} = 1$ cannot be interpreted as derived from a database of sample cases, simply because the total number $N'_{..}$ depends on the numbers of values n_C and n_A of the two attributes. However, this insight directly provides us with an idea how a likelihood equivalent variant of the Bayesian–Dirichlet metric can be constructed, which nevertheless uses an uninformative, i.e. uniform, prior distribution. We only have to choose $\forall i, j : N'_{ij} = \frac{N'_{..}}{n_A n_C}$ [Buntine 1991, Heckerman *et al.* 1995], where $N'_{..}$ is a parameter called the *equivalent sample size*, thus indicating that it represents the size of an imagined database. Intuitively, this parameter determines the strength of the uniform distribution assumption relative to the database to learn from. The result is the so-called *BDeu metric* (for Bayesian–Dirichlet, likelihood equivalent, and uniform).

Unfortunately, the BDeu metric is strongly biased towards many-valued attributes (cf. the discussion of Shannon information gain, which also has this property). This bias can be made plausible as follows: The larger the N'_{ij} are, the weaker is the influence of the database, simply because the distribution is "equalized" by the uniform prior distribution, and thus the more strongly dependent the attributes must be in order to achieve a high score. However, the more values there are for the attributes A and C, the smaller the N'_{ij} are and thus the more strongly dependent they appear to be.

[10]Note that the notion of Markov equivalence is useful only for directed acyclic graphs, since no two distinct undirected graphs can represent the same set of conditional independences statements.

In the following we discuss an extension of the Bayesian–Dirichlet metric, which we suggested in [Borgelt and Kruse 2001]. In order to arrive at this extension, it is useful to start by showing that the Shannon information gain can also be derived with a Bayesian approach. The idea is as follows: In the derivation of the K2 metric it is assumed that the density functions on the spaces of the conditional probabilities are uniform. However, after we have selected a graph as the basis of a model, we no longer integrate over all conditional probabilities. Rather we fix the structure and compute estimates of these probabilities using, for example, maximum likelihood estimation. Thus the idea suggests itself to reverse these steps. That is, we could estimate first for each graph the best assignments of conditional probabilities and then select the best graph based on these, then fixed, assignments. Formally, this can be done by choosing the density functions in such a way that the (maximum likelihood) estimated probabilities have probability 1 and all others have probability 0. Thus we get

$$f(\theta_{1j}, \dots, \theta_{n_C j}) = \prod_{i=1}^{n_C} \delta\left(\theta_{ij} - p_{i|j}\right) = \prod_{i=1}^{n_C} \delta\left(\theta_{ij} - \frac{N_{ij}}{N_{.j}}\right),$$

where δ is Dirac's δ-function (or, more precisely, δ-distribution, since it is not a classical function), which is defined to have the following properties:

$$\delta(x) = \begin{cases} +\infty, & \text{if } x = 0, \\ 0, & \text{if } x \neq 0, \end{cases} \qquad \int_{-\infty}^{+\infty} \delta(x)\,\mathrm{d}x = 1, \qquad \int_{-\infty}^{+\infty} \delta(x - y)\varphi(x)\,\mathrm{d}x = \varphi(y).$$

Inserting this density function into the function for $P(\vec{G}, D)$, we get as an evaluation measure [Borgelt and Kruse 2001]:

$$\begin{aligned} g_\infty(C, A) &= \prod_{j=1}^{n_A} \int \cdots \int_{\theta_{ij}} \left(\prod_{i=1}^{n_C} \theta_{ij}^{N_{ij}}\right) \left(\prod_{i=1}^{n_C} \delta(\theta_{ij} - p_{i|j})\right) \mathrm{d}\theta_{1j} \dots \mathrm{d}\theta_{n_C j} \\ &= \prod_{j=1}^{n_A} \left(\prod_{i=1}^{n_C} p_{i|j}^{N_{ij}}\right). \end{aligned}$$

(The name g_∞ for this measure is explained below.) Obviously,

$$\begin{aligned} \frac{1}{N_{..}} \log_2 \frac{g_\infty(C, A)}{g_\infty(C)} &= \frac{1}{N_{..}} \sum_{j=1}^{n_A} \sum_{i=1}^{n_C} N_{ij} \log_2 p_{i|j} - \sum_{i=1}^{n_C} N_{i.} \log_2 p_{i.} \\ &= \sum_{j=1}^{n_A} p_{.j} \sum_{i=1}^{n_C} p_{i|j} \log_2 p_{i|j} - \sum_{i=1}^{n_C} p_{i.} \log_2 p_{i.} \\ &= H(C) - H(C \mid A) \;=\; I_{\text{gain}}^{(\text{Shannon})}(C, A), \end{aligned}$$

where $g_\infty(C)$ is the assessment of a parentless attribute C, which is obtained formally by letting $n_A = 1$. That is, Shannon information gain can be seen as a Bayes factor obtained from the evaluation measure g_∞.

This derivation of the Shannon information gain may be doubted, because in it the database is, in a way, used twice, namely once directly and once indirectly through the estimation of the parameters of the conditional probability distribution. Formally this approach is not strictly correct, since the density function over the parameter space should be a prior distribution whereas the estimate we used clearly is a posterior distribution (since it is computed from the database). However, the fact that Shannon information gain results—a well-founded evaluation measure—is very suggestive evidence that this approach is worth to be examined.

The above derivation of the Shannon information gain assumed Dirac pulses at the maximum likelihood estimates of the conditional probabilities. However, we may also consider using the likelihood function directly, i.e.

$$f(\theta_{1j}, \ldots, \theta_{n_C j}) = \beta \cdot \prod_{i=1}^{n_C} \theta_{ij}^{N_{ij}}, \qquad \text{where} \qquad \beta = \frac{\Gamma(N_{.j} + n_C)}{\prod_{i=1}^{n_C} \Gamma(N_{ij} + 1)}.$$

With this consideration the idea suggests itself to derive a family of evaluation measures: First we normalize the likelihood function, so that the maximum of this function becomes 1. This is easily achieved by dividing it by the maximum likelihood estimate raised to the power N_{ij}. Then we introduce an exponent α, by which we can control the "width" of the density function around the maximum likelihood estimate. Hence, if the exponent is 0, we get a constant function, if it is 1, we get the likelihood function, and if it approaches infinity, the density approaches Dirac pulses at the maximum likelihood estimate. Thus we arrive at [Borgelt and Kruse 2001]

$$\begin{aligned} f_\alpha(\theta_{1j}, \ldots, \theta_{n_C j}) &= \gamma \cdot \left(\left(\prod_{i=1}^{n_C} p_{i|j}^{-N_{ij}} \right) \left(\prod_{i=1}^{n_C} \theta_{ij}^{N_{ij}} \right) \right)^\alpha \\ &= \gamma \cdot \left(\prod_{i=1}^{n_C} p_{i|j}^{-\alpha N_{ij}} \right) \left(\prod_{i=1}^{n_C} \theta_{ij}^{\alpha N_{ij}} \right) \\ &= \gamma' \cdot \prod_{i=1}^{n_C} \theta_{ij}^{\alpha N_{ij}}, \end{aligned}$$

where γ and γ' are normalization factors to be chosen in such a way that the integral over $\theta_{1j}, \ldots, \theta_{n_C j}$ is 1 (since f_α is a density function). Using the solution of Dirichlet's integral (see above) we find that

$$\gamma' = \frac{\Gamma(\alpha N_{.j} + n_C)}{\prod_{i=1}^{n_C} \Gamma(\alpha N_{ij} + 1)}.$$

Inserting the above parameterized density into the function for the probability $P(\vec{G}, D)$ and evaluating the resulting formula using Dirichlet's integral yields the family of evaluation measures

$$g_\alpha(C, A) = \prod_{j=1}^{n_A} \frac{\Gamma(\alpha N_{.j} + n_C)}{\Gamma((\alpha+1)N_{.j} + n_C)} \cdot \prod_{i=1}^{n_C} \frac{\Gamma((\alpha+1)N_{ij} + 1)}{\Gamma(\alpha N_{ij} + 1)}.$$

From the explanations given in the course of the derivation of this family of measures, it is clear that we have three interesting special cases:

$\alpha = 0$:	K2 metric
$\alpha = 1$:	"likelihood metric"
$\alpha \to \infty$:	expected entropy $N_{..}H(C \mid A)$

Note that, of course, we may also consider generalizing this family of evaluation measures in the same way as the K2 metric was generalized to the Bayesian–Dirichlet metric. This yields

$$\mathrm{BD}_\alpha(C, A) = \prod_{j=1}^{n_A} \frac{\Gamma(\alpha N_{.j} + N'_{.j})}{\Gamma((\alpha+1)N_{.j} + N'_{.j})} \cdot \prod_{i=1}^{n_C} \frac{\Gamma((\alpha+1)N_{ij} + N'_{ij})}{\Gamma(\alpha N_{ij} + N'_{ij})}.$$

Note also that both families of evaluation measures may be attacked on the grounds that, at least formally, the factor α can also be interpreted as the assumption that we observed the database $(\alpha + 1)$ times, which would be ridiculous from a strictly statistical point of view. However, in our derivation of these families of measures we emphasized that the factor α results from a choice of the prior distribution and it has to be admitted that the choice of the prior distribution is, to some degree, arbitrary.

Of course, there are strong arguments in favor of a uniform prior distribution, for instance, the insufficient reason principle (cf. Section 2.4.3). However, when learning Bayesian networks from data choosing a uniform prior distribution introduces a tendency to select simpler network structures. This tendency results from the fact that the size of the space of conditional probabilities is larger for more complex structures and thus, in a way, there are more "bad" choices of conditional probabilities (i.e. choices that make the database unlikely). Consequently, a more complex graph may be judged to be worse than a simpler graph, although there is a choice of probability parameters for which the database is much more likely than with any choice of parameters for the simpler graph. (Another explanation, which is based on the close formal relation of the K2 metric and a minimum description length measure, is given in the next section.) It has to be admitted, though, that such a tendency can be desirable in order to avoid overfitting the data. However, even then it is usually convenient to be able to control the strength of this tendency. With the above families of evaluation measures such control can be exerted via the parameter α: The greater the value of this parameter, the weaker the tendency to select simpler structures.

Reduction of Description Length

With Shannon information gain we already discussed a measure that can be interpreted as the reduction of the (per symbol) coding length of a sequence of symbols. In addition to these direct coding costs for the values, the *minimum description length principle* [Rissanen 1983, Rissanen 1987] takes into account the costs for the transmission of the coding scheme.

Intuitively, the basic idea is the following: A sender S wants to transmit a message to a receiver R. Since transmission is costly, it is tried to encode the message in as few bits as possible. It is assumed that the receiver R knows about the symbols that may appear in the message, but does not know anything about their probabilities. Therefore the sender S cannot use directly, for instance, a Huffman code for the transmission, because without the probability information the receiver R will not be able to decode it. Hence the sender must either use a simpler (and longer) code, for which this information is not required, or he must transmit first the coding scheme or the probabilities it is based on. If the message to be sent is long enough, transmitting the coding scheme can pay, since the total number of bits that have to be transmitted may be lower as with a standard coding that does not take into account the probability information.

For learning Bayesian networks the situation is imagined as follows: The aim is to transmit the database of sample cases. Both the sender S and the receiver R know the number of attributes, their domains, and the number of cases in the database,[11] but at the beginning only the sender knows the values the attributes are instantiated with in the sample cases. These values are transmitted attribute by attribute, i.e. in the first step the value of the first attribute is transmitted for all sample cases, then the value of the second attribute, and so on. Thus the transmission may exploit dependences between the next attribute to be transmitted and already transmitted attributes to code the values more efficiently.

This description already shows that it is especially suited for Bayesian networks. The attributes are simply transmitted in a topological order[12] and the dependence of an attribute on its parents is used for a more efficient coding. Of course, if we do so, we must indicate which attributes are the parents of the next attribute to be transmitted and we must transmit the conditional probabilities. With this information the transmitted sequence of bits can be decoded. Note that, in this respect, the costs for the transmission

[11] Note that a strict application of the minimum description length principle would assume that these numbers are unknown to the receiver. However, since they have to be transmitted in any case, they do not have an influence on the ranking and thus are usually neglected or assumed to be known.

[12] The notion of a topological order was defined in Definition 4.1.12 on page 98. Note that a strict application of the minimum description length principle requires a transmission of the topological order. However, since the order of the attributes must be agreed upon in any case, the costs for its transmission are usually neglected.

of the coding scheme can also be seen as a penalty for making the model more complex. The more parents there are, the more additional information has to be transmitted. If the reduction of the costs for the transmission of the values is less than the increase in the costs for the transmission of the parents and the conditional probabilities, the simpler model, i.e. the one with fewer parent attributes, is preferred. These considerations suggest that it is useful to compute the *reduction of the message/description length* that can be achieved by using a(nother) parent attribute.

Depending on the way the attribute values are coded, at least two measures can be distinguished. For the first measure it is assumed that the values are coded based on their *relative frequencies*, for the second that they are coded based on their ***absolute frequencies***. As usual, we state both measures w.r.t. a single parent attribute.

W.r.t. a relative frequency coding the reduction of the description length is computed as follows [Kononenko 1995]:

$$
\begin{aligned}
L^{(\mathrm{rel})}_{\mathrm{prior}}(C) &= \log_2 \frac{(N_{..} + n_C - 1)!}{N_{..}!\,(n_C - 1)!} + N_{..}\, H_C, \\
L^{(\mathrm{rel})}_{\mathrm{post}}(C, A) &= \log_2 k + \sum_{j=1}^{n_A} \log_2 \frac{(N_{.j} + n_C - 1)!}{N_{.j}!\,(n_C - 1)!} + \sum_{j=1}^{n_A} N_{.j}\, H_{C|a_j}, \\
L^{(\mathrm{rel})}_{\mathrm{red}}(C, A) &= L^{(\mathrm{rel})}_{\mathrm{prior}}(C) - L^{(\mathrm{rel})}_{\mathrm{post}}(C, A).
\end{aligned}
$$

$L^{(\mathrm{rel})}_{\mathrm{prior}}(C)$ is the length of a description based on coding the values of the attribute C for all sample cases without the aid of another attribute. The first term of this length describes the costs for transmitting the value frequencies, which are needed for the construction of the code. It is derived as follows: Suppose there is a code book that lists all possible divisions of the $N_{..}$ sample cases on n_C values, one division per page, so that we only have to transmit the number of the page the obtaining division is printed on. We learn from combinatorics that this code book must have $\frac{(N_{..}+n_C-1)!}{N_{..}!\,(n_C-1)!}$ pages. Consequently, if we assume that all divisions are equally likely, the Hartley information of the pages, i.e. the binary logarithm of the number of pages, is the number of bits to transmit. The costs of transmitting the actual data, i.e. the values of the attribute C, are described by the second term. It is computed with the help of Shannon entropy, which states the average number of bits per value (cf. the section on Shannon information gain).

The length of a description with the help of an attribute A is computed in a directly analogous way. The samples cases are divided into n_A subsets w.r.t. the value of the attribute A. For each subset the description length is determined in the same way as above and the results are summed. To this the term $\log_2 k$ is added, which describes the costs for identifying the attribute A or, if there may be several parents, the set of parent attributes. k is the number of possible choices of parent attributes, which are assumed to be equally likely.

The interpretation of the term $\log_2 k$ is as above: We imagine a code book that lists all possible choices of parent attributes, one per page, and transmit the number of the page the actual selection is printed on. Note, however, that this term is often neglected based on the following argument: It is clear that we have to indicate the parent attribute(s) if parents are used, but we also have to indicate if no parent attribute is used, because otherwise the message cannot be decoded. Hence we have to add this term also to the description length $L^{(\text{rel})}_{\text{prior}}(C)$, with k comprising the choice of an empty set of parents.

Finally, the reduction of the description length that results from using the attribute A to code the values instead of coding them directly is computed as the difference of the above two description lengths. Note that this difference is simply Shannon information gain times $N_{..}$ plus the difference in the costs of transmitting the value frequencies.

If the coding is based on the absolute frequency of the values the reduction of the description length is computed as follows [Kononenko 1995]:

$$\begin{aligned} L^{(\text{abs})}_{\text{prior}}(C) &= \log_2 \frac{(N_{..} + n_C - 1)!}{N_{..}!\,(n_C - 1)!} + \log_2 \frac{N_{..}!}{N_{1.}! \cdots N_{n_C.}!}, \\ L^{(\text{abs})}_{\text{post}}(C, A) &= \log_2 k + \sum_{j=1}^{n_A} \log_2 \frac{(N_{.j} + n_C - 1)!}{N_{.j}!\,(n_C - 1)!} + \sum_{j=1}^{n_A} \log_2 \frac{N_{.j}!}{N_{1j}! \cdots N_{n_C j}!}, \\ L^{(\text{abs})}_{\text{red}}(C, A) &= L^{(\text{abs})}_{\text{prior}}(C) - L^{(\text{abs})}_{\text{post}}(C, A). \end{aligned}$$

The first term of the description length $L^{(\text{abs})}_{\text{prior}}$ is the same as for a coding based on relative frequencies. It describes the costs for transmitting the frequency distribution of the values of the attribute C. In this measure, however, the second term is not based on Shannon entropy, but is derived with a similar consideration as the first. That is, we imagine a code book that lists all possible assignments of the values of the attribute C that are compatible with the transmitted frequency distribution of these values. It is clear that each such assignment can be obtained as follows: First $N_{1.}$ cases are selected and the value c_1 is assigned to them, then from the remaining cases $N_{2.}$ are selected and the value c_2 is assigned to them, and so on. Consequently, we learn from combinatorics that the code book must have $\frac{N_{..}!}{N_{1.}! \cdots N_{n_C.}!}$ pages. As above we assume that all possible assignments are equally likely. Thus we get the Hartley information of the pages, i.e. the binary logarithm of the number of pages, as the number of bits needed to transmit the values of the attribute C.

The length of a description with the help of an attribute A is computed in a directly analogous way. As above, the sample cases are divided w.r.t. the value of the attribute A and the description length for each of the subsets is computed and summed. $\log_2 k$ describes the costs for identifying the parent attribute(s), although, as above, this term is often neglected.

Finally, the reduction of the description length is computed, as above, as the difference of the two description lengths. Note that this difference is closely related to a Bayes factor for the K2 metric. To be more precise,

$$L_{\text{red}}^{(\text{abs})}(C, A) = \log_2 \frac{K_2(C, A)}{K_2(C)} + \text{const.}$$

The above explanations should have made it clear that the first term of a description length, which describes the costs for the transmission of the value frequencies and thus the coding scheme, can be seen as a penalty for making the model more complex. Clearly, this penalty introduces a bias towards simpler network structures. Therefore, in analogy to the extension of the Bayesian–Dirichlet metric discussed in the preceding section, the idea suggests itself to introduce a parameter by which we can control the strength of this bias. This can be achieved by defining, for example,

$$L_{\text{prior},\alpha}^{(\text{rel})}(C) = \frac{1}{\alpha + 1} \log_2 \frac{(N_{..} + n_C - 1)!}{N_{..}!\,(n_C - 1)!} + N_{..}\, H_C$$

and analogously for the other description length. The penalty term is weighted with the term $\frac{1}{\alpha+1}$ instead of a simple factor in order to achieve matching ranges of values for the parameter α and the corresponding parameter of the extended Bayesian–Dirichlet metric. For the Bayesian–Dirichlet metric the normal behavior results for $\alpha = 1$ and for $\alpha \to \infty$ the measure approaches Shannon information gain. With the above form of the weighting factor we have the same properties for the description length measures.

Information Criteria

The notion of an *information criterion* is well known in the statistical literature on model choice. It is defined generally as the log-likelihood of the data given the model to evaluate plus a term that depends on the number of parameters of the model. Thus this criterion takes into account both the statistical goodness of fit and the number of parameters that have to be estimated to achieve this particular degree of fit, by imposing a penalty for increasing the number of parameters [Everitt 1998]. For learning graphical models it can be defined as

$$\text{IC}_\kappa(G, \Theta \mid D) = -2 \ln P(D \mid G, \Theta) + \kappa |\Theta|,$$

where D is the database of sample cases, G is the (directed or undirected) graph underlying the model, Θ is the set of probability parameter associated with this graph, and $|\Theta|$ is the number of parameters. $P(D \mid G, \Theta)$ is the likelihood of the database given the graphical model that is described by G and Θ (the computation of this probability was discussed for a Bayesian network in Section 7.1.2). It is clear that for $\kappa = 0$ we get a measure that is equivalent to a maximum likelihood approach to model selection. However, pure maximum likelihood is usually a bad choice, as it does not take care of the number of parameters.

Important special cases of the above general form are the so-called *Akaike Information Criterion* (AIC) [Akaike 1974] and the *Bayesian Information Criterion* (BIC) [Schwarz 1978]. The former results for $\kappa = 2$ and is derived from asymptotic decision theoretic considerations. The latter has $\kappa = \ln N$, where N is the number of sample cases, and is derived from an asymptotic Bayesian argument [Heckerman 1998].

7.2.5 Possibilistic Evaluation Measures

There are much fewer possibilistic than probabilistic evaluation measures, mainly because possibility theory is a rather young theory. The first possibilistic evaluation measures were suggested in [Gebhardt and Kruse 1995, Gebhardt and Kruse 1996b]. Others followed in [Borgelt *et al.* 1996, Borgelt and Kruse 1997a]. All of them are derived either from relational or from probabilistic evaluation measures.

Specificity Gain

Specificity gain was already introduced in Definition 7.1.8 on page 183. Restated with the abbreviations introduced in Section 7.2.2 it reads

$$\begin{aligned} S_{\text{gain}}(C, A) \;=\; \int_0^{\sup(\pi_{ij})} \Bigg(& \log_2 \left(\sum_{i=1}^{n_C} [\pi_{i.}]_\alpha \right) + \log_2 \left(\sum_{j=1}^{n_A} [\pi_{.j}]_\alpha \right) \\ & - \log_2 \left(\sum_{i=1}^{n_C} \sum_{j=1}^{n_A} [\pi_{ij}]_\alpha \right) \Bigg) \, \mathrm{d}\alpha \end{aligned}$$

In Section 7.1.3 this measure was justified as a generalization of Hartley information gain drawing on the α-cut[13] view of a possibility distribution.

Another way of justifying this measure is via the notion of the *nonspecificity* of a possibility distribution, which was introduced in Definition 7.1.6 on page 178 and which plays a role in possibility theory that is similar to the role of Shannon entropy in probability theory. By using nonspecificity in the same way as we used Hartley entropy and Shannon entropy to derive the corresponding information gains, we get [Borgelt *et al.* 1996]:

$$S_{\text{gain}}(C, A) \;=\; \text{nonspec}(\pi_C) + \text{nonspec}(\pi_A) - \text{nonspec}(\pi_{CA}),$$

where π_C is the possibility distribution consisting of the degrees of possibility $\pi_{i.}$ for all values $i \in \{1, \ldots, n_C\}$ (cf. Section 7.1.3). This measure is identical to the evaluation measure suggested in [Gebhardt and Kruse 1996b], although it is not called *specificity gain* in that paper.

[13]The notion of an α-cut was introduced in Definition 7.1.5 on page 177.

In analogy to Hartley information gain and Shannon information gain there are several ways in which this measure may be normalized in order to eliminate or at least lessen a possible bias towards many-valued attributes. This leads to the *specificity gain ratio*

$$S_{\mathrm{gr}}(C, A) = \frac{S_{\mathrm{gain}}(C, A)}{\mathrm{nonspec}(\pi_A)} = \frac{\mathrm{nonspec}(\pi_C) + \mathrm{nonspec}(\pi_A) - \mathrm{nonspec}(\pi_{CA})}{\mathrm{nonspec}(\pi_A)}$$

and two *symmetric specificity gain ratios*, namely

$$S_{\mathrm{sgr}}^{(1)}(C, A) = \frac{S_{\mathrm{gain}}(C, A)}{\mathrm{nonspec}(\pi_{CA})} \quad \text{and}$$

$$S_{\mathrm{sgr}}^{(2)}(C, A) = \frac{S_{\mathrm{gain}}(C, A)}{\mathrm{nonspec}(\pi_A) + \mathrm{nonspec}(\pi_C)}.$$

A conditional version of specificity gain may also be defined by drawing on the conditional version of Hartley information gain (cf. Section 7.2.3):

$$S_{\mathrm{cgain}}(C, A) = \sum_{j=1}^{n_A} \int_0^{\pi_{.j}} \frac{[\pi_{.j}]_\alpha}{\sum_{j=1}^{n_A} [\pi_{.j}]_\alpha} \log_2 \frac{\sum_{i=1}^{n_C} [\pi_{i.}]_\alpha}{\sum_{j=1}^{n_A} [\pi_{i|j}]_\alpha} \, d\alpha.$$

Possibilistic Mutual Information

In Section 7.1.3 Shannon information gain was introduced under the name of mutual (Shannon) information as a measure that compares the actual joint distribution and an assumed independent distribution by computing their pointwise quotient. This idea can be transferred by defining

$$d_{\mathrm{mi}}(C, A) = -\sum_{i=1}^{n_C} \sum_{j=1}^{n_A} \pi_{ij} \log_2 \frac{\pi_{ij}}{\min\{\pi_{i.}, \pi_{.j}\}}$$

as a direct analog of mutual (Shannon) information [Borgelt and Kruse 1997a]. (The index "mi" stands for "mutual information".)

Possibilistic χ^2 measure

The χ^2 measure, as it was studied in Section 7.2.4, also compares the actual joint distribution and an assumed independent distribution. However, it does so by computing the pointwise squared difference. It is clear that this idea may as well be transferred, so that we get [Borgelt and Kruse 1997a]

$$d_{\chi^2}(C, A) = \sum_{i=1}^{n_C} \sum_{j=1}^{n_A} \frac{(\min\{\pi_{i.}, \pi_{.j}\} - \pi_{ij})^2}{\min\{\pi_{i.}, \pi_{.j}\}}.$$

Alternatively, one may compute the weighted sum of the squared differences of the individual degrees of possibility, i.e. one may compute

$$d_{\text{diff}}(C, A) = \sum_{i=1}^{n_C} \sum_{j=1}^{n_A} \pi_{ij} (\min\{\pi_{i.}, \pi_{.j}\} - \pi_{ij})^2.$$

As the corresponding alternative to the χ^2 measure in the probabilistic setting, this measure appears to be more natural.

Weighted Sum of Degrees of Possibility

As a global possibilistic evaluation measure we discussed in Section 7.1.3 the *weighted sum of degrees of possibility.* For a given graph G and a given database $D = (R, w_R)$ it is defined as [Borgelt and Kruse 1997b]

$$Q(G) = \sum_{t \in R} w_R(t) \cdot \pi_G(t)$$

where π_G is the possibility distribution represented by the graph G and its associated distribution functions. Since this measure was already studied in Section 7.1.3 we only mention it here for completeness.

Note that the weighted sum of possibility degrees may be penalized—in analogy to the information criteria in the possibilistic case (cf. Section 7.2.4)—by adding a term $\kappa|\Theta|$, where Θ is the number of parameters and κ is a constant. However, κ should be chosen by a factor of about 1000 smaller than in the probabilistic case.

7.3 Search Methods

Being provided by the preceding section with a variety of measures to evaluate a given graphical model, we turn to search methods in this section. As indicated above, a search method determines which graphs are considered in order to find a good model. In Section 7.3.1 we study an exhaustive search of all graphs, mainly to show that it is infeasible. Later we turn to heuristic approaches like random guided search (Section 7.3.2) and greedy search (Section 7.3.4). In addition, we consider the special case of a search based on conditional independence tests (Section 7.3.3).

7.3.1 Exhaustive Graph Search

The simplest search method is, of course, an exhaustive search of the space of possible graphs. That is, all possible candidate graphs are inspected in turn and evaluated. The graph with the best assessment is selected as the search result. As an illustration recall the examples of the preceding section, in which all eight possible candidate graphs were evaluated.

Table 7.3 Some examples for the number of undirected graphs (upper row) and the number of directed acyclic graphs (lower row) over n attributes.

n	2	3	4	5	6	7	8	10
$2^{\binom{n}{2}}$	2	8	64	1024	32768	$2.10 \cdot 10^6$	$2.68 \cdot 10^8$	$3.52 \cdot 10^{13}$
$f(n)$	3	25	543	29281	$3.78 \cdot 10^6$	$2.46 \cdot 10^8$	$7.84 \cdot 10^{11}$	$4.18 \cdot 10^{18}$

Clearly, this approach is guaranteed to find the "best" graphical model—at least w.r.t. the evaluation measure used. However, in applications it is infeasible, since the number of candidate graphs is huge unless the number of attributes used to describe the domain under consideration is *very* small. Therefore the main purpose of this section is not to discuss this search method as a reasonable alternative, but to bring out clearly that heuristic search methods are indispensable.

To see that the number of candidate graphs is huge, consider first the number of undirected graphs over n attributes. In an undirected graph any two attributes may either be connected by an edge or not (two possibilities) and from n attributes $\binom{n}{2}$ different pairs of attributes can be selected. Therefore the number of undirected graphs over n attributes is $2^{\binom{n}{2}}$. As an illustration recall that for the three-dimensional examples studied in the preceding section there were $2^{\binom{3}{2}} = 2^3 = 8$ possible undirected graphs (cf. Figures 7.1, 7.5, and 7.8 on pages 164, 173, and 180, respectively). Some examples for other values of n are shown in Table 7.3 (upper row). Obviously, for more than 6 or 7 attributes an exhaustive search is impossible.

Consider next the number of directed acyclic graphs over n attributes. This number is much more difficult to determine than the number of undirected graphs, because the requirement for acyclicity is a somewhat inconvenient constraint. However, a lower and an upper bound can easily be found: To determine an upper bound, we may simply drop the requirement for acyclicity and consider arbitrary directed graphs. In such a graph any two attributes A and B may be unconnected, or connected by an edge from A to B, or connected by an edge from B to A (three possibilities) and, as above, from n attributes $\binom{n}{2}$ pairs of attributes can be selected. Therefore the number of arbitrary directed graphs over n attributes is $3^{\binom{n}{2}}$. To determine a lower bound, suppose that a topological order[14] of the attributes has been fixed and consider the directed acyclic graphs that are compatible with this order. In such a graph any two attributes A and B are either unconnected or connected by a directed edge from the node having the lower rank in the topological order

[14]The notion of a topological order is defined in Definition 4.1.12 on page 98.

to the node having the higher rank (two possibilities). Again there are $\binom{n}{2}$ pairs of attributes and hence there are $2^{\binom{n}{2}}$ directed acyclic graphs over n attributes compatible with a given topological order. We conclude that the number of all directed acyclic graphs must be between $2^{\binom{n}{2}}$ and $3^{\binom{n}{2}}$. An exact recursive formula for the number of directed acyclic graphs with n nodes has been found by [Robinson 1977]

$$f(n) = \sum_{i=1}^{n} (-1)^{i+1} \binom{n}{i} 2^{i(n-i)} f(n-i).$$

Some examples of the value of this function are shown in Table 7.3 (lower row). Obviously, the situation is even worse than for undirected graphs, since the number of candidate graphs grows more rapidly, and thus an exhaustive search is clearly impossible for more than 5 or 6 attributes.

Of course, the mere fact that the space of candidate graphs is too large to be searched exhaustively does not imply that there is no feasible method to find the optimal result. For instance, the number of spanning trees over n attributes is n^{n-2} [Bodendiek and Lang 1995] and thus it is clearly impossible to search them exhaustively in order to find an optimum weight spanning tree w.r.t. given edge weights. Nevertheless, the well-known Kruskal algorithm [Kruskal 1956] is guaranteed to construct an optimum weight spanning tree and it is clearly efficient. (There is also an even more efficient algorithm for this task [Prim 1957], which, however, is more difficult to implement.)

However, no such efficient algorithm has been found yet for learning graphical models from data. Even worse, some special problems that occur in connection with learning graphical models are known to be NP-hard. For example, it is known from database theory that deciding whether a given relation is decomposable[15] w.r.t. a given family of attribute sets is NP-hard [Dechter and Pearl 1992]. Likewise, it is known to be NP-hard to find the minimal decomposition[16] of a given relation and it has been conjectured that it is NP-hard to determine whether a given relation can be decomposed even if the size of the attribute sets is restricted to some maximum number k [Dechter and Pearl 1992]. Finally, the specific task of learning a Bayesian network has been shown to be NP-hard if the Bayesian–Dirichlet metric (cf. Section 7.2.4) is used to evaluate the networks [Chickering *et al.* 1994, Chickering 1995]. As a consequence, it seems to be inevitable to accept suboptimal results.

7.3.2 Guided Random Graph Search

If an exhaustive search is infeasible or very costly, a standard solution is to use some heuristic search method, for example, a *guided random search.* The two best known examples of this class of search methods are *simulated*

[15]The notion of a relation being decomposable is defined in Definition 3.2.4 on page 65.
[16]The notion of a minimal decomposition is defined in Definition 3.2.5 on page 65.

annealing and *genetic* or *evolutionary algorithms*. We call these approaches "guided random search methods", because both involve an element of chance, but are also guided by an evaluation measure.

Simulated Annealing

The basic idea of *simulated annealing* [Metropolis *et al.* 1953, Kirkpatrick *et al.* 1983] is to start with a randomly generated candidate solution, which is evaluated. Then this candidate solution is modified randomly and the resulting new candidate solution is evaluated. If the new candidate solution is better than the original one, it is accepted and replaces the original one. If it worse, it is accepted only with a certain probability that depends on how much worse the new candidate solution is. In addition, this probability is lowered in the course of time, so that eventually only those new candidate solutions are accepted that are better than the current. Often the best solution found so far is recorded in parallel.

The reason for accepting a new candidate solution even though it is worse than the current is that without doing so the approach would be very similar to a gradient ascent (or descent). The only difference is that the direction of the gradient of the solution quality is not computed, but that the upward (or downward) direction is searched for by trial and error. However, it is well known that a gradient ascent (or descent) can easily get stuck in a local optimum. By accepting worse candidate solutions at the beginning of the process it is tried to overcome this undesired behavior. Intuitively, accepting worse candidate solutions makes it possible to cross the "barriers" that separate local optima from the global one, i.e. regions of the search space where the quality of the candidate solutions is worse. Later, however, when the probability for accepting worse candidate solutions is lowered, the quality function is optimized locally.

The name "simulated annealing" for this approach stems from the fact that it is similar to the physical minimization of the energy function (to be more precise: the atom lattice energy) when a heated piece of metal is cooled down very slowly. This process is usually called "annealing" and is used to soften a metal, relieve internal stresses and instabilities, and thus make it easier to work or machine. Physically, the thermal activity of the atoms prevents them from settling in a configuration that may be only a local minimum of the energy function. They "jump out" of this configuration. Of course, the "deeper" the (local) energy minimum, the harder it is for the atoms to "jump out" of the configuration. Hence, by this process they are likely to settle in a configuration of very low energy, the optimum of which is, in the case of a metal, a monocrystalline structure. It is clear, though, that it cannot be guaranteed that the global minimum of the energy function is reached. Especially if the piece of metal is not heated long enough, the atoms are likely to settle in a configuration that is only a local minimum (a polycrystalline structure in the case of a metal). Hence it is important to lower the temperature very slowly, so that there is a high probability that local minima, once reached, are left again.

This energy minimization can be visualized by imagining a ball rolling around on a curved landscape [Nauck *et al.* 1997]. The function to be minimized is the potential energy of the ball. At the beginning the ball is endowed with a certain kinetic energy which enables it to "climb" the slopes of the landscape. But due to friction this kinetic energy is diminished in the course of time and finally the ball comes to a rest in a valley (a minimum of the function to be optimized). Since it takes a higher kinetic energy to leave a deep valley than to leave a shallow one, the final resting point is likely to be in a rather deep valley and maybe in the deepest one (the global minimum).

The thermal energy of the atoms in the annealing process or the kinetic energy of the ball in the illustration is modeled by the decreasing probability for accepting a worse candidate solution. Often an explicit temperature parameter is introduced, from which the probability (parameterized by how much worse the new candidate solution is) is computed. Since the probability distribution of the velocities of atoms is often an exponential distribution (cf., for example, the Maxwell distribution, which describes the velocity distribution for an ideal gas [Greiner *et al.* 1987]), a function like $P(\text{accept}) = ce^{-\frac{dQ}{T}}$ is frequently used to compute the probability for accepting a worse solution, where dQ is the quality difference of the current and the new candidate solution, T is the temperature parameter and c is a normalization constant.

Genetic or Evolutionary Algorithms

The idea of of *genetic* or *evolutionary algorithms* [Nilsson 1998, Michalewicz 1996, Koza 1992], is to employ an analog of biological evolution [Darwin 1859, Dawkins 1976, Dawkins 1986] to optimize a given function (here: the quality of a graphical model w.r.t. the given data). In this approach the candidate solutions are coded into *chromosomes* with individual *genes* representing the components of a candidate solution. For example, for learning undirected graphical models a chromosome may be a simple bit-string in which each bit is a gene representing one edge. If the bit is set, the edge is present in the candidate solution described by the chromosome, otherwise it is absent. Thus each bit-string describes an undirected graph.

A genetic or evolutionary algorithm starts by generating a random initial *population* of individuals, each with its own chromosome. These individuals—or, to be more precise, the candidate solutions represented by their chromosomes[17]—are evaluated by a *fitness function*, which is the function to be optimized (or derived from it).

[17] As in biology one may distinguish between the *genotype* of a living being, which is its genetic constitution, and its *phenotype*, which denotes its physical appearance or, in the context of genetic algorithms, the represented candidate solution.

From the initial population a new population is generated by two means: The first is a simple *selection* process. A certain number of individuals is selected at random, with the probability that a given individual gets selected depending on its fitness. A simple method to achieve such a selection behavior is *tournament selection*: A certain number of individuals is picked at random from the population and the one with the highest fitness among them (the "winner of the tournament") is selected. It is clear that with this selection method individuals with a high fitness have a better chance to be passed into the new population than those with a low fitness and thus only the fittest individuals of a population "survive", illustrating the (somewhat simplistic) characterization of biological evolution as the *survival of the fittest.* Of course, the individuals are also randomly modified from time to time (as in simulated annealing), thus imitating *mutation*, which in living beings occurs due to errors in the chromosome copying process.

The second process that is involved in generating the new population imitates *sexual reproduction.* Two "parent" individuals are chosen from the population, again with a probability depending on their fitness (for example, using tournament selection). Then their chromosomes are *crossed over* in order to obtain two new individuals that differ from both "parents".[18] A very simple method to do so is to fix a breakage point on the chromosomes and then to exchange one of the parts. The idea underlying the crossing-over of chromosomes is that each of the "parent" chromosomes may already describe a good partial solution, which accounts for their high fitness (recall that the "parents" are selected with a probability depending on their fitness, so individuals with a high fitness are more likely to become "parents"). By crossing-over their chromosomes there is a good chance that these partial solutions are combined and that consequently an "offspring" chromosome is better than both of the "parents". This plausible argument is made formally precise by the *schema theorem* [Michalewicz 1996]. It explains why evolution is much faster with sexual reproduction than without it (i.e. with mutation being the only mechanism by which genetically new individuals can emerge).

Of course, the new population is then taken as a starting point for generating the next and so on, until a certain number of generations has been created or the fitness of the best member of the population has not increased in the last few generations. The result of a genetic algorithm is the fittest individual of the final generation or the fittest individual that emerged during the generations (if it is kept track of).

There are several variants of genetic or evolutionary algorithms, depending on whether only "offspring" is allowed into the next population or whether

[18]The term *crossing-over* was chosen in analogy to the biological process with the same name in which genetic material is exchanged between (homologous) chromosomes by breakage and reunion. This process happens during *meiosis* (reduction division), i.e. the division of (homologous) chromosome pairs so that each *gamete* (a sex cell, for example an egg) receives one chromosome.

"parents" are passed, too, whether the population is processed as a whole or split into subgroups with "mating" occurring only within subgroups and only rare "migrations" of individuals from one subpopulation to another etc. [Michalewicz 1996].

Application to Learning Graphical Models

As an illustration of how guided random search can be used to learn graphical models from data, we consider in the remainder of this section a simulated annealing approach to learn undirected graphs with hypertree structure[19] [Borgelt and Gebhardt 1997]. That is, we consider a method that tries to find a family of sets of attributes having the running intersection property, so that the corresponding undirected graph, i.e. the undirected graph, the maximal cliques of which are induced by these attribute sets, is optimal w.r.t. a given evaluation measure.

Trying to find directly the maximal cliques of an undirected graph instead of working with individual (directed or undirected) edges has several advantages: In the first place, compared to an approach based on directed acyclic conditional independence graphs the larger search space is avoided (cf. Section 7.3.1). Secondly, compared to an approach based on arbitrary undirected conditional independence graphs we need not determine the maximal cliques of graphs—which is necessary for evaluating them, since for this we have to compute the factor potentials on the maximal cliques (cf. Sections 3.3.3, 4.2.2, and 5.2) or an equivalent thereof—but have them readily available. In addition, for an undirected graph with hypertree structure it is much simpler to compute (an equivalent of) the factor potentials for the maximal cliques (see below). Furthermore, since not all undirected graphs have hypertree structure, the search space is smaller. Finally, and this may be the most important advantage in some applications, this approach allows us to control directly the complexity of the learned graphical model and, especially, the complexity of reasoning with this model, namely by limiting the size of the maximal cliques.

That the complexity of reasoning can be controlled in this way is obvious if join tree propagation (cf. Section 4.2.2) is used to update the represented distribution w.r.t. given evidence, but holds equally well for other propagation methods. To see that exerting such control is more difficult with other learning approaches recall that the preparations for join tree propagation involve a triangulation step (cf. Algorithm 4.2.1 on page 129), in which edges may be added to the graph. How many edges have to be added in this step depends in a complex way on the structure of the graph, which in practice makes it impossible to foresee and thus to control the size of the maximal cliques of the triangulated graph.

The main task when developing a simulated annealing approach to learn undirected graphs with hypertree structure is, obviously, to find an efficient

[19] The notion of a hypertree structure was defined in Definition 4.1.23 on page 110.

method to randomly generate and modify such graphs. In the following we consider two alternatives and discuss their respective merits and drawbacks. Both approaches exploit the fact that in order to ensure that a graph has hypertree structure we only have to make sure that the family of attribute sets underlying its maximal cliques has the running intersection property,[20] which guarantees acyclicity, and that there is a path between any pair of nodes, which guarantees connectedness (although in applications it is often useful and convenient to relax the latter condition).

The first approach relies directly on the defining condition of the running intersection property for a family $\mathcal{M}$ of sets, namely that there is an ordering $M_1, \ldots, M_m$ of the sets in $\mathcal{M}$, so that

$$\forall i \in \{2, \ldots, m\} : \exists k \in \{1, \ldots, i-1\} : \quad M_i \cap \Big(\bigcup_{1 \leq j < i} M_j \Big) \subseteq M_k.$$

W.r.t. graphs with hypertree structure such an ordering of the node sets underlying the maximal cliques is often called a *construction sequence* (cf. the proof of Theorem 4.1.24 in Section A.5 in the appendix).

The idea of the first approach is to select the node sets in exactly this order: We start with a random set M_1 of nodes. In step i, $i \geq 2$, a set M_k, $1 \leq k < i$, is selected at random and the set M_i, which is to be added in this step, is formed by randomly selecting nodes from $M_k \cup \big(U - \bigcup_{1 \leq j < i} M_j\big)$ making sure that at least one node from the set M_k and at least one node not in $\bigcup_{1 \leq j < i} M_j$ is selected. This process is repeated until all nodes are contained in at least one set M_j, $1 \leq j \leq i$. It is clear that the probabilities of the different set sizes and the probability with which a node in M_k or a node in $\bigcup_{1 \leq j < i} M_j$ is selected are parameters of this method. Convenient choices are uniform distributions on sizes as well as on nodes.

In order to randomly modify a given undirected graph with hypertree structure or, equivalently, the family $\mathcal{M}$ of node sets underlying its maximal cliques, this approach exploits the so-called *Graham reduction*, which is a simple method to test whether a family $\mathcal{M}$ of sets has the running intersection property [Kruse *et al.* 1994]:

Algorithm 7.3.1 *(Graham reduction)*

Input: *A finite family $\mathcal{M}$ of subsets of a finite set U of objects.*

Output: *Whether $\mathcal{M}$ has the running intersection property.*

The family $\mathcal{M}$ of sets of objects is reduced by iteratively applying one of the following two operations:

1. *Remove an object that is contained in only one set $M \in \mathcal{M}$.*
2. *Remove a set $M_1 \in \mathcal{M}$ that is a subset of another set $M_2 \in \mathcal{M}$.*

[20]The running intersection property was defined in Definition 4.1.23 on page 110.

The process stops if neither operation is applicable. If all objects appearing in the sets of the original family $\mathcal{M}$ could be removed, the original family $\mathcal{M}$ has the running intersection property, otherwise it does not have it.

Note that this algorithm is non-deterministic, since situations may arise in which both operations are applicable or in which one of them is applicable to more than one object or more than one set, respectively. Note also that the first operation need not be implemented explicitly if one maintains a counter for each object, which records the number of sets in the current family $\mathcal{M}$ the object is contained in. The subset test is then programmed in such a way that it takes into account only those objects for which the counter is greater than 1. Finally, note that this algorithm yields a construction sequence: The reverse of the order in which the sets $M \in \mathcal{M}$ were removed from $\mathcal{M}$ obviously provides us with such a sequence.

How Graham reduction works can easily be visualized by considering a *join tree* (cf. page 130) of the graph corresponding to the family $\mathcal{M}$, for example, the join tree for the Danish Jersey Cattle example (cf. Figure 4.14 on page 129). The node sets are removed by starting with those represented by the leaves of the join tree and then working inwards.

This illustration makes it clear that Graham reduction can also be seen as a hypertree pruning method and this provides us directly with an idea of how to exploit it for randomly modifying graphs with hypertree structure: We simply execute a few steps of the Graham reduction, randomly selecting the set to be removed if more than one can be removed at the same time, until only a certain percentage of the sets remain or only a certain percentage of the nodes is still covered. The reduced set $\mathcal{M}$ is then extended again in the same manner in which it was generated in the first place.

Unfortunately, although this approach is simple and clearly efficient, it has a serious drawback: Suppose that by accident the initial graph is a simple chain. Then in each step only the two sets corresponding to the edges at the ends of (the remainder of) the chain can be removed. Hence, if only a limited number of sets is removed, there is no or only a very small chance that the edges in the middle of the chain are removed. Obviously, this argument is not restricted to chain-like graphs: In general, the "inner cliques" are much less likely to be removed, since certain "outer cliques" have to be removed before an "inner clique" can be removed. Therefore the modification of candidate solutions with this method is severely biased, which renders it unsuited for most applications. Consequently, we have to look for less biased methods, although it was clearly necessary to consider this approach first, since it is the one that directly suggests itself.

The second approach which we are going to discuss is the method used in [Borgelt and Gebhardt 1997], although it was not described in detail in that paper. It is based on the insight that a family of node sets has the running intersection property if it is constructed by successively adding node sets M_i

to an initially empty family according to the following conditions:

1. M_i must contain at least one pair of nodes that are not connected in the graph represented by $\{M_1, \ldots, M_{i-1}\}$.
2. For each maximal subset S of nodes of M_i that are connected to each other in the graph represented by $\{M_1, \ldots, M_{i-1}\}$ there must be a set M_k, $1 \leq k < i$, so that $S \subset M_k$.

It is clear that the first condition ensures that all nodes are covered after a certain number of steps. It also provides us with a stopping criterion. The running intersection property is ensured by the second condition alone.

Theorem 7.3.2 *If a family $\mathcal{M}$ of subsets of objects of a given set U is constructed observing the two conditions stated above, then this family $\mathcal{M}$ has the running intersection property.*

Proof. The proof is carried out by a simple induction on the sets in $\mathcal{M}$. It can be found in Section A.9 in the appendix.

With this method, the family $\mathcal{M}$ is constructed by forming *subfamilies* of node sets, each of which represents a *connected component* of the graph, i.e. a subgraph, the nodes of which are connected to each other, but not to nodes in other subfamilies. Obviously, the main advantage of this approach is that we can connect subfamilies of node sets, whereas with the first approach we can only extend one family. This provides us with considerable freedom w.r.t. a random modification of the graph represented by a family of node sets. At first sight, it may even look as if we could select any subset of a given family of node sets and then fill it, respecting the two conditions stated above, with randomly generated sets to cover all nodes. However, an entirely unrestricted selection is not possible, because, unfortunately, if a family of node sets has the running intersection property, a subset of it need not have it. To see this, consider the family $\mathcal{M} = \{\{A_1, A_2, A_3\}, \{A_2, A_4, A_5\}, \{A_3, A_5, A_6\}, \{A_2, A_3, A_5\}\}$, which has the running intersection property, as can easily be verified by applying Graham reduction. Nevertheless, if the last set is removed, the property is lost. Therefore, since the running intersection property of the subfamilies is a prerequisite of the proof of Theorem 7.3.2, we have to be careful when choosing subsets of a given family of node sets.

Fortunately, there is a very simple selection method, which ensures that all resulting subfamilies have the running intersection property. It consists in shuffling the sets of the given family $\mathcal{M}$ into a random order and trying to add them in this order to an initially empty family, respecting the two conditions of the method, until a certain percentage of the sets of the original family has been added or a certain percentage of the nodes is covered. Clearly, the above theorem ensures that the subfamilies selected in this way have the running intersection property. The resulting family of node sets is then filled, again respecting the two conditions, with randomly generated node sets to cover all nodes, which yields a randomly modified graph.

It is clear that this method to modify randomly a given graph with hypertree structure is much less biased than the method of the first approach. It should be noted, though, that it is not completely unbiased due to the fact that the conditions a new set has to satisfy are, in a way, too strong. Situations can arise, in which a set is rejected, although adding it would not destroy the running intersection property of a subfamily of node sets. As an example consider the family $\{\{A_1, A_3, A_4\}, \{A_2, A_4, A_5\}\}$ and the new set $\{A_3, A_4, A_5, A_6\}$. Since the nodes A_3, A_4, and A_5 are already connected in the graph represented by the family, but are not contained in a single set of the family, the new set is rejected. However, as can easily be verified, if the family were enlarged by this set, it would still have the running intersection property.[21] It is evident, though, that this bias is negligible.

Having constructed a random graph with hypertree structure, either from scratch or by modifying a given graph, we must evaluate it. If the chosen evaluation measure (cf. Section 7.2) is defined on maximal cliques, or if the graphical model to be learned is an undirected possibilistic network that is to be evaluated by summing the weighted degrees of possibility for the tuples in the database (cf. Section 7.1.3), we only have to determine the marginal distributions on the maximal cliques. From these the network quality can easily be computed. If, however, the chosen evaluation measure is defined on conditional distributions, we have to transform the graph into a directed acyclic graph first, so that an appropriate set of conditional distributions can be determined. If we have a probabilistic graphical model that is to be evaluated by computing the log-likelihood of the database to learn from (cf. Section 7.1.2), it is convenient to carry out this transformation, because computing the log-likelihood of the dataset is much easier w.r.t. a Bayesian network.

A simple method to turn an undirected graph with hypertree structure into a directed acyclic graph is the following: First we obtain a construction sequence for the family of node sets underlying its maximal cliques. This can be achieved, for instance, by applying Graham reduction (cf. Algorithm 7.3.1 on page 230). Then we process the node sets in this order. For each set we divide the nodes contained in it into two subsets: Those nodes that are already processed, because they are contained in a node set preceding the current one in the construction sequence, and those that are unprocessed. We traverse the unprocessed nodes and assign to each of them as parents all already processed nodes of the current node set. Of course, after a node has been processed in this way, it is transferred to the set of processed nodes and must also be assigned as a parent to the next unprocessed node.

Finally, for a simulated annealing search, we must consider the probability function for accepting a solution that is worse than the current one. The problem here is that we usually do not know in advance the maximal quality

[21] A simple way to see this is to note that this family can be constructed if the sets are generated in a different order, for example if the set $\{A_3, A_4, A_5, A_6\}$ is added first.

difference of two graphical models and hence we cannot compute the normalization constant in the exponential distribution $P(\text{accept}) = ce^{-\frac{dQ}{T}}$. To cope with this problem one may use an adaptive approach, which estimates the maximal quality difference from the quality difference of the best and the worst graphical model inspected so far. A simple choice is the unbiased estimator for a uniform distribution [Larsen and Marx 1986], i.e.

$$\hat{dQ}_{\text{max}} = \frac{n+1}{n} |Q_{\text{best}} - Q_{\text{worst}}|,$$

where n is the number of graphical models evaluated so far, although the uniform distribution assumption is, of course, debatable. However, it is not very likely that the exact estimation function has a strong influence.

With these considerations we eventually have all components needed for a simulated annealing approach to learn a graphical model from data. It should be noted that, of course, the methods to randomly generate and modify an undirected graph with hypertree structure, which we studied in this section, can easily be adapted so that they can be used in a genetic or evolutionary algorithm. Therefore we do not discuss this alternative explicitly.

7.3.3 Conditional Independence Search

In Section 7.1 we considered the approach to learn a graphical model from data which is based on conditional independence test w.r.t. an exhaustive search of the possible graphs. Hence the considerations of Section 7.3.1, which showed that the search space is huge unless the number of attributes is *very* small, render this approach impossible in this most direct form. Unfortunately, in contrast to the approaches based on a direct test for decomposability or on the strengths or marginal dependences, it is much more difficult to use conditional independence tests in a heuristic search. The reason is that testing whether the conditional independence statements represented by a given graph hold yields only a binary result: Either the statements hold (at least w.r.t. to some evaluation measure and a given error bound) or they do not. However, for a heuristic search we need a gradual measure of how much better or worse one graph is compared to another.

Of course, one may try to construct such a measure, for example, by summing the assessments (computed with an evaluation measure) of all conditional independence statements represented by the graph. However, it is hard to see how such a measure can be normalized appropriately so that two graphs with a different structure can be compared with it. In addition, we face the problem that the number of conditional independence statements represented by a given graph can be very large. An extreme example is a simple undirected chain: The two attributes at the ends of the chain must be conditionally independent given any nonempty subset of the attributes between them (and these are, of course, not all conditional independence statements that have to be considered in this case). Although it may be possible to reduce the number of conditional independence statements that have to be considered by exploiting the equivalence of the different Markov properties of graphs (cf. Section 4.1.5) or by simply deciding to consider only pairwise conditional independences, it is clear that for a graph over a reasonable number of attributes there are still too many conditional independence statements that have to be checked. Therefore such an approach appears to be infeasible and, as far as we know, has not been tried yet.

As a consequence, conditional independence tests are usually employed in an entirely different manner to search for a suitable conditional independence graph. The rationale underlying the most well-known approach [Spirtes *et al.* 1993] is that if we knew that for the domain under consideration there is a *perfect map*[22], i.e. a graph that represents exactly those conditional independence statements that hold in the joint distribution on the domain, we could infer from a conditional independence $A \perp\!\!\!\perp B \mid S$, where A and B are attributes and S is a (possibly empty) set of attributes, that there cannot be an edge between A and B. The reason is that in a perfect map this conditional independence statement must be represented, but would not be, obviously, if there were an edge between A and B. Hence, provided that there is a perfect map of the conditional independence statements that hold in a given distribution, we can find this perfect map by the following algorithm [Spirtes *et al.* 1993]:

Algorithm 7.3.3 *(find a perfect map with conditional independence tests)*
Input: *A distribution δ over a set U of attributes.*
Output: *A perfect map $G = (U, E)$ for the distribution.*

1. *For each pair of attributes A and B, search for a set $S_{AB} \subseteq U - \{A, B\}$, so that $A \perp\!\!\!\perp_\delta B \mid S_{AB}$ holds, i.e. so that A and B are conditionally independent given S_{AB}. If there is no such set S_{AB}, connect the attributes by an undirected edge.*

If a directed acyclic perfect map $\vec{G} = (U, \vec{E})$ is to be found, the following two steps have to be carried out in order to direct the edges:

2. *For each pair of nonadjacent attributes A and B with a common neighbor $C \notin S_{AB}$, direct the edges towards C, i.e. $A \to C \leftarrow B$.*
3. *Recursively direct all undirected edges according to the rules:*
   - *If for two adjacent attributes A and B there is a strictly directed path from A to B not including $A - B$, then direct the edge towards B.*
   - *If there are three attributes A, B, and C with A and B not adjacent, $A \to C$, and $C - B$, then direct the edge $C \to B$.*
   - *If neither rule is applicable, arbitrarily direct an undirected edge.*

[22]The notion of a perfect map was defined in Definition 4.1.15 on page 101.

Obviously, the first step of this algorithm implements the idea stated above: Only those attributes are connected by an edge for which there is no set of attributes that renders them conditionally independent.

The reasons underlying the edge directing operations are as follows: If a set S_{AB} of attributes renders two attributes A and B conditionally independent, it must block all paths from A to B in the graph (otherwise this conditional independence statement would not be represented by the graph). Hence it must also block a path that runs via a common neighbor C of A and B. However, if C is not in S_{AB}, the only way in which this path can be blocked is by the fact that it has converging edges at C. This explains the second step of the algorithm.

The first rule of the third step simply exploits the fact that the resulting graph must be acyclic. If there already is a directed path from A to B not including the edge connecting A and B, directing the edge from B towards A would introduce a directed cycle. Hence it must be directed the other way. The second rule of the third step exploits the fact that C must be in the set S_{AB}, because otherwise step 2 of the algorithm would have directed the edge $C - B$ towards C. However, if C is in S_{AB}, the path from A to B via C can only be blocked by S_{AB}, if it does *not* have converging edges at C. Consequently, the edge $C - B$ cannot be directed towards C, but must be directed towards B. Finally, the by third rule of the third step deadlocks are avoided. For example, if the perfect map is a simple chain, neither step 2 nor the first two rules of step 3 will direct any edges. Hence there must be a default rule to ensure that all edges will eventually be directed.

Note that there can be at most one undirected perfect map for a given distribution. In contrast to this, there may be several directed acyclic perfect maps of a distribution. To see this, consider again a simple chain. Its edges may be directed away from an arbitrarily chosen attribute. All of these configurations represent the same set of conditional independence statements. Which of these alternatives is the result of the above algorithm depends, obviously, on the edge or edges chosen by the default rule and the direction they are endowed with.

Although this algorithm appears to be simple and convenient, there are some problems connected with it, which we are going to discuss, together with attempts at their solution, in the remainder of this section. The first problem is that in order to make sure that there is *no* set S_{AB} that renders two attributes A and B conditionally independent, it is, in principle, necessary to check all subsets of $U - \{A, B\}$, of which there are

$$s = \sum_{i=1}^{|U|-2} \binom{|U| - 2}{i}.$$

Even worse, some of these sets contain a large number of attributes (unless the number of attributes in U is small), so that the conditional independence tests to be carried out are of high order, where the order of a conditional independence test is simply the number of attributes in the conditioning set. The problem with a high order conditional independence test is that we have to execute it w.r.t. a database of sample cases by evaluating the joint distribution of the two attributes for each distinct instantiation of the conditioning attributes. However, unless the amount of available data is huge, the number of tuples with a given instantiation of a large number of conditioning attributes will be very small (if there are such tuples at all) and consequently the conditional independence test will not be reliable. In general: The higher the order of the test, the less reliable the result.

Usually this problem is dealt with by assuming the existence of a *sparse* perfect map, i.e. a perfect map with only a limited number of edges, so that any pair of attributes can be separated by an attribute set of limited size. With this assumption, we only have to test for conditional independence up the order that is fixed by the chosen size limit. If for two attributes A and B all conditional independence tests with an order up to this upper bound failed, we infer—from the assumption that the graph is sparse—that there is no set S_{AB} that renders them conditionally independent.

The sparsest graph, in which nevertheless each pair of attributes is connected, is, of course, a tree or its directed counterpart, a polytree. Thus it is not surprising that there is a special version of the above algorithm that is restricted to polytrees [Huete and de Campos 1993, de Campos 1996]. In this case the conditional independence tests, obviously, can be restricted to orders 0 and 1, because for any pair of attributes there is only one path connecting them, which can be blocked with at most one attribute. An overview of other specialized version that consider somewhat less restricted classes of graphs, but which are all, in one way or the other, based on the same principle, is given in [de Campos *et al.* 2000].

Note that the assumption of a sparse graph is admissible, because it can only lead to additional edges and thus the result of the algorithm must be at least an independence map (although it may be more complex than the type of graph assumed, for example more complex than a polytree).

The second problem connected with Algorithm 7.3.3 is that it is based on the assumption that there is a perfect map of the distribution. As already discussed in Section 4.1.4, there are distributions for which there is no perfect map, at least no perfect map of a given type. Hence the question suggests itself, what will be the result of this algorithm, if the assumption that there is a perfect map does not hold, especially, whether it yields at least an independence map in this case (which would be satisfactory).

Unfortunately, if the perfect map assumption does not hold, the graph induced by Algorithm 7.3.3 can be severely distorted: Consider first the induction of an undirected graph from the distribution shown in Figure 4.3 on page 102. Obviously, the result is the graph shown in the same figure—except that the edges are not directed. However, this makes a considerable

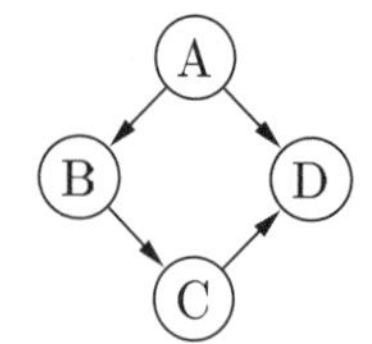

Figure 7.13 Directed acyclic graph constructed by Algorithm 7.3.3 for the probability distribution shown in Figure 4.4 on page 103.

difference. Whereas the directed graph represents the statements $A \perp\!\!\!\perp B \mid \emptyset$ and $A \not\perp\!\!\!\perp B \mid C$, the undirected one represents the statements $A \not\perp\!\!\!\perp B \mid \emptyset$ and $A \perp\!\!\!\perp B \mid C$. Hence, in this case the algorithm yields a result that is neither a dependence map nor an independence map.

A similar problem occurs w.r.t. directed acyclic graphs: Consider the induction of a directed acyclic graph from the distribution shown in Figure 4.4 on page 103. Algorithm 7.3.3 yields the undirected perfect map shown in this figure after its first step. However, since the set of conditional independence statements that hold in this distribution cannot be represented perfectly by a directed acyclic graph (cf. Section 4.1.4), it is not surprising that an attempt at directing the edges of this graph while preserving the represented conditional independence statements must fail. Indeed, since step 2 does not direct any edges, the default rule of the third step directs an arbitrary edge, say, $A \to B$. Then, by the other two rules of the third step, all other edges are directed and thus we finally obtain the graph shown in Figure 7.13. Since it represents the statements $A \perp\!\!\!\perp C \mid B$ and $A \not\perp\!\!\!\perp C \mid \{B, D\}$, whereas in the distribution it is $A \not\perp\!\!\!\perp C \mid B$ and $A \perp\!\!\!\perp C \mid \{B, D\}$, this graph is neither a dependence map nor an independence map.

In order to ensure that the resulting graph, whether directed or undirected, is at least a conditional independence graph, edges have to be reinserted. Fortunately, for directed graphs, it seems to be sufficient to consider nodes with converging edges, i.e. nodes towards which (at least) two edges are directed, and to add an edge (of appropriate directionality) between the nodes these edges start from. However, we do not have a proof for this conjecture. Yet if it holds, it also indicates a way to amend the failure of the algorithm in the case of undirected graphs. All we have to do is to check whether a common neighbor C of two nonadjacent attributes A and B is in the set S_{AB}. If it is not, a directed graph would have converging edges at C and thus we have to add an edge between A and B (compare the construction of a moral graph that was discussed in Section 4.2.2).

A problem that is closely connected to the one just discussed is that even if the distribution underlying a database of sample cases has a perfect map, the set of conditional independence statements that is determined with a chosen test criterion (i.e. an evaluation measure and an error bound) from the database may not have one. Even worse, this set of conditional independence statements may be inconsistent. The reason is that a conditional independence

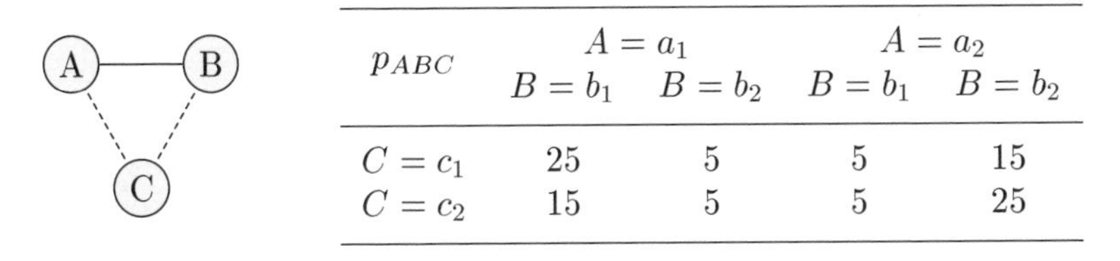

p_{ABC}	$A = a_1$		$A = a_2$	
	$B = b_1$	$B = b_2$	$B = b_1$	$B = b_2$
$C = c_1$	25	5	5	15
$C = c_2$	15	5	5	25

Figure 7.14 A database that suggests inconsistent independences.

test cannot be expected to be fully reliable due to possible random fluctuations in the data and thus may yield wrong results (that is, wrong w.r.t. what holds in the underlying distribution).

As an example consider the database shown in Figure 7.14 (each table entry indicates the number of occurrences of the respective value combination). From this database we can compute[23]

$$\begin{aligned} I_{\text{mut}}(A,B) &\approx 0.278, & I_{\text{mut}}(A,C) &= I_{\text{mut}}(B,C) &\approx 0.029,\\ I_{\text{mut}}(A,B \mid C) &\approx 0.256, & I_{\text{mut}}(A,C \mid B) &= I_{\text{mut}}(B,C \mid A) &\approx 0.007, \end{aligned}$$

if we compute the probabilities by maximum likelihood estimation. Clearly, there are no exact (conditional) independences and thus Algorithm 7.3.3 yields a complete graph. However, if we allow for some deviation from an exact conditional independence in order to cope with random fluctuations, we may decide to consider two attributes as (conditionally) independent if their (conditional) mutual information is less than 0.01 bits. If we do so, we have $A \perp\!\!\!\perp C \mid B$ and $B \perp\!\!\!\perp C \mid A$. Consequently, Algorithm 7.3.3 will remove the edge (A, C) as well as the edge (B, C), i.e. the edges indicated by dotted lines in Figure 7.14. Unfortunately, the resulting graph not only represents the above two conditional independences, but also $A \perp\!\!\!\perp C \mid \emptyset$ and $B \perp\!\!\!\perp C \mid \emptyset$, neither of which hold, at least according to the chosen criterion.

Note that the algorithm fails, because the set of conditional independence statements is inconsistent: The distribution underlying the database shown in Figure 7.14 must be strictly positive, since all value combinations occur. A strictly positive probability distribution satisfies the graphoid axioms (cf. Theorem 4.1.2 on page 93), which comprise the intersection axiom (cf. Definition 4.1.1 on page 92). With the intersection axiom $\{A, B\} \perp\!\!\!\perp C \mid \emptyset$ can be inferred from $A \perp\!\!\!\perp C \mid B$ and $B \perp\!\!\!\perp C \mid A$, and from this $A \perp\!\!\!\perp C \mid \emptyset$ and $B \perp\!\!\!\perp C \mid \emptyset$ follow with the decomposition axiom. Therefore it is impossible that the distribution underlying the database satisfies both conditional independence statements found, but not $A \perp\!\!\!\perp C \mid \emptyset$ and $B \perp\!\!\!\perp C \mid \emptyset$.

An approach to cope with the problem of inconsistent sets of conditional independence statements has been suggested by [Steck and Tresp 1999]. It is

[23]The mutual information I_{mut} of two attributes was defined in Definition 7.1.4 on page 174 and its conditional version was introduced on page 175.

based on the idea that the situations in which Algorithm 7.3.3 removes too many edges can be detected by associating each edge that may be removed with one or more paths in the graph, at least one of which must remain intact if the edge is actually to be removed (this is called the *causal path condition*—although we would prefer *dependence path condition*; cf. Chapter 9). From these associations pairs of edges can be determined, which alternatively may be removed, but one of which must be kept.

As an example reconsider the database of Figure 7.14. The result of the augmented algorithm would be the graph shown in the same figure, but with the dotted edges marked as alternatives, either of which may be removed, but not both: If the edge (A, C) is removed, the path $A - B - C$ must remain intact to ensure $A \not\perp\!\!\!\perp C \mid \emptyset$. Likewise, if the edge (B, C) is removed, the path $B - A - C$ must remain intact to ensure $B \not\perp\!\!\!\perp C \mid \emptyset$.

We close this section by pointing out that a variant of the directed version of Algorithm 7.3.3 has been suggested for finding causal relations between a set of attributes [Pearl and Verma 1991a, Pearl and Verma 1991b]. This algorithm and the assumptions underlying it are discussed in Chapter 9.

7.3.4 Greedy Search

Especially if a graphical model is to be learned by measuring the strengths of marginal dependences, it is not only possible to use a random guided search as discussed in Section 7.3.2. We may consider also the usually more efficient method of a *greedy search.* With a greedy search a graphical model is constructed from components, each of which is selected *locally*, i.e. relatively independent of other components and their interaction (although usually certain constraints have to be taken into account), so that a given (local) evaluation measure is optimized. Although it is clear that, in general, such an approach cannot be guaranteed to find the best solution, it is often reasonable to expect that it will find at least a good approximation.

The best-known greedy approach—and at the same time the oldest—is *optimum weight spanning tree construction* and was suggested by [Chow and Liu 1968]. All possible (undirected) edges over the set U of attributes used to describe the domain under consideration are evaluated with an evaluation measure ([Chow and Liu 1968] used *mutual information*[24]). Then an optimum weight spanning tree is constructed with either the (well-known) Kruskal algorithm [Kruskal 1956] or the (somewhat less well-known) Prim algorithm [Prim 1957] (or any other greedy algorithm for this task).

An interesting aspect of this approach is that if the probability distribution, for which a graphical model is desired, has a perfect map[25] that is a tree, optimum weight spanning tree construction is guaranteed to find the perfect map, provided the evaluation measure used has a certain property.

---

[24]Mutual information was defined in Definition 7.1.4 on page 174.

[25]The notion of a perfect map was defined in Definition 4.1.15 on page 101.

Theorem 7.3.4 *Let m be a symmetric evaluation measure satisfying*

$$\forall A, B, C: \quad m(C, AB) \geq m(C, B)$$

with equality obtaining only if the attributes A and C are conditionally independent given B. Let G be a singly connected undirected perfect map of a probability distribution p over a set U of attributes. Then constructing a maximum weight spanning tree for the attributes in U with m (computed from p) providing the edge weights uniquely identifies G.

Proof. The proof is based on the fact that with the property of the evaluation measure presupposed in the theorem, any edge between two attributes C and A that is not in the tree must have a weight less than the weight of all edges on the path connecting C and A in the tree. The details can be found in Section A.13 in the appendix.

From Lemma 7.2.2 and Lemma 7.2.6 we know that at least mutual information and the χ^2 measure have the property presupposed in the theorem.

It is clear that the above theorem holds also for directed trees, since any undirected conditional independence graph that is a tree can be turned into an equivalent directed tree by choosing an arbitrary root node and (recursively) directing the edges away from this node. However, with an additional requirement, it can also be extended to polytrees.

Theorem 7.3.5 *Let m be a symmetric evaluation measure satisfying*

$$\forall A, B, C: \quad m(C, AB) \geq m(C, B)$$

with equality obtaining only if the attributes A and C are conditionally independent given B and

$$\forall A, C: \quad m(C, A) \geq 0$$

with equality obtaining only if the attributes A and C are (marginally) independent. Let $\vec{G}$ be a singly connected directed perfect map of a probability distribution p over a set U of attributes. Then constructing a maximum weight spanning tree for the attributes in U with m (computed from p) providing the edge weights uniquely identifies the skeleton *of $\vec{G}$, i.e. the undirected graph that results if all edge directions are discarded.*

Proof. The proof is based on the same idea as the proof of the preceding theorem. The details can be found in Section A.14 in the appendix.

Note that the above theorem is an extension of a theorem shown in [Rebane and Pearl 1987, Pearl 1988], where it was proven only for mutual information providing the edge weights. However, from Lemma 7.2.6 we know that the χ^2 measure may also be used. Note also that the edges of the skeleton found with the above approach may be directed by applying steps 2 and 3 of

Algorithm 7.3.3 [Rebane and Pearl 1987]. It should be noted, though, that a directed perfect map need not be unique, as was discussed in connection with the notion of the likelihood equivalence of an evaluation measure (cf. page 212 in Section 7.2.4) and thus the directions of some edges may be arbitrary.

The above theorems only hold if there is a tree-structured perfect map of the distribution for which a graphical model is desired. It has to be admitted, though, that tree-structured perfect maps are not very frequent. Nevertheless the construction of an optimum weight spanning tree can be very useful. There are at least two reasons for this. The first is that an optimum weight spanning tree can serve as the starting point for another algorithm that is capable of constructing more complex graphs. For instance, a simulated annealing approach (cf. Section 7.3.2) may start its search with an optimum weight spanning tree. Another example is discussed below.

The second reason is that, since propagation in trees and polytrees is so very simple and efficient, we may be content with a good tree-structured approximation of the given distribution in order to be able to exploit this simplicity and efficiency. By a *tree-structured approximation* we mean a graphical model that is based on a (directed[26]) tree and which represents a distribution that is, in some sense, close to the original distribution. It is plausible that an optimum weight spanning tree may be a good tree-structured approximation. Indeed, we have the following theorem:

Theorem 7.3.6 [Chow and Liu 1968] *Let p be a strictly positive probability distribution over a set U of attributes. Then a best tree-structured approximation of p w.r.t. Kullback–Leibler information divergence[27] is obtained by constructing a maximum weight spanning undirected tree of U with mutual (Shannon) information[28] providing the edge weights, then directing the edges away from an arbitrarily chosen root node, and finally computing the (conditional) probability distributions associated with the edges of the tree from the given distribution p.*

Proof. The proof exploits mainly the properties of the Kullback–Leibler information divergence and is somewhat technical. It can be found in Section A.15 in the appendix.

Because of this theorem a tree-structured graphical model that is constructed as described in this theorem (and which is often called a Chow–Liu tree) is frequently used as some kind of baseline to assess the quality of more complex graphs. For the extension of an optimum weight spanning tree skeleton to a polytree some limiting results for the possible improvement of fit have been obtained by [Dasgupta 1999].

---

[26] One may confine oneself to directed trees, because, as already mentioned above, any undirected tree can be turned into an equivalent directed tree.

[27] Kullback–Leibler information divergence was defined in Definition 7.1.3 on page 171.

[28] Mutual (Shannon) information was defined in Definition 7.1.4 on page 174.

A straightforward generalization of the above theorem can be obtained by using its principle to augment naive Bayes classifiers (cf. Section 6.1), so that the strong conditional independence assumptions of naive Bayes classifiers are mitigated [Geiger 1992]. The idea is to start from a naive Bayes classifier (i.e. a star-like Bayesian network with the class at the center) and add edges between the descriptive attributes that form a (directed) tree. The resulting classifier is often called a *tree-augmented naive Bayes classifier (TAN)* [Friedman and Goldszmidt 1996]. Not surprisingly, the tree-augmented naive Bayes classifier that best approximates a given probability distribution w.r.t. Kullback–Leibler information divergence can be obtained in basically the same way as the best tree-structured approximation:

Theorem 7.3.7 [Geiger 1992] *Let p be a strictly positive probability distribution over a set $U \cup \{C\}$ of attributes, where C is a class attribute. Then a best tree-augmented naive Bayes classifier approximation of p w.r.t. Kullback–Leibler information divergence is obtained by augmenting naive Bayes classifier for the attribute C with a maximum weight spanning undirected tree of U with conditional mutual (Shannon) information*

$$\begin{aligned} I_{\text{mutual}}^{(\text{Shannon})}(A, B \mid C) &= H(A \mid C) + H(B \mid C) - H(A, B \mid C) \\ &= \sum_{j=1}^{n_A} \sum_{k=1}^{n_B} \sum_{i=1}^{n_C} P(a_j, b_k, c_i) \log_2 \frac{P(a_j, b_k \mid c_i)}{P(a_j \mid c_i) \cdot P(b_k \mid c_i)} \end{aligned}$$

providing the edge weights, then directing the edges away from an arbitrarily chosen root node, and finally computing the (conditional) probability distributions associated with the resulting graph structure (the star of the naive Bayes classifier plus the constructed tree) from the given distribution p.

Proof. The proof of this theorem is obtained in direct analogy to the proof of Theorem 7.3.6: All one has to do is to add the class attribute as an additional condition in all expressions referring to the tree considered in Theorem 7.3.6. This changes the occurring instances of mutual information to conditional mutual information and thus provides the proof of this theorem.

Note that, unfortunately, neither of the above Theorems 7.3.4–7.3.7 can be transferred to possibilistic networks, as can be seen from the simple relational example discussed in Section 7.1.3 (cf. Figure 7.4 on page 170). Nevertheless, the construction of a maximum weight spanning tree is a valuable heuristic method for learning possibilistic networks.

As already mentioned above, an optimum weight spanning tree may also serve as the starting point for another algorithm. In the following we suggest an algorithm the can be seen as a modification of an algorithm mentioned above, namely the one by [Rebane and Pearl 1987] for constructing a polytree. The basic idea of this algorithm is as follows: First an (undirected) maximum weight spanning tree is constructed. Then this tree is enhanced by edges

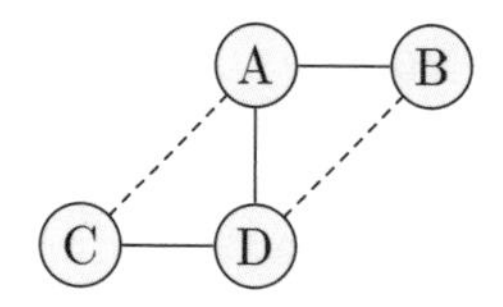

Figure 7.15 The dotted edges cannot both be the result of "marrying" parents in a directed graph.

where a conditional independence statement implied by the tree does not hold (cf. also Section 7.1.2). The main advantage of such an approach is that by introducing restrictions w.r.t. to which edges may be added, we can easily control the complexity of the resulting graph—a consideration that is usually important for applications (cf. also Section 7.3.2). For example, we may allow adding edges only between nodes that have a common neighbor in the maximum weight spanning tree. With this restriction the algorithm is very closely related to the algorithm by [Rebane and Pearl 1987]: Directing two edges so that they converge at a node is equivalent to adding an edge between the source nodes of these edges to the corresponding moral graph.[29] However, our approach is slightly more general as can be seen from Figure 7.15. The two dotted edges cannot both be the result of "marrying" parents.

An interesting further restriction of the edges that may be added is the following requirement: If all edges of the optimum weight spanning tree are removed, the remaining graph must be acyclic. This condition is interesting, because it guarantees that the resulting graph has hypertree structure and that its maximal cliques comprise at most three nodes.

Theorem 7.3.8 *If an undirected tree is extended by adding edges only between nodes with a common neighbor in the tree and if the added edges do not form a cycle, then the resulting graph has hypertree structure and its maximal cliques contain at most three nodes.*

Proof. The proof, which exploits a simple observation about cliques with more than three nodes and the fact that a triangulated graph has hypertree structure, can be found in Section A.16 in the appendix.

It should be noted this approach cannot be guaranteed to find the best possible graph with the stated properties. This can be seen clearly from the counterexample studied in Section 7.1.2 (cf. Figure 7.7 on page 176). Note also that it is difficult to generalize this approach to graphs with larger maximal cliques, since there seem to be no simple conditions by which it can be ensured that the resulting graph has hypertree structure and that the maximal cliques have a size at most n, $n > 3$. However, this approach may be used to augment a naive Bayes classifier beyond a tree-augmented version. Then the cliques are restricted to four nodes, but it can no longer be guaranteed that the result is optimal w.r.t. Kullback–Leibler information divergence.

[29] The notion of a moral graph was introduced on page 128.

Another extension of optimum weight spanning tree construction is the so-called K2 algorithm [Cooper and Herskovits 1992], which was already mentioned in Section 7.2.4. It is an algorithm for learning directed acyclic graphs by greedily selecting parent attributes. The basic idea is as follows: In order to narrow the search space and to ensure the acyclicity of the resulting graph a topological order[30] of the attributes is fixed. Fixing a topological order restricts the permissible graphs, because the parents of an attribute can only be selected from the attributes preceding it in the topological order. The topological order can either be stated by a domain expert or can be derived automatically with the help of conditional independence tests [Singh and Valtorta 1993] (compare also Section 7.3.3).

As already indicated, the parent attributes are selected greedily: At the beginning the value of an evaluation measure is computed for an attribute, treating it as a parentless child attribute. Then in turn each of the parent candidates (the attributes preceding an attribute in the topological order) is temporarily added to the child attribute and the evaluation measure is recomputed. The parent candidate that yields the highest value of the evaluation measure is selected as a first parent and is added permanently. In the third step each remaining parent candidate is added temporarily as a second parent and again the evaluation measure is recomputed. As before, the parent candidate that yields the highest value of the evaluation measure is permanently added. The process stops if either no more parent candidates are available, a given maximal number of parents is reached, or none of the parent candidates, if added, yields a value of the evaluation measure that exceeds the best value obtained in the preceding step.

It is clear that this algorithm is equivalent to the construction of an optimum weight spanning tree if only one parent is selected per attribute—provided, of course, a suitable topological order has been fixed. If more than one parent may be selected, polytrees or general directed acyclic graphs may be learned. From the above theorems and their proofs it is clear that if there is a perfect map of the given distribution that is a directed tree or a polytree, then the K2 algorithm will yield a perfect map, provided a topological order compatible with a perfect map is fixed and an evaluation measure satisfying the presuppositions of the above theorems is used.

For more general graphs, however, the greedy character of the algorithm can lead to a selection of wrong parent attributes. An example illustrating this is shown in Figure 7.16. Independent of the topological order used, a wrong parent attribute is selected, namely either C as a parent for D or D as a parent for C, although these two attributes are conditionally independent given A and B. Note that from the distributions shown in Figure 7.16 and the symmetries of the example it is immediately clear that this behavior is independent of the evaluation measure used.

---

[30]The notion of a topological order was introduced in Definition 4.1.12 on page 98.

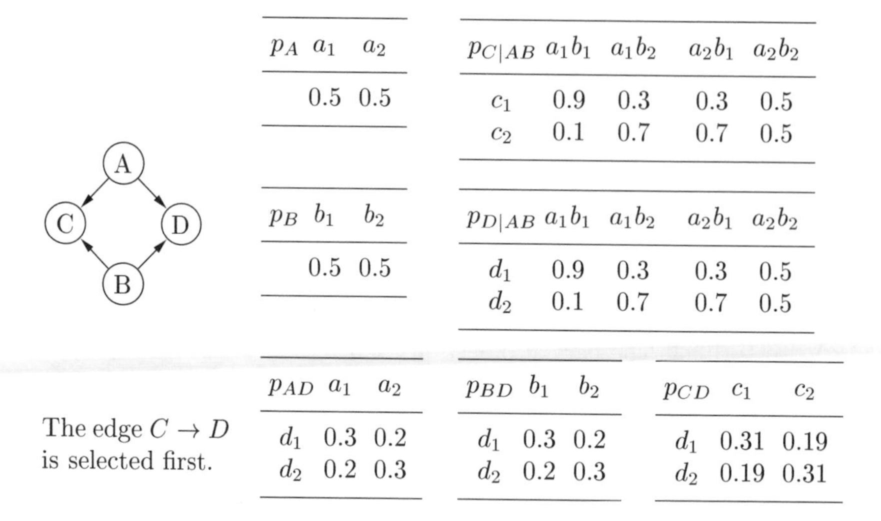

p_A	a_1	a_2
	0.5	0.5

$p_{C\|AB}$	a_1b_1	a_1b_2	a_2b_1	a_2b_2
c_1	0.9	0.3	0.3	0.5
c_2	0.1	0.7	0.7	0.5

p_B	b_1	b_2
	0.5	0.5

$p_{D\|AB}$	a_1b_1	a_1b_2	a_2b_1	a_2b_2
d_1	0.9	0.3	0.3	0.5
d_2	0.1	0.7	0.7	0.5

p_{AD}	a_1	a_2
d_1	0.3	0.2
d_2	0.2	0.3

p_{BD}	b_1	b_2
d_1	0.3	0.2
d_2	0.2	0.3

p_{CD}	c_1	c_2
d_1	0.31	0.19
d_2	0.19	0.31

Figure 7.16 Greedy parent selection can lead to suboptimal results, if there is more than one path connecting two attributes.

In order to cope with this drawback one may consider adding a step in which parent attributes are removed, again in a greedy fashion. With this approach the failure of the K2 algorithm in the above example is amended. If, for instance, A, B, and C have been selected as parents for the attribute D, this second step would remove the attribute C and thus the correct graph would be found. However, it should be noted that, in general, incorrect selections are still possible, so the result may still be suboptimal.

Another drawback of the K2 algorithm is the requirement of a topological order of the attributes, although this drawback is mitigated, as already mentioned, by methods that automatically construct such an order [Singh and Valtorta 1993]. An alternative is the following generalized version: All possible directed edges (each of which represents the selection of a parent attribute) are evaluated and the one receiving the highest score is selected. In each step all candidate edges are eliminated that would lead to a directed cycle and only the remaining ones are evaluated. The process stops if no edge that may be added leads to an improvement of the evaluation measure. However, such an unrestricted search suffers from the drawback that in a directed acyclic independence graph an attribute is, obviously, strongly dependent not only on its parents, but also on its children. In such a case the properties of the evaluation measure—for instance, whether it yields a higher value for an edge from an attribute having many values towards an attribute having only few than for an edge having the opposite direction—determine the direction of the edge, which is not always desirable.

7.4 Experimental Results

In order to illustrate the capabilities of (some of) the evaluation measures and search methods discussed in the preceding sections we report in the following some experimental results we obtained with a prototype implementation called INES (Induction of NEtwork Structures), which was developed by the first author of this book [Borgelt and Kruse 1997a, Borgelt and Kruse 1997b]. The list of results is not complete, since we have not yet implemented all search methods discussed in Section 7.3. In addition, we selected a subset of the large variety of evaluation measures discussed in Section 7.2. As a basis for the experiments we chose the Danish Jersey cattle blood type determination example [Rasmussen 1992], which was discussed in Section 4.2.2.

7.4.1 Learning Probabilistic Networks

The probabilistic learning methods were tested on ten pairs of databases with 1000 tuples each. These databases were generated by Monte Carlo simulation from the human expert designed Bayesian network for the Danish Jersey cattle blood type determination example (cf. Figure 4.13 on page 127). The first database of each pair was used to induce a graphical model, the second to test this model. The results were then averaged over all ten pairs. As a baseline for comparisons we used the original graph the databases were generated from and a graph without any edges, i.e. with independent attributes. All networks are assessed by computing the log-likelihood of the data (natural logarithm; cf., for example, Section 7.1.2). In order to avoid problems with impossible tuples, the probabilities were estimated with a Laplace correction[31] of 1.

The results are shown in Table 7.4. In addition to the network quality this table shows the total number of edges (parents/conditions) and the number of (probability) parameters as a measure of the complexity of the network. For the K2 algorithm [Cooper and Herskovits 1992], i.e. the greedy selection of parent attributes w.r.t. a topological order, the learned network is also compared to the original one by counting the number of additional and missing edges. This is possible here, because any edge selected must have the same direction as in the original network. Analogous numbers are not shown for the optimum weight spanning trees or the results of simulated annealing, because here the edges may have different directions and thus a comparison to the original network is not that straightforward.

The results obtained for optimum weight spanning tree construction show that this simple method already achieves a very good fit of the data, which is not surprising, since the original graph is rather sparse. These results also show the bias of Shannon information gain $I_{\text{gain}}^{(\text{Shannon})}$ and the χ^2 measure towards many-valued attributes. Although the number of edges is (necessarily) identical for all trees, for these two measures the number of parameters is

[31] The notion of Laplace correction was introduced on page 137 in Section 5.2.

Table 7.4 Results of probabilistic network learning.

Baseline

network	edges	params.	train	test
indep.	0	59	−19921.2	−20087.2
orig.	22	219	−11391.0	−11506.1

Optimum Weight Spanning Tree Construction

measure	edges	params.	train	test
$I_{\text{gain}}^{(\text{Shannon})}$	20.0	285.9	−12122.6	−12339.6
$I_{\text{sgr1}}^{(\text{Shannon})}$	20.0	169.5	−12149.2	−12292.5
χ^2	20.0	282.9	−12122.6	−12336.2

Greedy Parent Selection w.r.t. a Topological Order

measure	edges	add.	miss.	params.	train	test
$I_{\text{gain}}^{(\text{Shannon})}$	35.0	17.1	4.1	1342.2	−11229.3	−11817.6
$I_{\text{gr}}^{(\text{Shannon})}$	24.0	6.7	4.7	208.6	−11614.8	−11736.5
$I_{\text{sgr1}}^{(\text{Shannon})}$	32.0	11.3	1.3	316.6	−11387.8	−11574.9
Gini	35.0	17.1	4.1	1341.6	−11233.1	−11813.4
χ^2	35.0	17.3	4.3	1300.8	−11234.9	−11805.2
K2	23.3	1.4	0.1	229.9	−11385.4	−11511.5
BDeu	39.0	24.1	7.1	1412.0	−11387.3	−11978.0
$L_{\text{red}}^{(\text{rel})}$	22.5	0.6	0.1	219.9	−11389.5	−11508.2

Simulated Annealing

penalty	edges	params.	train	test
no	28.6	596.4	−12492.5	−12832.5
yes	27.9	391.0	−12696.2	−12960.0

considerably higher than for the less biased symmetric Shannon information gain ratio $I_{\text{sgr1}}^{(\text{Shannon})}$. This behavior leads to some overfitting, as can be seen from the fact that both Shannon information gain and the χ^2 measure lead to a better fit of the training data, but a worse fit of the test data.

For greedy parent selection, for which the maximum number of conditions was set to 2, $L_{\text{red}}^{(\text{rel})}$, i.e. the reduction of the description length based on relative frequency coding, and the K2 metric, which is equivalent to $L_{\text{red}}^{(\text{abs})}$, i.e. the reduction of the description length based on absolute frequency coding (cf. Section 7.2.4), lead to almost perfect results. Both recover almost exactly the original structure, as can be seen from the very small numbers of additional and missing edges. Other measures, especially the Bayesian–Dirichlet likelihood equivalent uniform metric (BDeu), the Shannon information gain $I_{\text{gain}}^{(\text{Shannon})}$ the χ^2 measure and the Gini index suffer considerably from overfitting effects: The better fit of the training data is more than outweighed by the worse fit of the test data. Shannon information gain ratio $I_{\text{gr}}^{(\text{Shannon})}$ seems to select the "wrong" parents, since not even the fit of the training data is good. A notable alternative is the symmetric Shannon information gain ratio $I_{\text{sgr1}}^{(\text{Shannon})}$. Although it tends to select too many conditions, this does not lead to strong overfitting.

At first sight, the simulated annealing results (clique size restricted to three attributes; for the penalized version the Akaike Information Criterion was used to evaluate networks instead of the maximum likelihood criterion, cf. Section 7.2.4) are somewhat disappointing, since they are worse than the results of the simple optimum weight spanning tree construction. However, one has to take into account that they were obtained with a "pure" simulated annealing approach, i.e. by starting from a randomly generated initial network, and not from, for example a maximum weight spanning tree. If such a starting point is chosen, better results are obtained. As indicated in Section 7.3.2 an important advantage of random guided search approaches like simulated annealing is that they can be used to tune a solution that has been generated with another algorithm. Therefore these results may be somewhat misleading w.r.t. the true powers of simulated annealing.

7.4.2 Learning Possibilistic Networks

For evaluating the learning methods for possibilistic networks we used a real-world database for the Danish Jersey cattle blood type determination example with 500 sample cases. This database contains a considerable number of missing values and thus is well suited for a possibilistic approach. As a baseline for comparisons we chose, as in the probabilistic case, a graph without any edges and the human expert designed network. However, the results obtained with the latter are not very expressive, since it captures a different kind of dependence, as it is based on a different uncertainty calculus.

All possibilistic networks were assessed by computing the weighted sum of the degrees of possibility for the tuples in the database, which should be as small as possible (cf. Section 7.1.3). However, since the database contains several tuples with missing values, a precise degree of possibility cannot always be computed. To cope with this problem, we computed for a tuple with missing values the minimum, the maximum, and the average degree of possibility of all precise tuples compatible with it. The results are then summed separately for all tuples in the database.

We did not divide the dataset into training and test data. The main reason is that it is not clear how to evaluate a possibilistic network w.r.t. test data, since, obviously, the measure used to evaluate it w.r.t. the training data cannot be used to evaluate it w.r.t. the test data: If the marginal possibility distributions do not fit the test data, the weighted sum of the degrees of possibility for the test data tuples will be small, although in this case this clearly indicates that the network is bad. This is also the reason why we did not use artificially generated datasets. Such datasets are of little use if the learning results cannot be evaluated on test data.

The results of learning possibilistic networks are shown in Table 7.5. Clearly, as already mentioned above, the original network is not well suited as a baseline for comparisons. As in the probabilistic case, the optimum weight spanning tree construction yields very good results. The possibilistic analog of mutual information d_{mi} seems to provide the best results in this case, since it achieves the best fit with the smallest number of parameters. For greedy parent selection the situation is similar. The specificity gain S_{gain} and d_{χ^2}, the possibilistic analog of the χ^2-measure, lead to graphical models that are too complex as can be seen from the high number of parameters. Despite this high number of parameters the model generated with specificity gain not even fits the data well. The specificity gain ratio seems to be too reluctant to select parents, and thus leads to a model that is simple, but does not fit the data well. In all, as for the optimum weight spanning tree construction, d_{mi} provides the best results.

In contrast to the somewhat disappointing results of simulated annealing in the probabilistic case, this approach seems to work very well in the possibilistic case (clique size again restricted to three attributes; for the penalized version $\kappa = 0.001$ was chosen; cf. Section 7.2.5). Actually, it yields the best results of all approaches, although it was also "pure", i.e. the search started from a randomly generated initial network. There are two possible interpretations of this result. In the first place, there may be much more "good" solutions in the possibilistic case, so that it is simpler to find one of them, even with a guided random search. Alternatively, one may conjecture that the possibilistic evaluation measures are bad and do not lead to a construction of appropriate networks. We guess that the former is the case. However, this problem needs further investigation.

Table 7.5 Results of possibilistic network learning.

Baseline

network	edges	params.	min.	avg.	max.
indep.	0	80	10.064	10.160	11.390
orig.	22	308	9.888	9.917	11.318

Optimum Weight Spanning Tree Construction

measure	edges	params.	min.	avg.	max.
S_{gain}	20	438	8.878	8.990	10.714
S_{sgr1}	20	442	8.716	8.916	10.680
d_{χ^2}	20	472	8.662	8.820	10.334
d_{mi}	20	404	8.466	8.598	10.386

Greedy Parent Selection w.r.t. a Topological Order

measure	edges	params.	min.	avg.	max.
S_{gain}	31	1630	8.524	8.621	10.292
S_{gr}	18	196	9.390	9.553	11.100
S_{sgr1}	28	496	8.946	9.057	10.740
d_{χ^2}	35	1486	8.154	8.329	10.200
d_{mi}	33	774	8.206	8.344	10.416

Simulated Annealing

penalty	edges	params.	min.	avg.	max.
no	22.7	829.0	7.955	8.255	9.974
yes	20.2	379.9	8.170	8.465	10.126

Chapter 8

Learning Local Structure

In contrast to the *global structure* of a graphical model, which is the structure of the conditional independence graph underlying it, the term *local structure* refers to regularities in the (conditional) probability or possibility distributions associated with this graph. In this chapter we consider a decision graph representation of the local structure of directed graphical models and how to learn such local structure from data. The decision graph representation shows that the induction of a decision tree from data can be seen as a special case of learning a Bayesian network with local structure.

8.1 Local Network Structure

As already indicated, the term *local structure* refers to regularities in the (conditional) distributions associated with the conditional independence graph underlying a graphical model. Particularly for Bayesian networks several approaches to exploit such regularities have been studied in order to capture additional (i.e. context-specific) independences and, as a consequence, to (potentially) enhance evidence propagation. Among these are similarity networks [Heckerman 1991] and the related multinets [Geiger and Heckerman 1991], the use of asymmetric representations for decision making [Smith *et al.* 1993], probabilistic Horn rules [Poole 1993], and also decision trees [Boutilier *et al.* 1996] and decision graphs [Chickering *et al.* 1997]. In this chapter we focus on the decision tree/decision graph approach and review it in the following for Bayesian networks. In doing so we confine ourselves, as usual, to attributes that have a finite set of possible values.

As a simple example, consider the section of a Bayesian network shown on the left in Figure 8.1 (and assume that in this network the attribute C has no other parents than the attributes A and B). The probabilities that have to be stored with attribute C in this case are $P(C = c_i \mid A = a_j, B = b_k)$

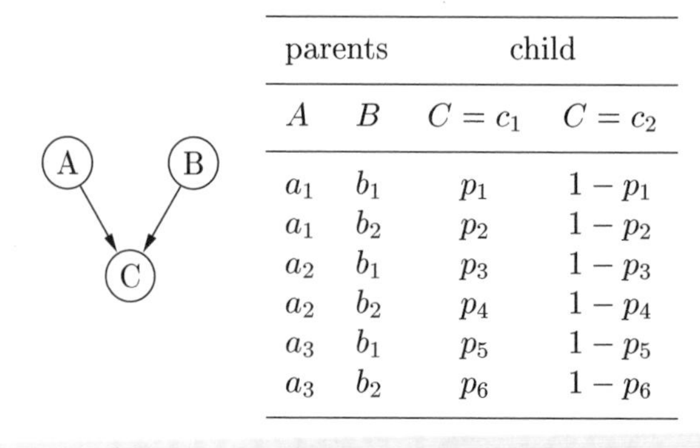

parents		child	
A	B	$C = c_1$	$C = c_2$
a_1	b_1	p_1	$1 - p_1$
a_1	b_2	p_2	$1 - p_2$
a_2	b_1	p_3	$1 - p_3$
a_2	b_2	p_4	$1 - p_4$
a_3	b_1	p_5	$1 - p_5$
a_3	b_2	p_6	$1 - p_6$

Figure 8.1 A section of a Bayesian network and the corresponding conditional probability table, which states a probability distribution for each possible instantiation of the parent attributes.

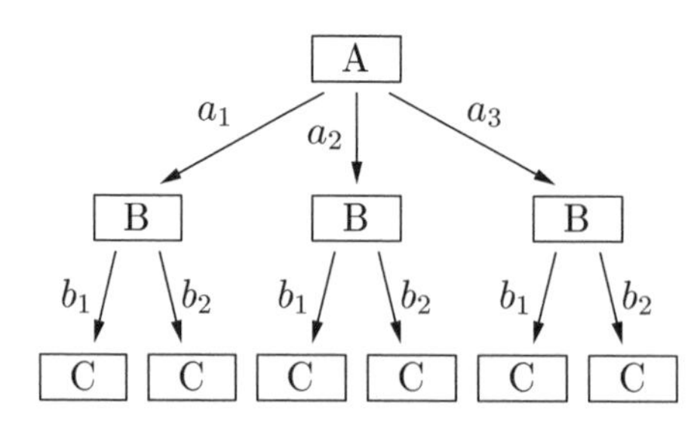

Figure 8.2 A full decision tree for the child attribute C that can be used to store the conditional probability table shown in Figure 8.1.

for all values c_i, a_j, and b_k. A very simple way to encode these conditional probabilities is a table in which for each possible instantiation of the conditioning attributes A and B there is a line stating the corresponding conditional probability distribution on the values of the attribute C. If we assume that $\text{dom}(A) = \{a_1, a_2, a_3\}$, $\text{dom}(B) = \{b_1, b_2\}$, and $\text{dom}(C) = \{c_1, c_2\}$, this table may look like the one shown in Figure 8.1.

However, the same conditional probabilities can also be stored in a tree in which the leaves hold the conditional probability distributions and each level of inner nodes corresponds to a conditioning attribute (cf. Figure 8.2). The branches of this tree are labeled with the values of the conditioning attributes and thus each path from the root to a leaf corresponds to a possible instantiation of the conditioning attributes. Obviously such a tree is equivalent to a *decision tree* [Breiman *et al.* 1984, Quinlan 1993] for the attribute C with the following restrictions: All leaves have to lie on the same level and in each level of the tree the same attribute has to be tested on all paths. If these restrictions hold, we call the tree a *full decision tree*, because it contains a complete set of test nodes.

Consider now a situation in which there are some regularities in the conditional probability table, namely those indicated in the table on the left of Figure 8.3. From this table it is clear that the value of the attribute B matters only if attribute A has the value a_2, i.e. C is independent of B given $A = a_1$ or $A = a_3$. This is usually called *context-specific independence*. Hence the tests

parents		child	
A	B	$C = c_1$	$C = c_2$
a_1	b_1	p_1	$1 - p_1$
a_1	b_2	p_1	$1 - p_1$
a_2	b_1	p_2	$1 - p_2$
a_2	b_2	p_3	$1 - p_3$
a_3	b_1	p_4	$1 - p_4$
a_3	b_2	p_4	$1 - p_4$

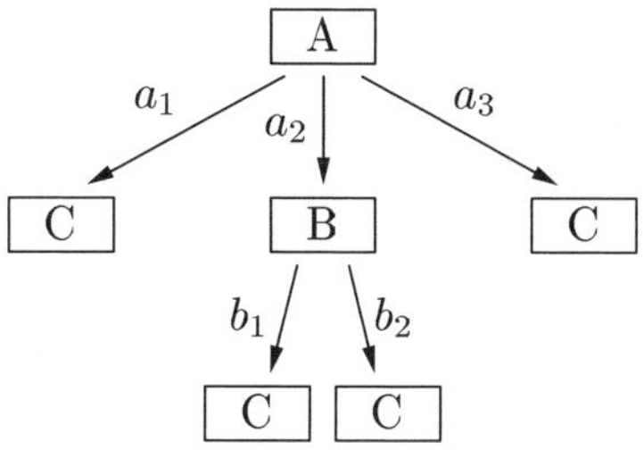

Figure 8.3 A conditional probability table with some regularities and a partial decision tree for the child attribute C that captures them.

of attribute B can be removed from the branches for the values a_1 and a_3, as illustrated with the tree on the right in Figure 8.3. This tree we call a *partial decision tree*, because it has a reduced set of nodes.

Unfortunately, however, decision trees are not powerful enough to capture all possible regularities. Although a lot can be achieved by accepting changes of the order in which the attributes are tested and by allowing binary splits and multiple tests of the same attribute (in this case, for example, the regularities in the table on the left of Figure 8.4 can be represented by the decision tree shown on the right in the same figure), the regularities shown in the table on the left of Figure 8.5 cannot be represented by a decision tree.

The problem is that in a decision tree a test of an attribute splits the lines of a conditional probability table into disjoint subsets that cannot be brought together again. In the table shown on the left in Figure 8.5 a test of attribute B separates lines 1 and 2 and a test of attribute A separates lines 4 and 5. Hence either test prevents us from exploiting one of the two cases in which distributions coincide. Fortunately, this drawback can easily be overcome by allowing a node of the tree to have more than one parent, i.e. by using *decision graphs* [Chickering *et al.* 1997]. Thus the regularities can easily be captured, as can be seen on the right in Figure 8.5.

8.2 Learning Local Structure

To learn a decision tree/decision graph representation from data, a simple top-down method may be used as for standard decision tree induction: First the split attribute at the root node is chosen based on an evaluation measure. Then the sample cases to learn from are split w.r.t. the value they have for this attribute and the algorithm is applied recursively to each subset. For learning decision graphs, we must also be able to merge nodes. Thus we arrive at the

parents		child	
A	B	$C = c_1$	$C = c_2$
a_1	b_1	p_1	$1 - p_1$
a_1	b_2	p_1	$1 - p_1$
a_2	b_1	p_2	$1 - p_2$
a_2	b_2	p_3	$1 - p_3$
a_3	b_1	p_2	$1 - p_2$
a_3	b_2	p_4	$1 - p_4$

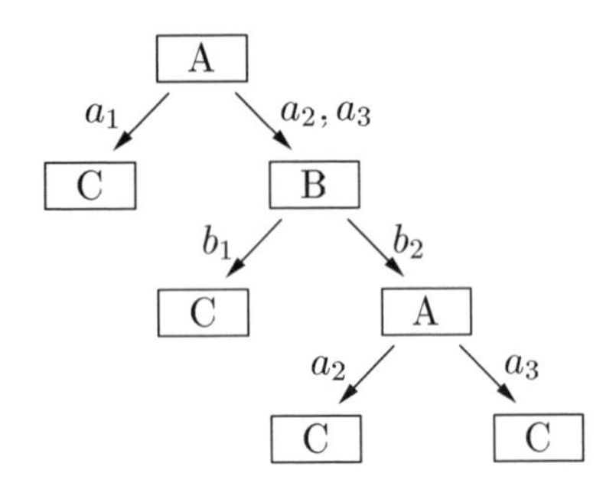

Figure 8.4 A conditional probability table with a second kind of regularities and a decision tree with two tests of attribute A that captures them.

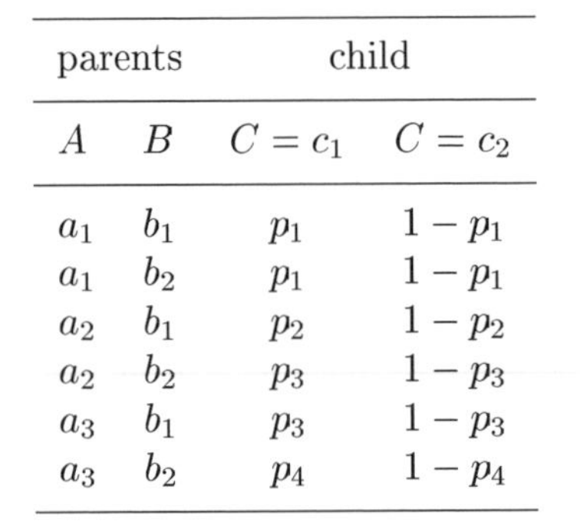

parents		child	
A	B	$C = c_1$	$C = c_2$
a_1	b_1	p_1	$1 - p_1$
a_1	b_2	p_1	$1 - p_1$
a_2	b_1	p_2	$1 - p_2$
a_2	b_2	p_3	$1 - p_3$
a_3	b_1	p_3	$1 - p_3$
a_3	b_2	p_4	$1 - p_4$

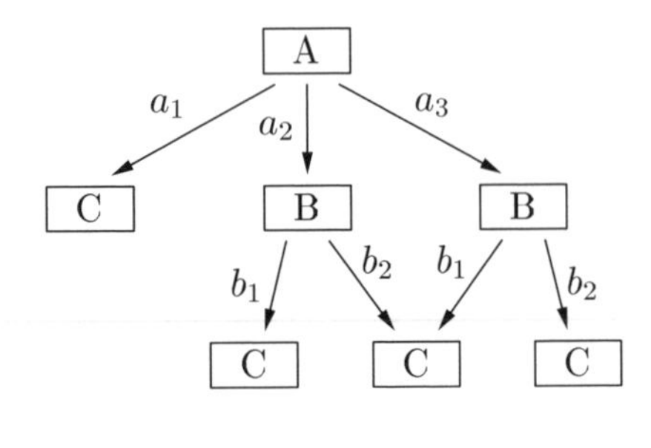

Figure 8.5 A conditional probability table with a third kind of regularities and a *decision graph* that captures them.

following set of operations [Chickering *et al.* 1997]:

- *full split*: Split a leaf node w.r.t. to the values of an attribute.
- *binary split*: Split a leaf node so that one child corresponds to one value of an attribute and the other child to all other values.
- *merge*: Merge two distinct leaf nodes.

As for decision tree induction the operation to execute is chosen greedily: All possible operations of the types listed above are temporarily applied (wherever possible) to an (initially empty) decision graph and the results are evaluated. The operation that yields the greatest improvement of the evaluation measure is carried out permanently. This greedy search is carried out until none of the operations listed above leads to an improvement.

It is obvious that without the merge operation this algorithm is equivalent to the well-known top-down induction algorithm for decision trees [Breiman *et al.* 1984, Quinlan 1993]. Hence decision tree induction can be seen as a special case of Bayesian network learning, namely as learning the local structure of a Bayesian network, in which there is only one child attribute, namely the class attribute. This view provides us with another justification for using the attribute selection measures of decision tree induction also as evaluation measures for learning Bayesian networks (cf. Section 7.2.4).

It should be noted that the greedy algorithm described above can be seen as a generalization of the K2 algorithm, which was discussed in Section 7.3.4. The only difference is that the K2 algorithm is restricted to one operation, namely to split in parallel all leaf nodes w.r.t. to the values of an attribute, which, in addition, must be the same for all leaves. Hence the K2 algorithm always yields a full decision tree as the local structure. In this respect it is important that the topological order presupposed by the K2 algorithm enables us to construct each decision graph independently of any other, since the topological order already ensures the acyclicity. If we use the generalized version of the K2 algorithm (cf. Section 7.3.4), the operations have to be applied in parallel to all decision graphs that are constructed in order to ensure that the global structure of the resulting Bayesian network, i.e. the conditional independence graph, is acyclic. All operations that may lead to a cycle have to be eliminated.

Our own approach to learning local network structure, which we suggested in [Borgelt and Kruse 1998d], is a slight modification of the above algorithm. It is based on the view explained in the preceding paragraph, namely that the above algorithm can be seen as a generalization of the K2 algorithm. This view suggests the idea to exploit the additional degree of freedom of decision graphs compared to decision trees, namely that a node in a decision graph can have more than one parent, not only to capture a larger set of regularities, but also to improve the learning process. The basic idea is as follows: With decision graphs, we can always work with the complete set of inner nodes of a full decision tree and let only leaves have more than one parent. Even if we do not care about the order of the conditioning attributes in the decision graph and if we require that any attribute may be tested only once on each path, such a structure can capture all regularities in the examples examined in the preceding section. For instance, the regularities of the table shown on the left in Figure 8.4 are captured by the decision graph shown in Figure 8.6. Note that the test of attribute B in the leftmost node is without effect, since both branches lead to the same leaf node.

It is easy to see that such an approach can capture any regularities that may be present in conditional probability tables: Basically, it consists in merging the leaves of a full decision tree and this is the same as merging lines of a conditional probability table. The decision graph structure only makes it simpler to keep track of the different instantiations of the conditioning attributes, for which the same probability distribution on the values of the conditioned attribute has to be adopted.

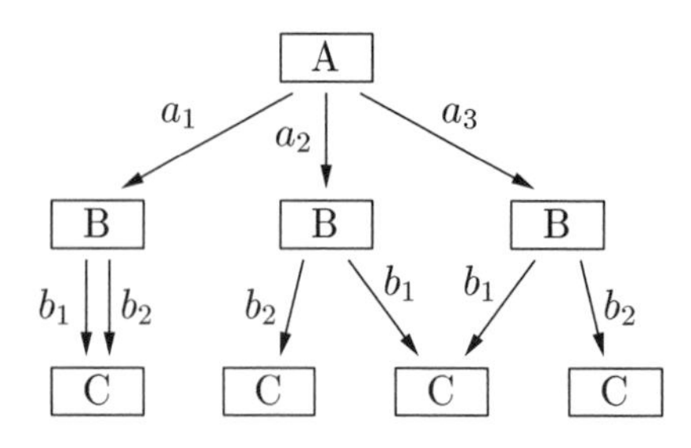

Figure 8.6 A decision graph with a full set of inner nodes that captures the regularities in the table shown in Figure 8.4. Note that the test of attribute B in the leftmost node is without effect.

Consequently we use only two operations [Borgelt and Kruse 1998d]:

- *split*: Add a new level to a decision graph, i.e. split all leaves according to the values of a new parent attribute.
- *merge*: Merge two distinct leaf nodes.

These operations are executed in turn. First a level is added to a decision graph and then leaves are merged. When a new level is added, one may either split the merged leaves of the previous step or start over from a full decision tree. To find a good set of mergers of leaf nodes, a greedy approach suggests itself. That is, in each step all mergers of two leaves are evaluated and then that merger is carried out that yields the largest improvement of the evaluation measure. Merging leaves stops, when no merger improves the value of the evaluation measure. If a mechanism for re-splitting leaf nodes is provided, simulated annealing may also be used.

At first sight this algorithm may appear to be worse than the algorithm reviewed above, because the operation that splits all leaves seems to be more costly than the split operations for single leaves. However, it is clear that the algorithm by [Chickering *et al.* 1997] has to check all splits of leaf nodes in order to find the best split and thus actually carries out the same operation. Note also that our algorithm needs to access the database of sample cases only as often as an algorithm for learning a Bayesian network without local structure, for instance, the K2 algorithm: The conditional frequency distributions have to be recomputed only after a split operation has been executed. The next step, i.e. the step in which leaves are merged, can be carried out without accessing the database, since all necessary information is already available in the leaf nodes. In contrast to this, the algorithm by [Chickering *et al.* 1997] has to access the database whenever two leaf nodes are merged in order to evaluate the possible splits of the resulting node.

It should be noted that for merging leaves both algorithms can exploit the fact that most evaluation measures (cf. Section 7.2.4) are computed from terms that can be computed leaf by leaf. Hence, if two leaves are merged, the decision graph need not be re-evaluated completely, but the change of the evaluation measure can be computed locally from the distributions in the merged leaves and the distribution in the resulting leaf.

Table 8.1 Results of learning Bayesian networks with local structure.

measure	edges	add.	miss.	params.	train	test
$I_{\text{gain}}^{(\text{Shannon})}$	35.0	17.1	4.1	1169.1	−11229.3	−11817.6
$I_{\text{gr}}^{(\text{Shannon})}$	21.9	15.3	15.4	284.4	−16392.0	−16671.1
$I_{\text{sgr1}}^{(\text{Shannon})}$	34.7	14.5	1.8	741.3	−11298.8	−11692.1
Gini	35.0	17.1	4.1	1172.3	−11233.1	−11813.4
χ^2	35.0	17.3	4.3	1143.8	−11234.9	−11805.2
K2	27.0	6.3	1.3	612.1	−11318.3	−11639.9
BDeu	39.0	24.1	7.1	1251.0	−11387.3	−11978.0
$L_{\text{red}}^{(\text{rel})}$	25.1	4.2	1.1	448.7	−11354.4	−11571.8

A drawback of our algorithm, but also of the algorithm by [Chickering *et al.* 1997] reviewed above, is that it can lead to a complicated structure that may hide a simple structure of context-specific independences. Therefore some post-processing to simplify the structure by changing the order of the attributes and by introducing multiple tests along a path is advisable.

8.3 Experimental Results

We incorporated our algorithm for learning the local structure of a Bayesian network into the INES program (Induction of NEtwork Structures) [Borgelt and Kruse 1997a, Borgelt and Kruse 1997b], which was also used to obtain the experimental results reported in Section 7.4. As a test case we chose again the Danish Jersey cattle blood group determination example [Rasmussen 1992] (cf. Section 4.2.2), so that the results of this section can easily be compared to those reported in Section 7.4. The set of databases used is identical and the networks were evaluated by the same means.

The results, which were obtained with greedy parent selection w.r.t. a topological order, are shown in Table 8.1. The meaning of the columns is the same as in Section 7.4. It is worth noting that the Shannon information gain ratio $I_{\text{gr}}^{(\text{Shannon})}$ fails completely and that the K2 metric and the reduction of description length measures again perform best.

It can also be seen that some measures tend to select more conditions (parents), thus leading to overfitting. At first sight it is surprising that allowing local structure to be learned can make the global structure more complex, although the number of parameters can be reduced (and actually is for some measures). But a second thought (and a closer inspection of the learned networks) reveals that this could have been foreseen. In a frequency distribution determined from a database of sample cases random fluctuations are to be expected. Usually these do not lead to additional conditions (except for measures like Shannon information gain or the χ^2 measure), since the "costs" of an additional level with several (approximately) equivalent leaves prevents the selection of such a condition. But the disadvantage of (approximately) equivalent leaves is removed by the possibility to merge these leaves, and thus those fluctuations that show a higher deviation from the true (independent) probability distribution are filtered out and become significant to the measure.

This effect is less pronounced for a larger dataset, but does not vanish completely. We guess that this is a general problem any learning algorithm for local structure has to cope with. Therefore it may be advisable not to combine learning global and local network structure, but to learn the global structure first, relying on the score for a full decision tree, and to simplify this structure afterwards by learning the local structure.

Chapter 9

Inductive Causation

If A causes B, an occurrence of A should be accompanied or (closely) followed by an occurrence of B. That causation implies conjunction or correlation is the basis of all reasoning about causation in statistics. But is this enough to infer causal relations from statistical data, and, if not, are there additional assumptions that provide reasonable grounds for such an inference? These are the questions we are going to discuss in this chapter.

In Section 9.1 we consider the connection of correlation and causation for two variables. Since this connection turns out to be unreliable, we extend our considerations in Section 9.2 to the connection of causal and probabilistic structures, especially Bayesian networks. Section 9.3 is concerned with two presuppositions of inductive causation, namely the stability assumption and the treatment of latent variables. In Section 9.4 we describe the inductive causation algorithm [Pearl and Verma 1991a], and criticize the assumptions underlying it in Section 9.5. Finally, in Section 9.6, we evaluate our discussion.

9.1 Correlation and Causation

Correlation is perhaps the most frequently used concept in applied statistics. Its standard measure is the correlation coefficient, which assesses what can be called the intensity of linear relationship between two measures [Everitt 1998]. Correlation is closely related to probabilistic dependence, although the two concepts are not identical, because zero correlation does not imply independence. However, since this difference is of no importance for our discussion, we use the term "correlation" in the vernacular sense, i.e. as a synonym for (probabilistic) dependence.

Note that neither in the narrower statistical nor in the wider vernacular sense correlation is connected directly to causal relation. We usually do not know *why* a correlation exists or does not exist, only *that* it is present or not.

Nevertheless such erroneous interpretation is tempting [Gould 1981]:

> Much of the fascination of statistics lies embedded in a gut feeling—and never trust a gut feeling—that abstract measures summarizing large tables of data must express something more real and fundamental than the data itself. (Much professional training in statistics involves a conscious effort to counteract this gut feeling.) The technique of *correlation* has been particularly subject to such misuse because it seems to provide a path for inferences about causality. [...] [But t]he inference of cause must come from somewhere else, not from the simple fact of correlation—though an unexpected correlation may lead us to search for causes so long as we remember that we may not find them. [...] The invalid assumption that correlation implies cause is probably among the two or three most serious and common errors of human reasoning.

It is easily demonstrated that indeed the vast majority of all correlations are, without doubt, noncausal. Consider, for example, the distance between the continents America and Europe over the past twenty years. Due to continental drift this distance increases a few centimeters every year. Consider also the average price of Swiss cheese in the United States over the same period.[1] The correlation coefficient of these two measures is close to 1, i.e. even in the narrow statistical sense they are strongly correlated. But obviously there is no causal relation whatsoever between them.

Of course, we could have used also any other measure that increased over the past years, for example, the distance of Halley's comet (since its last visit in 1986) or the reader's age. The same can be achieved with pairs of measures that *de*creased. Therefore, causality may neither be inferred from correlation with certainty (since there are counterexamples), nor even inferred with a high probability (since causal correlations themselves are fairly rare).

According to these arguments it seems to be a futile effort to try to infer causation from observed statistical dependences. Indeed, there is no way to causation from a *single* correlation (i.e. a dependence between two variables). But this does not exclude immediately the possibility to infer from a set of (conditional) dependences and independences between *several* variables something about the underlying causal influences. There could be connections between the causal and the probabilistic *structure*, which enable us to discover the former at least partly.

9.2 Causal and Probabilistic Structure

Our intuition of causation is perhaps best captured by a binary predicate "X (directly) causes Y" or "X has a (direct) causal influence on Y", where X

[1] We do not know much about the average price of Swiss cheese in the United States over the past twenty years, but we assume that it has risen. If it has not, substitute the price of any other consumer good that has.

is the *cause* and Y the *effect*. This predicate is usually seen as antisymmetric, i.e. if "X (directly) causes Y" holds, then "Y (directly) causes X" does *not* hold. Thus there is an inherent direction in causal influence, which seems to be a characteristic property of causation. For the most part it is due to our intuition that a cause precedes its effect in time.

Another formal interpretation, which was advocated by Bertrand Russell [Vollmer 1981], is that an effect is a function of its cause. But we reject this interpretation for several reasons. The first is that it brings in an assumption through the back door, which we want to make explicit (cf. Section 9.5). Secondly, a function is not necessarily antisymmetric and thus cannot always represent the direction of causation. Thirdly, if one variable is a function of another, then there need not be a causal connection (cf. Section 9.1). Hence functional dependence and causal influence should not be identified.

Because of the inherent direction, we can use a *directed graph* to represent causal influences, which we call the *causal structure*. In principle directed cycles, i.e. circular causal influences, are possible. (Such cycles are often exploited for control mechanisms, for example Watt's conical pendulum governor of the steam engine.) Nevertheless we do not consider circular causal structures here, but assume that the causal influences form a *directed acyclic graph* in order to make them comparable to a probabilistic structure.

We need not say much about the notion of the probabilistic structure of a domain here, because it is simply the conditional independence graph underlying a graphical model. Since we defined the causal structure of a domain as a directed acyclic graph, the theory of Bayesian networks, which are also based on directed acyclic graphs, suggests itself as an appropriate framework for a discussion of the connection between the causal and the probabilistic structure of a domain. Indeed, Bayesian networks are not only studied on purely statistical grounds as we did in the earlier chapters of this book, but they are often also used to describe a causal structure. Sometimes this is emphasized by calling them *probabilistic causal networks*. The reason for this is that, since Bayesian networks are based on directed graphs, the idea suggests itself to direct their edges in such a way that they represent causal influences. Actually human domain experts, who want to built a Bayesian network model, often start from a causal model of the underlying domain and simply enhance it by conditional probability distributions.

Of course, it is perfectly acceptable to call a Bayesian network a *causal* network as long as the term *causal* is only meant to indicate that a human expert, who "manually" constructed the network, did so by starting from a causal model of the domain under consideration. In this case the knowledge of the human expert about the causal relations in the modeled domain ensures that the Bayesian network represents not only statistical (in)dependences, but also causal influences. If, however, the conditional independence graph of a Bayesian network is learned automatically from data, calling it a *causal* network seems to be questionable. What could it be that ensures in an automatically generated network that directed edges show the directions of causal influences (from cause to effect)? Obviously we need to establish a relation between causal dependence and statistical dependence, since it is only statistical dependence we can test for automatically. The most direct approach is, of course, to identify the two structures, i.e. to use the causal structure as a conditional independence graph of a given domain (see above). If the causal and a probabilistic structure of a domain are identified in this way, we should be able to read from the causal structure, using the d-separation criterion,[2] certain conditional independences that hold in the domain. This suggests the idea to invert the procedure, i.e. to identify the causal structure or at least a part of it by conditional independence tests.

9.3 Stability and Latent Variables

A fundamental problem of an approach that tries to discover the causal structure with conditional independence tests is that the d-separation criterion does not says anything about the dependence or independence of two sets X and Y of attributes given a third set Z if X and Y are *not* d-separated by Z. This suffices if a Bayesian network is constructed for a given domain, since for applications it is not essential to find and represent *all* independences (cf. Chapter 4). However, we need more to identify a causal structure, because we must be able to infer something about the causal structure from a conditional independence statement, which we cannot do if we do not know whether this statement is represented by d-separation in the causal structure or not.

In order to cope with this problem it is assumed that in a (sampled) probability distribution p of the domain under consideration there exist exactly those (conditional) independences that can be read from the causal structure using d-separation. This assumption is called *stability* [Pearl and Verma 1991a] and can be formalized as $(X \perp\!\!\!\perp_p Y \mid Z) \Leftrightarrow$ (Z d-separates X and Y in the causal structure). Obviously, the stability assumption is equivalent to the assumption that the causal structure of the domain is a *perfect map*[3] of the (conditional) independence statements that hold in the domain. Note that the stability assumption asserts that there is "no correlation without causation" (also known as Reichenbach's dictum), because it assumes a direct causal influence between any two variables that are dependent given any set of other variables. In addition, any correlation between two variables is explained by a (direct or indirect) causal connection.

An important property of d-separation together with the stability assumption is that they distinguish a common effect of two causes on the one hand from the mediating variable in a causal chain and the common cause of two effects on the other. In the two structures shown on the left in Figure 9.1

---

[2]The notion of d-separation was defined in Definition 4.1.14 on page 98.

[3]The notion of a perfect map was introduced in Definition 4.1.15 on page 101.

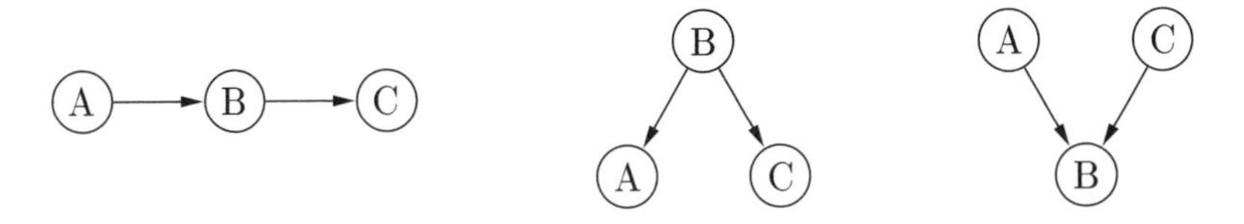

Figure 9.1 Possible causal connections of three attributes.

A and C are independent given B, but dependent if B is not given. In contrast to this, in the structure on the right A and B are independent unconditionally, but dependent if B is given. It is this (alleged) fundamental asymmetry of the basic causal structures, which was studied first in [Reichenbach 1956],[4] that makes statistical inferences about causal relations possible.

However, even with d-separation and the stability assumption there are usually several causal structures that are compatible with the observed (conditional) dependences and independences. The main reason is that d-separation and stability cannot distinguish between causal chains and common causes. But in certain situations all compatible causal structures have a common substructure. The aim of inductive causation is to find these substructures.

Furthermore, if we want to find causal relations in real world problems, we have to take care of *hidden* or *latent* variables. To handle latent variables, the notion of a *latent structure* and of a *projection* of a latent structure are introduced [Pearl and Verma 1991a]. The idea is to restrict the number and influence of latent variables while preserving all dependences and independences. A latent structure is simply a causal structure in which some variables are unobservable. A projection is defined as follows:

Definition 9.3.1 *A latent structure L_1 is a* **projection** *of another latent structure L_2, if and only if*

1. *Every unobservable variable in L_1 is a parentless common cause of exactly two nonadjacent observable variables.*
2. *For every stable distribution p_2 that can be generated by L_2, there exists a stable distribution p_1 generated by L_1 such that*

$$\begin{aligned}&\forall X, Y \in O, S \subseteq O \backslash \{X, Y\}:\\ &\qquad (X \perp\!\!\!\perp Y \mid S \textit{ holds in } p_2|_O) \Rightarrow (X \perp\!\!\!\perp Y \mid S \textit{ holds in } p_1|_O),\end{aligned}$$

where O is the set of observable variables and $p|_O$ denotes the marginal probability distribution on these variables.

(A stable distribution satisfies the stability assumption, i.e. exhibits only those independences identifiable by the d-separation criterion.)

[4]However, [Reichenbach 1956] did not consider this asymmetry as a means to discover causal structure, but as a means to define the direction of time.

It can be shown that for every latent structure there is at least one projection [Pearl and Verma 1991a]. Note that a projection must exhibit only the same (in)dependence structure (w.r.t. d-separation), but need not be able to generate the same distribution.[5] In essence, the notion of a projection is only a technical trick to be able to represent dependences that are due to latent variables by bidirected edges (which are an intuitive representation of a hidden common cause of exactly two variables).

9.4 The Inductive Causation Algorithm

With the ingredients listed above, we finally arrive at the following algorithm [Pearl and Verma 1991a], which is very closely related to Algorithm 7.3.3 on page 235 (the first two steps are actually identical, the third is similar—it only does not try to direct *all* edges).

Algorithm 9.4.1 (Inductive Causation Algorithm)

Input: *A (sampled) distribution p over a set U of attributes.*

Output: *A marked hybrid acyclic graph* core(p).

1. *For each pair of attributes A and B, search for a set $S_{AB} \subseteq U - \{A, B\}$, so that $A \perp\!\!\!\perp_p B \mid S_{AB}$ holds, i.e. so that A and B are conditionally independent given S_{AB}. If there is no such set S_{AB}, connect the attributes by an undirected edge.*

2. *For each pair of nonadjacent attributes A and B with a common neighbor C (i.e. C is adjacent to A as well as to B), check whether $C \in S_{AB}$. If it is not, direct the edges towards C, i.e. $A \to C \leftarrow B$.*

3. *Form* core(p) *by recursively directing edges according to the following two rules:*

   - *If for two adjacent attributes A and B there is a strictly directed path from A to B not including the edge connecting A and B, then direct the edge towards B.*

   - *If there are three attributes A, B, and C with A and B not adjacent, either $A \to C$ or $A \leftarrow C$, and $C - B$, then direct the edge $C \to B$.*

4. *For each triplet of attributes A, B, and C: If A and B are not adjacent, $C \to B$, and either $A \to C$ or $A \leftarrow C$, then mark the edge $C \to B$.*

[5] Otherwise a counterexample could easily be found: Consider seven binary variables A, B, C, D, E, F, and G, i.e. $\text{dom}(A) = \text{dom}(B) = \ldots = \text{dom}(G) = \{0, 1\}$. Let A be hidden and $E = A \cdot B$, $F = A \cdot C$, and $G = A \cdot D$ (where $\cdot$ denotes multiplication). A projection of this structure contains three latent variables connecting E and F, E and G, and F and G, respectively. It is easy to prove that such a structure cannot generate the distribution resulting from the functional dependences given above.

Step 1 determines the attribute pairs between which there must exist a direct causal influence or a hidden common cause, because an indirect influence should enable us to find a set S_{AB} that renders the two attributes independent. In step 2 the asymmetry inherent in the *d*-separation criterion is exploited to direct edges towards a common effect. Part 1 of step 3 ensures that the resulting structure is acyclic. Part 2 uses the fact that $B \rightarrow C$ is impossible, since otherwise step 2 would have already directed the edge in this way. Finally, step 4 marks those unidirected links that cannot be replaced by a hidden common cause. The reason is that if the attributes B and C named in this step were connected by a hidden common cause, A and B would not be independent given C, which they must, because otherwise step 2 would have directed both edges towards C.

The output graph core(p) has four kinds of edges:

1. marked unidirected edges representing *genuine causal influences* (which must be direct causal influences in a projection),
2. unmarked unidirected edges representing *potential causal influences* (which may be direct causal influences or indirect influences brought about by a hidden common cause),
3. bidirected edges representing *spurious associations* (which are due to a hidden common cause in a projection), and
4. undirected edges representing unclassifiable relations.

9.5 Critique of the Underlying Assumptions

In this section we discuss the assumptions underlying *d*-separation and stability by considering some special cases with only few variables [Borgelt and Kruse 1999]. The simplest case are causal chains, like the one shown on the left in Figure 9.1. If a variable has a direct causal influence on another, they should be dependent at least unconditionally, i.e. $A \not\perp\!\!\!\perp B \mid \emptyset$ and $B \not\perp\!\!\!\perp C \mid \emptyset$. It is also obvious, that $A \perp\!\!\!\perp C \mid B$. A direct cause, if fixed, should shield the effect from any change in an indirect cause, since a change in the indirect cause can influence the effect only by changing the direct cause. But to decide whether B and C are dependent given A or not, we need to know the causal influences in more detail. For instance, if $B = f(A)$ and $C = g(B)$, then $B \perp\!\!\!\perp C \mid A$. But if the value of A does not completely determine the value of B, then B and C will usually be dependent. Although the former is not uncommon, the stability assumption excludes it.

The next cases are diverging or converging causal influences, as shown in the middle and on the right in Figure 9.1 on page 265. The main problems with these structures are whether the statements $B \perp\!\!\!\perp C \mid A$ (middle) and $A \perp\!\!\!\perp B \mid C$ (right) hold or not. The assumptions by which *d*-separation and the stability assumption handle these problems are as follows:

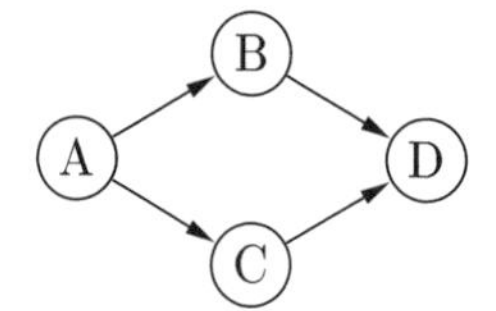

Figure 9.2 Interaction of common cause and common effect assumption.

Common Cause Assumption (Causal Markov Assumption)

Given *all* of their (direct or indirect) common causes, two effects are independent, i.e. in the structure in the middle of Figure 9.1 the attributes B and C are independent given A. If B and C are still dependent given A, it is postulated that either B has a causal influence on C or vice versa or there is another (hidden) common cause of B and C (apart from A). That is, the causal structure is considered to be incomplete.

Common Effect Assumption

Given *one* of their (direct or indirect) common effects, two causes are dependent, i.e. in the structure on the right of Figure 9.1 the attributes A and B are dependent given C. For applications of Bayesian networks this assumption is not very important, since little is lost if it is assumed that A and B are dependent given C though they are not—only the storage savings resulting from a possible decomposition cannot be exploited. However, for inferring causal relations this assumption is very important.

Note that the common cause assumption necessarily holds, if causation is interpreted as functional dependence. Then it only says that fixing all the arguments that (directly or indirectly) enter both functions associated with the two effects renders the effects independent. But this is obvious, since any variation still possible has to be due to independent arguments that enter only one function. This is the main reason why we rejected the interpretation of causation as functional dependence. It is not at all obvious that causation should satisfy the common cause assumption.

A situation with diverging causal influences also poses another problem: Are B and C independent unconditionally? In most situation they are not, but if, for example, $\text{dom}(A) = \{0, 1, 2, 3\}$, $\text{dom}(B) = \text{dom}(C) = \{0, 1\}$ and $B = A \bmod 2$, $C = A \operatorname{div} 2$, then they will be. The stability assumption rules out this not very unlikely possibility.

The two assumptions also interact and this can lead to a priority problem. Consider, for example, the structure shown in Figure 9.2: Given A as well as D, are B and C independent? The common cause assumption affirms this, the common effect assumption denies it. Since the stability assumption requires B and C to be dependent, it contains the assumption that in case of a tie

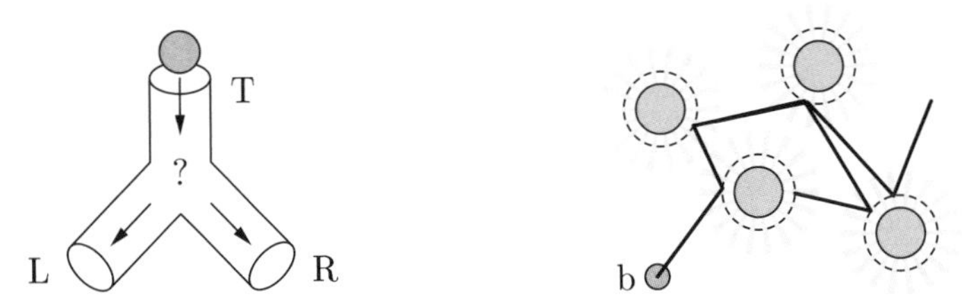

Figure 9.3 Left: Y-shaped tube arrangement into which a ball is dropped. Since it can reappear only at L *or* at R, but not at both, the corresponding variables are dependent. Right: Billiard with round obstacles exhibits sensitive dependence on the initial conditions.

the common effect assumption has the upper hand. Note that from strict functional dependence $B \perp\!\!\!\perp C \mid \{A, D\}$ follows.

In the following we examine some of the assumptions identified above in more detail, especially the common cause and the common effect assumption, which are at the heart of the alleged asymmetry of causal relations.

Common Cause Assumption (Causal Markov Assumption)

Consider a Y-shaped arrangement of tubes like the one shown on the left in Figure 9.3. If a ball is dropped into this arrangement, it will reappear shortly afterwards at one of the two outlets. If we neglect the time it takes the ball to travel through the tubes, we can define three binary variables T, L, and R indicating whether there is a ball at the top T, at the left outlet L or at the right outlet R. Obviously, whether there is a ball at T or not has a causal influence on L and on R. But L and R are dependent given T, because the ball can reappear only at one outlet.

At first sight the common cause assumption seems to fail in this situation. However, we can always assume that there is a hidden common cause, for instance, an imperfectness of the ball or the tubes, which influences its course. If we knew the state of this cause, the outlet at which the ball will reappear could be determined and hence the common cause assumption would hold. Obviously, if there is a dependence between two effects, we can always say that there must be another hidden common cause. We only did not find it, because we did not look hard enough. Since this is a statement of existence, it cannot be disproven. Although using statements that cannot be falsified is bad scientific methodology [Popper 1934], we have even better grounds on which to reject such an explanation.

The idea that, in principle, we could discover the causes that determine the course of the ball is deeply rooted in the mechanistic paradigm of physics, which is perhaps best symbolized by Laplace's demon.[6] But quantum theory suggests that such a view is wrong [Feynman *et al.* 1965, von Neumann 1932]: It may very well be that even if we look hard enough, we will not find a hidden common cause to explain the dependence.

To elaborate a little: Among the fundamental statements of quantum mechanics are Heisenberg's uncertainty relations. One of these states that $\Delta x \cdot \Delta p_x \geq \frac{\hbar}{2}$. That is, we cannot measure both the location x and the momentum p_x of a particle with arbitrary precision in such a way that we can predict its exact trajectory. There is a finite upper bound due to the unavoidable interaction with the observed particle. However, in our example we may need to predict the exact trajectory of the ball in order to determine the outlet with certainty.

The objection may be raised that $\frac{\hbar}{2}$ is too small to have any observable influence. To refute this, we could add to our example an "uncertainty amplifier" based on the ideas studied in chaos theory, i.e. a system that exhibits a sensitive dependence on the initial conditions. A simple example is billiard with round obstacles [Ruelle 1993], as shown on the right in Figure 9.3. The two trajectories of the billiard ball b, which at the beginning differ only by about $\frac{1}{100}$ degree, differ by about 100 degrees after only four collisions. (This is a precisely computed example, not a sketch.) Therefore, if we add a wider tube containing spheres or semi-spheres in front of the inlet T, it is plausible that even a tiny change of the position or the momentum of the ball at the new inlet may change the outlet at which the ball will reappear. Therefore quantum mechanical uncertainty cannot be neglected.

Another objection is that there could be "hidden parameters", which, if discovered, would remove the statistical nature of quantum mechanics. However, as [von Neumann 1932] showed,[7] this is tantamount to claiming that quantum mechanics is false—a claim for which we do not have any convincing evidence, since quantum mechanics is a well-confirmed theory and there is currently no alternative theory with the same explanatory power.

---

[6] Laplace wrote [Kline 1980]: "We may regard the present state of the universe as the effect of its past and the cause of its future. An intellect which at any given moment knew all the forces that animate nature and the mutual positions of the beings that compose it, if this intellect were vast enough to submit the data to analysis, could condense into a single formula the movement of the greatest bodies of the universe and that of the lightest atom: for such an intellect nothing would be uncertain; and the future just like the past would be present before its eyes."

[7] von Neumann wrote: "[...] the established results of quantum mechanics can never be re-derived with their [the hidden parameters'] help. In fact, we have even ascertained that it is impossible that the same physical quantities exist with the same function connections [...], if other variables (i.e. "hidden parameters") should exist in addition to the wave functions. Nor would it help if there existed other, as yet undiscovered physical quantities, [...], because the relations assumed by quantum mechanics [...] would have to fail already for the known quantities [...] It is therefore not, as often assumed, a question of a re-interpretation of quantum mechanics, — the present system of quantum mechanics would have to be objectively false, in order that another description of the elementary processes than the statistical one be possible."

murderer	sentence death	other	$\sum$
black	59	2448	2507
white	72	2185	2257
$\sum$	131	4633	4764

Table 9.1 Death sentencing and race in Florida 1973–1979. The hypothesis that the two attributes are independent can be rejected only with an error probability greater than 7.8% (according to a χ^2 test).

victim	murderer	death	other
black	black	11	2209
	white	0	111
white	black	48	239
	white	72	2074

Table 9.2 Death sentencing and race in Florida 1973–1979, full table. For white victims the hypothesis that the two other attributes are independent can be rejected with an error probability less than 0.01% (according to a χ^2 test).

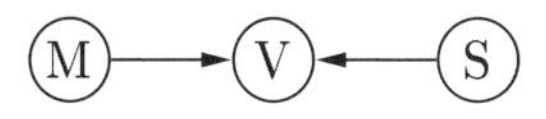

Figure 9.4 Core inferred by the inductive causation algorithm for the above data.

Common Effect Assumption

According to [Salmon 1984], it seems to be difficult to come up with an example in which the common effect assumption does not hold. Part of the problem seems to be that most macroscopic phenomena are described by continuous real-valued functions, but there is no continuous n-ary function, $n \geq 2$, which is injective. (Such a function would be a simple, though not the only possible counterexample).

However, there are real-world examples that come close, for instance, statistical data concerning death sentencing and race in Florida 1973–1979 (according to [Krippendorf 1986] as cited in [Whittaker 1990]). From Table 9.1 it is plausible to assume that *murderer* and *sentence* are independent. Splitting the data w.r.t. *victim* shows that they are strongly dependent given this variable (see Table 9.2). Hence the inductive causation algorithm yields the causal structure shown in Figure 9.4. But this is not acceptable: A direct causal influence of *sentence* on *victim* is obviously impossible (since the sentence follows the murder in time), while a common cause is hardly imaginable. The most natural explanation of the data, namely that *victim* has a causal influence on *sentence*, is explicitly ruled out by the algorithm.

This example shows that an argument mentioned in [Pearl and Verma 1991a] in favor of the stability assumption is not convincing. It refers to [Spirtes *et al.* 1989], where it is shown that, if the parameters of a distribution

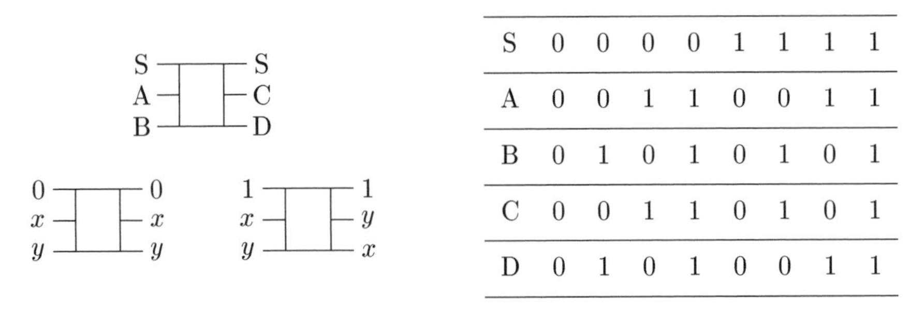

S	0	0	0	0	1	1	1	1
A	0	0	1	1	0	0	1	1
B	0	1	0	1	0	1	0	1
C	0	0	1	1	0	1	0	1
D	0	1	0	1	0	0	1	1

Figure 9.5 The Fredkin gate [Fredkin and Toffoli 1982].

are chosen at random from any reasonable distribution, then any unstable distribution has measure zero. But the problem is that this is not the correct set of distributions to look at. When trying to infer causal influence, we have to take into account all distributions that *could be mistaken for an unstable distribution.* Indeed, the true probability distribution in our example may very well be stable, i.e. *murderer* and *sentence* may actually be marginally dependent. But the distribution in the sample is so close to an independent distribution that it may very well be confused with one.

In addition, the special parameter assignments leading to unstable distributions may have high probability. For example, it would be reasonable to assume that two variables are governed by the same probability distribution, if they were the results of structurally equivalent processes. Yet such an assumption can lead to an unstable distribution, especially in a situation, in which common cause and common effect assumption interact. For instance, for a Fredkin gate [Fredkin and Toffoli 1982] (a universal gate for computations in conservative logic, see Figure 9.5), the two outputs C and D are independent if the two inputs A and B assume the value 1 with the same probability. In this case, as one can easily verify, the causal direction assigned to the connection A—C depends on whether the variables A, B, and C or the variables A, C, and D are observed.

9.6 Evaluation

The discussion of the assumptions underlying the inductive causation algorithm showed that at least some of them can be reasonably doubted. In addition, the inductive causation algorithm cannot deal adequately with accidental correlations. But we saw in Section 9.1 that we sometimes reject a causal explanation in spite of the statistical data supporting such a claim. In our opinion it is very important for an adequate theory of causation to explain such a rejection.[8] In summary, when planning to apply this algorithm, one should carefully check whether the assumptions can be accepted and whether the underlying interpretation of causality is adequate for the problem at hand.

A related question is: Given a causal relation between two variables, we are usually much more confident in an inference from the state of one of them to the state of the other than we would be if our reasoning was based only on a number of similar cases we observed in the past. But the inductive causation algorithm infers causation from a set of past observations, namely a sampled probability distribution. If the result is not substantiated by other means, in particular, by a model of the underlying mechanism, can we be any more confident in our reasoning than we would be if we based it directly on the observed correlations? It seems to be obvious that we can not. Hence the question arises whether the inductive causation algorithm is more than a heuristic method to point out possible causal connections, which then have to be further investigated. Of course, this does not discredit the inductive causation algorithm, since good heuristics are a very valuable thing to have. However, it warns against high expectations and emphasizes that this algorithm should not be seen as the *ultima ratio* for inferences about causality.

[8] An approach to causation that does not suffer from this deficiency was suggested by K. Lorenz and later developed, for example, in [Vollmer 1981]. It models causal connections as a transfer of energy. [Lemmer 1996] suggests a closely related model.

Chapter 10

Applications

In this chapter we conclude our discussion of graphical models and how to learn them from data by pointing out their practical value with some applications. In Section 10.1 we review an approach to fraud detection in telecommunications developed at AT&T [Ezawa and Norton 1995, Ezawa *et al.* 1996], which is based on learning Bayesian network classifiers from large phone call databases. In Section 10.2 we study the construction of a Markov network to predict the supply of automobile parts at the Volkswagen corporation, which is derived from a relational graphical model built from technical and marketing rules [Detmer and Gebhardt 2001]. In Section 10.3 we describe an application of learning Bayesian networks from data for fault analysis in automobiles, which was part of a cooperation between the Otto-von-Guericke-University of Magdeburg and the DaimlerChrysler corporation [Borgelt *et al.* 1998b].

10.1 Application in Telecommunications

Telecommunication is a fast growing business with a high turnover and a high competitive pressure. A severe problem in this area are fraud and uncollectible debt, from which the telecommunications industry incurs considerable losses every year. In this section we review, following [Ezawa and Norton 1995, Ezawa *et al.* 1996], an approach to identify potentially fraudulent phone calls, which is based on learning Bayesian network classifiers from large databases of already settled phone calls, i.e. of phone calls for which it is known whether the fee for them could be collected or not.

At first sight this problem may seem to be a simple classification task, which could be solved with any type of classifier. However, there are several aspects that make it difficult. In the first place, the interesting cases, i.e. the non-paying customers, are rare: They comprise only 1% or 2% of the population. Nevertheless they cannot be neglected: Due to the high turnover in the telecommunications industry (more than 100 billion dollars per year) this small percentage causes losses of several billion dollars. Unfortunately, most classifier learners have difficulties characterizing a minority class that small, because in such extreme situations they tend to predict the majority class for all cases, thus minimizing the error rate (decision tree learners are especially prone to this problem). Even though in this application the databases to learn from have been enriched to 9–12% non-paying customers, thus alleviating the task, the unequal class distribution still causes problems.

Secondly, the misclassification costs are highly unequal. It is less important to identify non-paying customers: If they go undetected initially, they will be identified after a couple of billing cycles nevertheless and thus the expected loss is moderate. However, if a valuable paying customer is classified as fraudulent and, consequently, some action is taken against him, the company may loose this customer forever. Due to the high competition in the market, a customer has a wide range of alternatives to choose from and may select a different company in the future. In this case the loss of potential revenue may be considerable. Even though some state-of-the-art classifiers offer ways to take misclassification costs into account (for instance C5.0, the successor of C4.5 [Quinlan 1993]), the costs often have to be fixed in advance, and a change in costs enforces a reconstruction of the classifier. It would be more convenient if a classification threshold could be fixed later.

Thirdly, some of the attributes describing a phone call have large unordered sets of values, such as telephone exchange and city name. Such attributes also pose problems for standard classifiers. Decision tree inducers, for instance, cannot handle them adequately, because such attributes lead to a very fine-grained division of the training data, thus limiting the number of attributes that can reasonably be tested in a decision tree. As a consequence it may not be possible to exploit the information contained in some attributes.

Finally, the phone call databases to learn from are very large (they typically have 4–6 million records and 600–800 million bytes) and missing values are very common. Some learning methods cannot handle such volumes of data in reasonable time, because they may have to process the data several times to reach a solution (for example artificial neural networks). Missing values also pose severe problems for some classifier induction methods.

As a possible solution, which nicely handles most of the difficulties pointed out above, a naive Bayes classifier suggests itself (cf. Section 6.1). A naive Bayes classifier can cope with the unequal class distribution and the unequal misclassification costs, since it only computes class probabilities. The threshold probability at which an action is taken w.r.t. a possibly fraudulent costumer can be fixed later, depending, for instance, on an analysis of expected loss based on test data. A naive Bayes classifier can also exploit at least part of the information conveyed by any of the attributes, since the attributes are treated independently and thus using one attribute does not restrict using any other. However, the main drawback of a naive Bayes classifier, namely the strong independence assumptions, leads to suboptimal performance in this application. Therefore it is advisable not to rely on a pure naive Bayes classifier, but to augment it with edges between some of the descriptive attributes. If appropriate heuristics are used to select these edges, it is possible to retain the efficiency with which a naive Bayes classifier can be constructed, thus enabling this approach to handle large volumes of data.

The APRI (Advanced Pattern Recognition and Identification) system developed at AT&T [Ezawa and Norton 1995, Ezawa *et al.* 1996] was designed according to the general scheme outlined above. It builds a Bayesian network classifier starting from a naive Bayes structure and employs heuristics based on mutual information (cf. Sections 7.2.4 and 7.3.4) to select the attributes used in the classifier and the edges between them.

The Bayesian network construction in APRI has four phases. In the first phase the database is scanned to determine the domains of all attributes. This phase may involve a discretization of continuous attributes using an information-based approach. In the second phase the descriptive attributes to be used in the classifier are selected. To this end, the mutual information of the class attribute C and each descriptive attribute A is computed. Then a minimal set of attributes $M \in U$ (where U is the set of all descriptive attributes) is determined, which satisfies

$$\sum_{A \in M} I_{\text{mutual}}(C, A) \ \geq \ \alpha \cdot \sum_{A \in U} I_{\text{mutual}}(C, A),$$

where α is a parameter (specified as a percentage) that governs how many attributes get selected. If α is set to 100%, all attributes are used. If α is less than 100%, the attributes conveying the least information about the class are discarded (and are not used in the consecutive steps either). By proceeding in this way it can be hoped that a simplification of the classifier can be achieved without having to pay too much in terms of reduced prediction accuracy. Note, however, that this selection scheme differs from the one studied in Section 6.3, because it does not rely on the classification error to select the attributes.

The third phase is similar to the second. Again mutual information is employed to determine which attributes should be connected with an edge (in addition to the naive Bayes structure, in which there is a directed edge from the class to each of the attributes selected in the second step). The selection criterion is analogous to the one used in the second step. The *conditional mutual information* (cf. Theorem 7.3.7 on page 243) of all pairs of descriptive attributes is computed and the possible edges are ordered w.r.t. to this value. Then a minimal set E_{add} of edges $(A, B) \in U' \times U'$ (where U' is the reduced set of attributes resulting from the second phase) is selected, which satisfies

$$\sum_{(A,B) \in E_{\text{add}}} I_{\text{mutual}}(A, B \mid C) \ \geq \ \beta \cdot \sum_{(A,B) \in U' \times U'} I_{\text{mutual}}(A, B \mid C),$$

where β is a parameter (specified as a percentage) that governs the density of the resulting Bayesian network. If β is set to 100% (which is not advisable, though), all edges are selected and the result is a complete network. If β is less than 100% the edges corresponding to the weakest dependences are discarded. The selected edges are directed according to the mutual information ranking computed in the second phase.

Note that this selection scheme can lead to more general networks than the *tree-augmented naive Bayes classifiers* studied in Section 7.3.4, because the set of additional edges is not restricted to a tree. A descriptive attribute may have an arbitrary number of parents, whereas in a tree-augmented naive Bayes classifier each descriptive attribute has at most two: The class attribute and maybe one other descriptive attribute. Note also that it is not checked whether an edge is unnecessary, because the attributes it connects share a parent (other than the class attribute) that renders them conditionally independent.

Finally, in the fourth phase of the APRI construction algorithm, the conditional probability distributions associated with the edges of the constructed network are determined. The result is a Bayesian network classifier.

It is clear that each phase of the APRI algorithm requires exactly one traversal of the database to learn from. Hence the classifier can be learned with four traversals of the database, which make this approach feasible even if the database has to be read from secondary storage (because it is too large to be loaded into main memory). This is important, because in the application considered here efficiency is an important issue. The patterns shown by fraudulent phone calls are likely to change over time depending on the policy adopted to identify and to prevent them. Hence, in order to be able to react quickly to the changing patterns, an efficient method to reinduce a classifier from newer training data is highly desirable.

[Ezawa and Norton 1995] report the result of an experiment, in which APRI was applied to a phone call database of about 4 million records, each described by 33 attributes. This database contains about 10% uncollectible calls. The algorithm described above was parameterized by $\alpha = 95\%$ and either $\beta = 25\%$ or $\beta = 45\%$, which, judging from the empirical evaluation, seem to be reasonable values. As a probability threshold for the classification as an uncollectible call a more aggressive 50% and a more conservative 70% were tested. The approach was compared to a naive Bayes classifier, either with all attributes or a simplified version induced by greedily adding attributes (cf. Section 6.3). The results show that the APRI approach is clearly successful, being able to outperform the naive Bayes classifiers by a considerable margin. The best results were achieved with $\beta = 45\%$ and a probability threshold of 70%. With these parameters over 20% of the uncollectible calls are correctly identified while less than 3% of the collectible calls are misclassified, which translates to virtually identical absolute numbers. The pure naive Bayes classifier either misclassifies a lot more collectible calls or identifies considerably fewer uncollectible calls (depending on the parameters).

10.2 Application at Volkswagen

In order to satisfy the individual needs and wishes of their customers, it is the marketing policy of the Volkswagen corporation to offer their customers the possibility to configure their cars as individually as possible. That is, a customer may choose—in addition to the model and a certain line of basic equipment—from a wide variety of options, including engines with different horsepower, manual or automatic gearshift, different types of tires, several special equipment items and so on [Detmer and Gebhardt 2001].

As convenient as this may be for a customer, who is not forced to adapt to some standard configuration of a car, this freedom of choice leads to high demands on logistics. The reason is that in todays highly competitive market any company is forced to reduce production costs wherever possible. In particular, it is tried to reduce the amount of stock in components and raw material that has to be kept in a production plant, because stockkeeping is an important cost factor. That is, it is tried to achieve ideally a situation in which the needed components arrive "just in time" for assembly so that no stock is necessary at all. Of course, for such a scheme to be feasible, it is necessary to plan the production process with high precision. This, in turn, requires highly accurate prediction of the supply of components, because otherwise it would not be possible to order the correct quantities from suppliers early enough.

As Niels Bohr remarked: "Prediction is very difficult, especially of the future." To predict the future supply of components is, of course, no exception, but it is manageable if the range of products is narrow enough. If a company produces, say, only a dozen different goods, then for each of these goods there will be enough samples, so that time series can be set up for each of them. Based on an extrapolation of these time series into the future (done, for instance, with standard statistical techniques) a reasonable prediction of the supply of components can be achieved by simply computing the quantities needed to produce the predicted number of goods.

However, such an approach is no longer feasible with the wide range of possible configurations of cars offered by the Volkswagen corporation, each of which has, in principle, to be treated as a different product. For its most successful car, for example, the VW Golf, a configuration is described by fixing a value for each of about 165 different families of properties, which comprise engine type, gearshift, tires, color, special equipment items etc. Each of these families has typically 3–5 values, but may also have up to 150, leading to billions of configurations. Even though not all of these properties are chosen directly by a customer (often several values are fixed by a single choice of a certain line of equipment) and a large number of these configurations cannot be built due to technical and marketing restrictions, there remain hundreds of millions of possible products. Of course, not all of the possible configurations are actually selected by some customer (in the year 2000 about 650 000 cars of type VW Golf were produced in Europe, thus restricting the number of existing configurations), but that customers indeed take advantage of the freedom of choice offered to them can be seen from the average number of identical cars. For the VW Golf this average number is less than five, even though it is computed *without* distinguishing between cars of different color (i.e. cars that only differ in color are counted as identical; however, the sets of identical cars that are ordered for the fleets of large corporations or by car rental companies have been discounted first) [Detmer and Gebhardt 2001]. As a consequence extrapolating time series for individual products is not feasible, since the available samples are much too small.

As an alternative one may consider predicting the supply of components directly from time series for individual components. However, this approach, although it appears to be a convenient solution at first sight, does not handle the problem adequately. In the first place, there are dependences between components, which are imposed by technical or marketing restrictions. If such dependences exist, the correct trend of the demand of a component may not be visible from the time series for that component. In addition, such an approach cannot handle well the restriction that the percentage of alternative components has to add up to 100%. That is, every car needs a steering wheel, an engine, four wheels, etc. But there are different choices of, say, steering wheels, and the sum of the supply of these has to add up to the total number of cars. These restrictions are especially hard to satisfy in predictions if there are dependences between components. For instance, a certain type of steering wheel is only installed if an airbag has been selected, and leather coated steering wheels may not be combined with a certain type of multi-function display. For the VW Golf, for example, such dependences result in about 70 000 combinations of properties that are relevant for the exact set of components that are installed in a car [Detmer and Gebhardt 2001].

In order to cope with the prediction problem describe above the Volkswagen corporation is currently developing a system for car configuration prediction that is based on a Markov network (cf. Definition 4.1.28 on page 114) over the set of property families, which describes a probability distribution on the space of possible car configurations [Detmer and Gebhardt 2001]. With this network it is tried to capture important dependences between car properties, while exploiting at the same time (conditional) independences to reduce the complexity of the prediction task. The basic idea is that given certain properties others may be selected independent of each other, thus making a decomposition of the joint probability distribution possible. Based on this decomposition, the frequency of relevant combinations of properties can be estimated, relying on all cars in the sales database having these combinations. That is, the decomposition allows us to remove the restriction to individual configurations and thus to enlarge the samples that can be evaluated. Finally the supply of individual components is computed from these estimates.

In order to actually build the graphical model, a relational network is constructed first based on a catalog of technical and marketing rules that describes the constructible cars, i.e. that distinguishes the possible from the impossible configurations. For the VW Golf this catalog, which is formulated in a kind of a logical calculus, comprises about 10 000 entries. The idea of starting from this catalog is that impossible configurations should be clearly represented as impossible, because the numbers predicted for such configurations must necessarily be zero. If impossible configurations were not explicitly marked as impossible, this requirement could not be enforced, thus making the subsequent prediction worse than necessary.

The construction of the relational model is carried out in three steps. In the first step the catalog of technical and marketing rules is translated into a relational representation. That is, all rules that refer to the same set of attributes (i.e. the same set of property families) is represented by a relation over this set of attributes: Starting from the Cartesian product of the attribute domains all tuples are discarded that are incompatible with one of the rules. In a second step these relations are organized in a lattice structure defined by the subset relation on the domains of the relations (i.e. the sets of attributes underlying them). The maximal elements of the resulting lattice are used as the maximal cliques of the graph of a relational graphical model. In order to make inferences possible this graph is finally triangulated. (It should be noted that it is actually necessary to draw inferences, because the sets of attributes that are relevant for predicting the supply of certain components need not be contained in the same clique of the graph.)

Unfortunately, this straightforward approach does not lead directly to a usable model, because the cliques of the resulting graph are too large to be handled efficiently. Hence it is necessary to simplify the graphical model, accepting approximations to make inferences feasible. This done by computing for each relation in the lattice a measure that makes it possible to assess which relations can be dispensed with without deteriorating the model quality too much. The idea of this measure is closely related to the Hartley information gain and the relative number of possible value combinations studied in Section 7.1.1. It is based on the notion of the *restrictivity* of a relation, which is the relative number of tuples that are excluded by this relation from the Cartesian product of the attribute domains underlying it. That is, it measures the percentage of combinations that are excluded by the relation. From the considerations in Section 7.1.1 it is clear that a high value of this measure indicates that a relation is very important for the graphical model. One has to take care, though, because the restrictivity of the maximal sets of the lattice is not the correct measure to look at. If one of them is discarded one or more subsets of it will take its place, thus making the loss in restrictivity less severe than indicated by this measure. Therefore a modified measure, the *additional restrictivity* is used, which is the percentage of tuples that is excluded by a relation relative to the natural join of its subsets in the lattice [Gebhardt 1999]. For this measure a limiting value may be specified, which governs the removal of relations. However, since it is difficult to specify an optimal limit, the process of constructing the relational model must be iterated, increasing the limit in each step, until a model of feasible complexity is reached.

Subsequently the relational model is turned into a Markov network by simply taking the graph structure of the relational model and enhancing it with probabilistic distribution functions, which are estimated from a sales database. However, this model is still not the final one from which the supply of components is predicted, because it does not respect the production capacities of different plants and suppliers. In order to take these into account the parameters of the Markov network are modified so that the predicted distribution changes as slightly as possible, but the external requirements are met. Unfortunately, however, the details of the algorithm underlying this step have not been made publicly accessible yet [Gebhardt 1999].

Test runs of the system have shown that it is successful, although details are confidential. Currently the Volkswagen corporation is working on making the developed system operative.

10.3 Application at DaimlerChrysler

Even high quality products like Mercedes-Benz vehicles sometimes show an undesired behavior. Since its is a major concern of the DaimlerChrysler corporation to further improve the quality of their products, a lot of effort is dedicated to finding the causes of these faults in order to be able to prevent similar faults from occurring in the future. To support these efforts DaimlerChryler maintains a quality information database to control the quality of produced vehicles. In this database for every produced vehicle (car, van, lorry or truck) its configuration (product line, engine type, special equipment etc.) and any faults detected during production or maintenance are recorded.

The problems one faces in this application are similar to those discussed in connection with the supply prediction at Volkswagen. Like Volkswagen, DaimlerChrysler offers its customers the possibility to configure their cars individually. As a consequence there are millions of different configurations, each of which is bought only very few times (the average numbers of identical cars are comparable to those of the Volkswagen corporation). Therefore it is not possible to monitor the behavior of individual configurations, simply because there are much too few example cases for each of them.

Unfortunately, it is not uncommon that an aggregate fails only when installed in combination with specific other components. Therefore a simple check whether individual aggregates show an unusually high frequency of failure (although such a check is, of course, done) cannot uncover all weaknesses: Their normal behavior in other configurations may hide the problem. However, since the space of possible combinations is extremely large (for some vehicles there are more than 100 special equipment items that may be chosen), it is clear that this space cannot be searched manually.

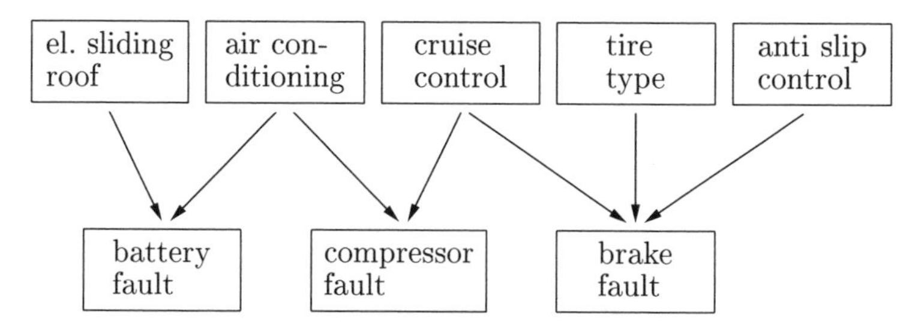

Figure 10.1 A section of a fictitious two-layered network for the dependences of faults (bottom) on vehicle properties (top).

Table 10.1 A fictitious dependence of battery faults on the presence or absence of an electrical sliding roof and an air conditioning system.

(fictitious) frequency of battery faults		air conditioning with	air conditioning without
electrical sliding roof	with	9 %	3 %
	without	3 %	2 %

In order to find a solution for this problem a cooperation between our research group at the Otto-von-Guericke-University of Magdeburg and the DaimlerChrysler Research and Technology Center Ulm was established. In this cooperation an approach based on learning Bayesian networks from data was tried in order to detect dependences between faults and vehicle properties. The idea underlying it was to exploit the search methods and evaluation measures developed for learning Bayesian networks from data to search automatically for sets of attributes that are strongly dependent on each other. In the course of this cooperation the first author of this book wrote a first version of the INES program (Induction of NEtwork Structures), a prototype implementation of several learning algorithms for probabilistic and possibilistic networks. This program was already mentioned in Sections 7.4 and 8.3, where it was used to produce the reported experimental results.

The idea underlying the application of INES to the vehicle database is very simple. Since we are interested in causes of faults, a two-layered network is learned. The top layer of this network contains attributes describing the vehicle configuration, whereas the bottom layer contains attributes describing possible vehicle faults. This is illustrated in Figure 10.1 and Table 10.1. (Since real dependences and numbers are, of course, highly confidential, these figures show fictitious examples. Any resemblance to actual dependences and numbers is purely coincidental.[1]) Figure 10.1 shows a possible learned two-layered network, Table 10.1 the frequency distribution associated with the first of its subnets. Since in this example the fault rate for cars with an air conditioning system and an electrical sliding roof is considerably higher than that of cars without one or both of these items, we can conjecture that the increased consumption of electrical energy due to installed air conditioning and electrical sliding roof is a cause of increased battery faults. Of course, such a conjecture has to be substantiated by a technical analysis.

It should be noted that this approach is preferable to the alternative that naturally comes to mind first, namely learning, for instance, a decision tree for each of the faults. The first reason is that faults (like fraudulent calls in the telecommunications application) are rare. However, decision trees have to achieve a classification rate higher than 50% for a fault in a leaf node in order to actually predict a fault. Otherwise the decision tree is simplified and—in the worst case—simply predicts the majority class, which is, of course, that no fault occurs. One may try to mitigate this problem by enriching the database to learn from with examples of faulty behavior. However, this distorts the frequency information, making the results difficult to interpret, but nevertheless cannot completely remove the problem. Secondly, the distributions learned in a Bayesian network are symmetric w.r.t. to the different parent attributes. Hence it is possible to remove any of them in order to study their individual influence, by which evaluating the result is simplified considerable. Finally, with the described approach it easier to detect common causes of different faults, which show as common parents in the resulting network.

Although specific results are confidential, we can remark here that when applied to a truck database INES easily found a dependence pointing to a possible cause of a unusual fault, which was already known to the domain experts, but had taken them considerable effort to discover "manually" (querying the database and studying the blueprints describing the construction of the trucks). Even if one takes into account the time needed to select and preprocess the data, so that INES becomes applicable, it is evident that with the help of this program the solution would have been found much faster. Other dependences detected by INES were welcomed by the domain experts as valuable starting points for further technical investigations.

In a second application INES was applied to a problem that had occurred in the area of utility vehicles, but where it was not known at the time of the application whether there existed a dependence on a certain configuration of the vehicle. In this case INES could not find any dependence that pointed to a possible cause. Although this sounds disappointing, it was partly due to this result that investigations where directed elsewhere and, indeed, the cause

[1] Actually it can already be seen from the items used in Table 10.1 that this is a fictitious example, because a car with both an electrical sliding roof and air conditioning is a very uncommon choice: There is no need for an electrical sliding roof if there is air conditioning, and air conditioning does not work properly if the sliding roof is open.

was finally found at a supplier, who had changed the quality of a lubricant. This information was not contained in the database presented to INES and thus there was no chance of detecting it automatically. Even though the result obtained by INES was still useful in this case (because it prevented a fruitless search for dependences on vehicle properties), it warns against too optimistic expectations. If the database to learn from does not contain the relevant information, no automatic search can find the solution. Nevertheless we can conclude that learning probabilistic networks is a very useful method to support the detection of product weaknesses [Borgelt *et al.* 1998b].

It should be noted that this application also provided some important impulses for our research. For example, the extension of the K2 metric to a family of evaluation measures by introducing a parameter α, which was described in Section 7.2.4, was in part triggered by practical problems, which turned up in the cooperation with DaimlerChrysler. In tests of the learning algorithm against expert knowledge sometimes dependences of faults on the vehicle equipment, which were known to the domain experts, could not be discovered if the K2 metric was used. If Shannon information gain was used instead, the desired dependences could be detected, though.

A closer inspection revealed that the dependences that were difficult to discover were rather weak. This made it plausible why they were not detected with the K2 metric: As discussed in Section 7.2.4 the K2 metric has a tendency to select simpler models. Therefore another parent attribute is selected only if the improvement of fit outweighs the penalty for a more complex model. However, for this to be the case the dependence has to be fairly strong. Therefore, due to the penalty, the mentioned measures are sometimes not sensitive enough (at least for the application described above). On the other hand, Shannon information gain has its drawbacks, too. As discussed in Section 7.2.4 this measure is biased towards many-valued attributes. Therefore, if the parent attributes are combined into one pseudo-attribute (as described in Section 7.2.1), Shannon information gain tends to "overfit" the data, because it is over-sensitive to random fluctuations in the data. As a consequence it sometimes indicates pseudo-dependences.

Not surprisingly, these results created the desire to have a parameter, by which the sensitivity of the evaluation measure w.r.t. dependences between attributes can be controlled. Since by adjusting the value of the parameter α of the extended K2 metric (or the reduction of description length measures), a smooth transition from this (less sensitive) measure to the (more sensitive) Shannon information gain can be achieved, it is such a parameter. It allows us to control the sensitivity of the evaluation measure in a convenient way.

Appendix A

Proofs of Theorems

In this appendix the proofs of all theorems of this book can be found. It also contains a few additional lemmata and their proofs, which are needed in the proofs of some of the theorems. For reading convenience each section restates the theorem proven in it.

A.1 Proof of Theorem 4.1.2

Theorem 4.1.2 *Conditional probabilistic independence as well as conditional possibilistic independence satisfy the semi-graphoid axioms.*[1] *If the considered joint probability distribution is strictly positive, conditional probabilistic independence satisfies the graphoid axioms.*

Proof. Since the proof of the semi-graphoid axioms is very similar for the two calculi, we only provide a proof for the possibilistic case. A proof for the probabilistic case can be derived from it very easily.[2]

That conditional possibilistic independence satisfies the semi-graphoid axioms can easily be demonstrated by drawing on the axioms of Definition 2.4.5 on page 39, the notion of a conditional degree of possibility as defined in Definition 2.4.9 on page 47 and the notion of conditional possibilistic independence as defined in Definition 3.4.2 on page 86 (extended to sets of attributes). In the following W, X, Y, and Z are four disjoint sets of attributes. Expressions like $W = w$, $X = x$, $Y = y$, and $Z = z$ are used as abbreviations for $\bigwedge_{A_i \in W} A_i = a_i$ etc. Although this is somewhat sloppy, it simplifies the notation considerably.

[1] The semi-graphoid and the graphoid axioms are defined in Definition 4.1.1 on page 92.

[2] We chose the possibilistic case, because a spelled out proof for the probabilistic case is likely to be found in introductory books on probabilistic graphical models, for example, in [Castillo *et al.* 1997]. On the other hand, [Pearl 1988] and [Lauritzen 1996] do not provide a proof, but leave it as an exercise for the reader.

1. symmetry:

$$
\begin{aligned}
& X \perp\!\!\!\perp_{\Pi} Y \mid Z \\
& \quad \Rightarrow \quad \forall x, y, z, \Pi(Z = z) > 0 : \\
& \qquad \Pi(X = x, Y = y \mid Z = z) \\
& \qquad\quad = \min\{\Pi(X = x \mid Z = z), \Pi(Y = y \mid Z = z)\} \\
& \quad \Rightarrow \quad \forall x, y, z, R(Z = z) > 0 : \\
& \qquad \Pi(Y = y, X = x \mid Z = z) \\
& \qquad\quad = \min\{\Pi(Y = y \mid Z = z), \Pi(X = x \mid Z = z)\} \\
& \quad \Rightarrow \quad Y \perp\!\!\!\perp_{\Pi} X \mid Z
\end{aligned}
$$

2. decomposition:

$$
\begin{aligned}
& W \cup X \perp\!\!\!\perp_{\Pi} Y \mid Z \\
& \quad \Rightarrow \quad \forall w, x, y, z, \Pi(Z = z) > 0 : \\
& \qquad \Pi(W = w, X = x, Y = y \mid Z = z) \\
& \qquad\quad = \min\{\Pi(W = w, X = x \mid Z = z), \Pi(Y = y \mid Z = z)\} \\
& \quad \Rightarrow \quad \forall w, x, y, z : \\
& \qquad \Pi(W = w, X = x, Y = y, Z = z) \\
& \qquad\quad = \min\{\Pi(W = w, X = x, Z = z), \Pi(Y = y, Z = z)\} \\
& \quad \Rightarrow \quad \forall w, x, y, z : \\
& \qquad \Pi(X = x, Y = y, Z = z) \\
& \qquad\quad = \Pi(\textstyle\bigvee_w W = w, X = x, Y = y, Z = z) \\
& \qquad\quad = \max_w \Pi(W = w, X = x, Y = y, Z = z) \\
& \qquad\quad = \max_w \min\{\Pi(W = w, X = x, Z = z), \Pi(Y = y, Z = z)\} \\
& \qquad\quad = \min\{\max_w \Pi(W = w, X = x, Z = z), \Pi(Y = y, Z = z)\} \\
& \qquad\quad = \min\{\Pi(\textstyle\bigvee_w W = w, X = x, Z = z), \Pi(Y = y, Z = z)\} \\
& \qquad\quad = \min\{\Pi(X = x, Z = z), R(Y = y, Z = z)\} \\
& \quad \Rightarrow \quad \forall x, y, z, \Pi(Z = z) > 0 : \\
& \qquad \Pi(X = x, Y = y \mid Z = z) \\
& \qquad\quad = \min\{\Pi(X = x \mid Z = z), \Pi(Y = y \mid Z = z)\} \\
& \quad \Rightarrow \quad X \perp\!\!\!\perp_{\Pi} Y \mid Z
\end{aligned}
$$

3. weak union:

$$
\begin{aligned}
& W \cup X \perp\!\!\!\perp_{\Pi} Y \mid Z \\
& \quad \Rightarrow \quad \forall w, x, y, z, \Pi(Z = z) > 0 : \\
& \qquad \Pi(W = w, X = x, Y = y \mid Z = z) \\
& \qquad\quad = \min\{\Pi(W = w, X = x \mid Z = z), \Pi(Y = y \mid Z = z)\}
\end{aligned}
$$

$$\begin{aligned}
\Rightarrow\quad & \forall w, x, y, z: \\
& \quad \Pi(W = w, X = x, Y = y, Z = z) \\
& \qquad = \min\{\Pi(W = w, X = x, Z = z), \Pi(Y = y, Z = z)\} \\
& \qquad = \min\{\Pi(W = w, X = x, Z = z), \\
& \qquad\qquad \Pi(\textstyle\bigvee_w W = w, Y = y, Z = z)\} \\
& \qquad = \min\{\Pi(W = w, X = x, Z = z), \\
& \qquad\qquad \max_{w'} \Pi(W = w', Y = y, Z = z)\} \\
& \qquad \geq \min\{\Pi(W = w, X = x, Z = z), \Pi(W = w, Y = y, Z = z)\} \\
& \qquad \geq \Pi(W = w, X = x, Y = y, Z = z) \\
\Rightarrow\quad & \forall w, x, y, z: \\
& \quad \Pi(W = w, X = x, Y = y, Z = z) \\
& \qquad = \min\{\Pi(W = w, X = x, Z = z), \Pi(W = w, Y = y, Z = z)\} \\
\Rightarrow\quad & \forall w, x, y, z, \Pi(Z = z, W = w) > 0: \\
& \quad \Pi(W = w, X = x, Y = y \mid Z = z) \\
& \qquad = \min\{\Pi(X = x \mid Z = z \cup W = w), \\
& \qquad\qquad \Pi(Y = y \mid Z = z \cup W = w)\} \\
\Rightarrow\quad & X \perp\!\!\!\perp_\Pi Y \mid Z \cup W
\end{aligned}$$

4. contraction:

$$\begin{aligned}
& (W \perp\!\!\!\perp_\Pi X \mid Z) \quad \wedge \quad (W \perp\!\!\!\perp_\Pi Y \mid Z \cup X) \\
\Rightarrow\quad & \forall w, x, y, z, \Pi(Z = z) > 0: \\
& \quad \Pi(W = w, X = x \mid Z = z) \\
& \qquad = \min\{\Pi(W = w \mid Z = z), \Pi(X = x \mid Z = z)\} \\
\wedge\quad & \forall w, x, y, z, \Pi(X = x, Z = z) > 0: \\
& \quad \Pi(W = w, Y = y \mid X = x, Z = z) \\
& \qquad = \min\{\Pi(W = w \mid X = x, Z = z), \Pi(Y = y \mid X = x, Z = z)\} \\
\Rightarrow\quad & \forall w, x, y, z: \\
& \quad \Pi(W = w, X = x, Z = z) \\
& \qquad = \min\{\Pi(W = w, Z = z), \Pi(X = x, Z = z)\} \\
\wedge\quad & \forall w, x, y, z: \\
& \quad \Pi(W = w, Y = y, X = x, Z = z) \\
& \qquad = \min\{\Pi(W = w, X = x, Z = z), \Pi(Y = y, X = x, Z = z)\} \\
\Rightarrow\quad & \forall w, x, y, z: \\
& \quad \Pi(W = w, Y = y, X = x, Z = z) \\
& \qquad = \min\{\Pi(W = w, Z = z), \Pi(X = x, Z = z), \\
& \qquad\qquad \Pi(Y = y, X = x, Z = z)\}
\end{aligned}$$

$$
\begin{aligned}
&= \min\{\Pi(W = w, Z = z), \Pi(Y = y, X = x, Z = z)\} \\
\Rightarrow \quad & \forall w, x, y, z, \Pi(Z = z) > 0 : \\
& \Pi(W = w, X = x, Y = y \mid Z = z) \\
& \quad = \min\{\Pi(W = w \mid Z = z), \Pi(X = x, Y = y \mid Z = z)\} \\
\Rightarrow \quad & W \perp\!\!\!\perp_{\Pi} X \cup Y \mid Z
\end{aligned}
$$

Note that in these derivations we require $\Pi(Z = z) > 0$ whenever $Z = z$ is the condition of a conditional degree of possibility (although there is no such requirement in Definition 2.4.9 on page 47), in order to strengthen the analogy to the probabilistic case. Note also that this requirement is dropped for unconditional degrees of possibility, since the relations for them trivially also hold for $P(Z = z) = 0$ (as in the probabilistic case).

The intersection axiom can be demonstrated to hold for strictly positive probability distributions by drawing on the basic axioms of probability theory and the definition of a conditional probability and conditional probabilistic independence as follows:

$$
\begin{aligned}
& (W \perp\!\!\!\perp_P Y \mid Z \cup X) \quad \wedge \quad (X \perp\!\!\!\perp_P Y \mid Z \cup W) \\
\Rightarrow \quad & \forall w, x, y, z : \\
& P(W = w, Y = y \mid Z = z, X = x) \\
& \quad = P(W = w \mid Z = z, X = x) \cdot P(Y = y \mid Z = z, X = x) \\
\wedge \quad & \forall w, x, y, z : \\
& P(X = y, Y = y \mid Z = z, W = w) \\
& \quad = P(X = x \mid Z = z, W = w) \cdot P(Y = y \mid Z = z, W = w) \\
\Rightarrow \quad & \forall w, x, y, z : \\
& P(W = w, X = y, Y = y, Z = z) \\
& \quad = P(W = w, X = x, Z = z) \cdot P(Y = y \mid Z = z, X = x) \\
\wedge \quad & \forall w, x, y, z : \\
& P(W = w, X = y, Y = y, Z = z) \\
& \quad = P(W = w, X = x, Z = z) \cdot P(Y = y \mid Z = z, W = w) \\
\Rightarrow \quad & \forall w, x, y, z : \\
& P(Y = y \mid Z = z, X = x) = P(Y = y \mid Z = z, W = w).
\end{aligned}
$$

This equation can only hold for all possible assignments x and w, if

$$
\begin{aligned}
\forall w, x, y, z : \quad & P(Y = y \mid Z = z, X = x) = P(Y = y \mid Z = z) \quad \text{and} \\
& P(Y = y \mid Z = z, W = w) = P(Y = y \mid Z = z).
\end{aligned}
$$

Note that this argument can fail if the joint probability distribution is not strictly positive. For example, it may be $P(X = x, Z = z) = 0$ and thus $P(Y = y \mid Z = z, X = x)$ may not be defined, although $P(Y = y \mid Z = z)$ is.

Substituting this result back into the first two equations we get

$$\begin{aligned}
& \forall w, x, y, z: \\
& \quad P(W = w, Y = y \mid Z = z, X = x) \\
& \quad\quad = P(W = w \mid Z = z, X = x) \cdot P(Y = y \mid Z = z) \\
\wedge\ & \forall w, x, y, z: \\
& \quad P(X = y, Y = y \mid Z = z, W = w) \\
& \quad\quad = P(X = x \mid Z = z, W = w) \cdot P(Y = y \mid Z = z) \\
\Rightarrow\ & \forall w, x, y, z: \\
& \quad P(W = w, X = y, Y = y, Z = z) \\
& \quad\quad = P(Y = y \mid Z = z) \cdot P(W = w, X = x, Z = z) \\
\Rightarrow\ & \forall w, x, y, z: \\
& \quad P(W = w, X = y, Y = y \mid Z = z) \\
& \quad\quad = P(Y = y \mid Z = z) \cdot P(W = w, X = x \mid Z = z) \\
\Rightarrow\ & W \cup X \perp\!\!\!\perp_P Y \mid Z.
\end{aligned}$$

Note that no requirements $P(Z = z) > 0$, $P(Y = y, Z = z) > 0$ etc. are needed, because the probability distribution is presupposed to be strictly positive and therefore these requirements necessarily hold.

A.2 Proof of Theorem 4.1.18

Theorem 4.1.18 *If a joint distribution δ over a set of attributes U satisfies the graphoid axioms w.r.t. a given notion of conditional independence, then the pairwise, the local, and the global Markov property of an undirected graph $G = (U, E)$ are equivalent.*[3]

Proof. From the observations made in the paragraph following the definition of the Markov properties for undirected graphs (Definition 4.1.17 on page 103) we know that the global Markov property implies the local and that, if the weak union axiom holds, the local Markov property implies the pairwise. So all that is left to show is that, given the graphoid axioms, the pairwise Markov property implies the global.

The idea of the proof is very simple, as already pointed out on page 105. Consider three arbitrary disjoint subsets X, Y, and Z of nodes such that $\langle X \mid Z \mid Y \rangle_G$. We have to show that $X \perp\!\!\!\perp_\delta Y \mid Z$ follows from the pairwise conditional independence statements that correspond to non-adjacent attributes. To do so we start from an arbitrary statement $A \perp\!\!\!\perp_\delta B \mid U - \{A, B\}$

[3]The graphoid axioms are defined in Definition 4.1.1 on page 92 and the Markov properties of undirected graphs are defined in Definition 4.1.17 on page 103.

with $A \in X$ and $B \in Y$, and then shifting nodes from the separating set to the separated sets, thus extending A to (a superset of) X and B to (a superset of) Y and shrinking $U - \{A, B\}$ to Z. The shifting is done by applying the intersection axiom, drawing on other pairwise conditional independence statements. Finally, any excess attributes in the separated sets are cut away with the help of the decomposition axiom.

Formally, the proof is carried out by *backward* or *descending induction* [Pearl 1988, Lauritzen 1996] over the number n of nodes in the separating set Z. If $n = |U| - 2$ then X and Y both contain exactly one node and thus the conditional independence w.r.t. δ follows directly from the pairwise Markov property (induction anchor).

So assume that $|Z| = n < |V| - 2$ and that separation implies conditional independence for all separating sets S with more than n elements (induction hypothesis). We first assume that $U = X \cup Y \cup Z$, implying that at least one of X and Y has more than one element, say, X. Consider an attribute $A \in X$: $Z \cup \{A\}$ separates $X - \{A\}$ from Y and $Z \cup (X - \{A\})$ separates A from Y. Thus by the induction hypothesis we have

$$X - \{A\} \perp\!\!\!\perp_\delta Y \mid Z \cup \{A\} \quad \text{and} \quad \{A\} \perp\!\!\!\perp_\delta Y \mid Z \cup (X - \{A\}).$$

Applying the intersection axiom yields $X \perp\!\!\!\perp_\delta Y \mid Z$.

If $X \cup Y \cup Z \subset U$, we choose an attribute $A \in U - (X \cup Y \cup Z)$. We know that $Z \cup \{A\}$ separates X and Y (due to the monotony of u-separation, cf. page 99, expressed by the strong union axiom, cf. page 99), implying $X \perp\!\!\!\perp_\delta Y \mid Z \cup \{A\}$ (by the induction hypothesis). Furthermore, either $X \cup Z$ separates Y from A or $Y \cup Z$ separates X from A (due to the transitivity of u-separation, cf. page 99, together with the strong union axiom, cf. page 99). The former case yields $\{A\} \perp\!\!\!\perp_\delta Y \mid X \cup Z$ (by the induction hypothesis) and by applying the intersection and the decomposition axiom we derive $X \perp\!\!\!\perp_\delta Y \mid Z$. The latter case is analogous.

A.3 Proof of Theorem 4.1.20

Theorem 4.1.20 *If a three-place relation $(\cdot \perp\!\!\!\perp_\delta \cdot \mid \cdot)$ representing the set of conditional independence statements that hold in a given joint distribution δ over a set U of attributes satisfies the semi-graphoid axioms, then the local and the global Markov property of a directed acyclic graph $\vec{G} = (U, \vec{E})$ are equivalent. If it satisfies the graphoid axioms, then the pairwise, the local, and the global Markov property are equivalent.*[4]

Proof. As already mentioned on page 107, a proof for this theorem can be found, for example, in [Verma and Pearl 1990], although this may be a little

[4]The graphoid and the semi-graphoid axioms are defined in Definition 4.1.1 on page 92 and the Markov properties of directed graphs are defined in Definition 4.1.19 on page 106.

difficult to recognize, because the definition of certain notions has changed since the publication of that paper. (Surprisingly enough, [Lauritzen 1996] only contains a specialized proof for the probabilistic case.) Here we present our own proof (which is similar to the one in [Verma and Pearl 1990], though).

From the observations made in the paragraph following the definition of the Markov properties for directed graphs (Definition 4.1.19 on page 106), namely that the set of parents of a node obviously d-separates it from all its other non-descendants, we know that the global Markov property implies the local. Therefore, for the first part of the theorem, we only have to show that the local Markov property implies the global.

So let X, Y, and Z be three disjoint subsets of attributes, such that X and Y are d-separated by Z in the graph $\vec{G}$. We have to show that $X \perp\!\!\!\perp_\delta Y \mid Z$ follows from the local conditional independence statements $A \perp\!\!\!\perp_\delta \text{nondescs}(A) - \text{parents}(A) \mid \text{parents}(A)$ that can be derived from the local Markov property of the graph $\vec{G}$.

The proof consists of two phases. In the first phase the graph is simplified by removing nodes that do not have a bearing on the conditional independence statement to be derived. In the second phase the remaining attributes of the graph are added step by step to a set of three conditional independence statements. After all nodes have been added, the desired conditional independence statement can be derived from one of these statements by applying the decomposition axiom.

In the first phase a sequence of graphs $\vec{G}_0 = \vec{G}, \vec{G}_1, \ldots, \vec{G}_k$ is formed. Each graph $\vec{G}_{i+1}$ is constructed from $\vec{G}_i$ by removing an arbitrary childless node not in $X \cup Y \cup Z$ (and, of course, all edges leading to it). The process stops when no more nodes can be removed. In the following we show by induction on the graph sequence that all local conditional independence statements derivable from the resulting graph $\vec{G}_k$ are implied by those derivable from the original graph $\vec{G}$. Clearly, all such statements derivable from $\vec{G}_0$ are implied by those derivable from $\vec{G}$, because the two graphs are identical (induction anchor). So suppose that all local conditional independence statements derivable from $\vec{G}_i$ are implied by those derivable from $\vec{G}$ (induction hypothesis). Let B be the node that is removed from $\vec{G}_i$ to construct $\vec{G}_{i+1}$ and let A be an arbitrary node of $\vec{G}_{i+1}$. Note first that B cannot be a parent of A in $\vec{G}_i$, since then B would not be childless in $\vec{G}_i$, contradicting the construction. If B is a descendant of A in $\vec{G}_i$, then the local conditional independence statement $A \perp\!\!\!\perp_\delta \text{nondescs}(A) - \text{parents}(A) \mid \text{parents}(A)$ is identical in $\vec{G}_i$ and in $\vec{G}_{i+1}$ and thus must be implied by the statements derivable from $\vec{G}$ due to the induction hypothesis. If B is a non-descendant of A in $\vec{G}_i$ (but not a parent, see above), then

$$\begin{aligned} & A \perp\!\!\!\perp_\delta \text{nondescs}_{\vec{G}_i}(A) - \text{parents}_{\vec{G}_i}(A) \mid \text{parents}_{\vec{G}_i}(A) \\ \equiv\ & A \perp\!\!\!\perp_\delta (\text{nondescs}_{\vec{G}_{i+1}}(A) - \text{parents}_{\vec{G}_{i+1}}(A)) \cup \{B\} \mid \text{parents}_{\vec{G}_{i+1}}(A) \end{aligned}$$

and therefore the local conditional independence statement

$$A \perp\!\!\!\perp_\delta \text{nondescs}_{\vec{G}_{i+1}}(A) - \text{parents}_{\vec{G}_{i+1}}(A) \mid \text{parents}_{\vec{G}_{i+1}}(A)$$

can be derived by applying the decomposition axiom. Together with the induction hypothesis we have that it is implied by the local conditional independence statements derivable from $\vec{G}$. This completes the induction.

The set of nodes of the resulting graph $\vec{G}_k$ is often called the *smallest ancestral set* of $X \cup Y \cup Z$. An ancestral set is a set of nodes of a directed graph that is closed under the ancestor relation, i.e., for each node in the set all its ancestors are also in the set. A set S of nodes is called the smallest ancestral set of a set W of nodes, if it is an ancestral set containing W and there is no true subset of S that is also an ancestral set containing W. (We do not prove, though, that the set of nodes of the resulting graph $\vec{G}_k$ is indeed a smallest ancestral set, because this is not relevant for the proof.)

In the second phase of the proof the desired conditional independence statement is derived from the local conditional independence statements that can be read from $\vec{G}_k$. This is done by forming four sequences $(W_i)_{i \in I}$, $(X_i)_{i \in I}$, $(Y_i)_{i \in I}$, and $(Z_i)_{i \in I}$, $I = \{0, \ldots, |U_k|\}$, of disjoint subsets of nodes, such that certain conditional independence statements hold for all i. These sequences are constructed as follows: The first elements are all empty, i.e., $W_0 = X_0 = Y_0 = Z_0 = \emptyset$. Let $o : U_k \to \{1, \ldots, |U_k|\}$ be a topological order[5] of the nodes in $\vec{G}_k$ and let A_{i+1}, $i \geq 0$, be the $(i+1)$-th node of this topological order, i.e., $o(A_{i+1}) = i + 1$. In the step from i to $(i+1)$ the node A_{i+1} is added to exactly one of the sets W_i, X_i, Y_i, and Z_i according to

$$W_{i+1} = \begin{cases} W_i \cup \{A_{i+1}\}, & \text{if } A_{i+1} \notin X \cup Y \cup Z \wedge \langle A_{i+1} \mid Z \mid X \rangle_{\vec{G}_k} \\ & \qquad \wedge \langle A_{i+1} \mid Z \mid Y \rangle_{\vec{G}_k}, \\ W_i, & \text{otherwise,} \end{cases}$$

$$X_{i+1} = \begin{cases} X_i \cup \{A_{i+1}\}, & \text{if } A_{i+1} \in X \vee \\ & \quad (A_{i+1} \notin Z \wedge \neg \langle A_{i+1} \mid Z \mid X \rangle_{\vec{G}_k}), \\ X_i, & \text{otherwise,} \end{cases}$$

$$Y_{i+1} = \begin{cases} Y_i \cup \{A_{i+1}\}, & \text{if } A_{i+1} \in Y \vee \\ & \quad (A_{i+1} \notin Z \wedge \neg \langle A_{i+1} \mid Z \mid Y \rangle_{\vec{G}_k}), \\ Y_i, & \text{otherwise,} \end{cases}$$

$$Z_{i+1} = \begin{cases} Z_i \cup \{A_{i+1}\}, & \text{if } A_{i+1} \in Z, \\ Z_i, & \text{otherwise.} \end{cases}$$

From this construction it is clear that $\forall A \in W_i \cup X_i \cup Y_i \cup Z_i : o(A) \leq i$, that W_i, X_i, Y_i, and Z_i are disjoint for all i (recall that X and Y are d-separated by Z in $\vec{G}$ and thus in $\vec{G}_k$), and that $X \subseteq X_{|U_k|}$, $Y \subseteq Y_{|U_k|}$, and $Z_{|U_k|} = Z$.

[5]The notion of a topological order is defined in Definition 4.1.12 on page 98.

We show now by induction on the four sequences that the following three conditional independence statements hold for all i: $W_i \perp\!\!\!\perp_\delta X_i \cup Y_i \mid Z_i$, $X_i \perp\!\!\!\perp_\delta W_i \cup Y_i \mid Z_i$, and $Y_i \perp\!\!\!\perp_\delta W_i \cup X_i \mid Z_i$. While proving this it is helpful to keep in mind the picture shown on the right which illustrates the above set of conditional independence statements.

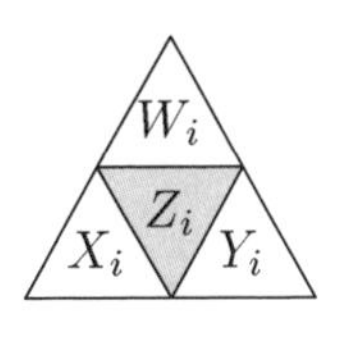

Since $W_0 = X_0 = Y_0 = Z_0 = \emptyset$, it is trivially $W_0 \perp\!\!\!\perp_\delta X_0 \cup Y_0 \mid Z_0$, $X_0 \perp\!\!\!\perp_\delta W_0 \cup Y_0 \mid Z_0$, and $Y_0 \perp\!\!\!\perp_\delta W_0 \cup X_0 \mid Z_0$ (induction anchor). So suppose that for some i the statements $W_i \perp\!\!\!\perp_\delta X_i \cup Y_i \mid Z_i$, $X_i \perp\!\!\!\perp_\delta W_i \cup Y_i \mid Z_i$, and $Y_i \perp\!\!\!\perp_\delta W_i \cup X_i \mid Z_i$ hold (induction hypothesis).

For the induction step, since $\forall A \in W_i \cup X_i \cup Y_i \cup Z_i : o(A) \leq i$, we can conclude from the properties of a topological order that parents$(A_{i+1}) \subseteq W_i \cup X_i \cup Y_i \cup Z_i$ (since all parents of a node must precede it in a topological order) and $W_i \cup X_i \cup Y_i \cup Z_i \subseteq$ nondescs(A_{i+1}) (since no child of a node and thus no descendant can precede it in a topological order). Therefore, by applying the decomposition axiom to the local conditional independence statement involving A_{i+1} that can be read from the graph $\vec{G}_k$, we have

$$A_{i+1} \perp\!\!\!\perp_\delta (W_i \cup X_i \cup Y_i \cup Z_i) - \text{parents}(A_{i+1}) \mid \text{parents}(A_{i+1}). \tag{A.1}$$

To derive from this statement the three desired conditional independence statements for $(i + 1)$, one has to show two things:

1. If the node A_{i+1} is added to Z_i, then the parents of A_{i+1} must be either all in $W_i \cup Z_i$, or all in $X_i \cup Z_i$ or all in $Y_i \cup Z_i$.
2. If the node A_{i+1} is added to X_i, then all parents of A_{i+1} must be in $X_i \cup Z_i$. (Analogously, if A_{i+1} is added to W_i or Y_i.)

However, this is easily achieved:

1. Let A_{i+1} be added to Z_i. Suppose that it has parents in both X_i and Y_i and let these parents be B_X and B_Y, respectively. From the construction of the sequences we know that $B_X \in X$ or $\neg\langle B_X \mid Z \mid X\rangle_{\vec{G}_k}$. Similarly, we know that $B_Y \in Y$ or $\neg\langle B_Y \mid Z \mid Y\rangle_{\vec{G}_k}$. From the facts that the path $B_X \to A_{i+1} \leftarrow B_Y$ has converging edges at A_{i+1} (since B_X and B_Y are parents of A_{i+1}) and that $A_{i+1} \in Z_{i+1} \subseteq Z$ we know that there is an active path connecting B_X and B_Y. Combining these observations, we arrive at the conclusion that given Z there must be an active path from a node in X to a node in Y in the graph $\vec{G}_k$ (and thus in $\vec{G}$, since the same path exists in $\vec{G}$) contradicting the assumption that X and Y are d-separated by Z in $\vec{G}$. In a similar way we see that A_{i+1} cannot have parents in both W_i and X_i or in both W_i and Y_i, only that here the contradiction follows from the construction of the sequences alone.
2. Let A_{i+1} be added to W_i, X_i, or Y_i, say, X_i (the three cases are analogous). Then A_{i+1} cannot have a parent in W_i or Y_i, because this parent together with A_{i+1} forms an active path. Combining this observation, as in the

first case, with statements derivable from the construction of sequences, we conclude that given Z there is an active path from a node in W_i to a node in X (if the parent is in W_i) or from Y to X (if the parent is in Y_i). However, the former contradicts the construction of the sequences, the latter the assumption that X and Y are d-separated by Z in $\vec{G}$. It is clear that the other two cases, i.e., that A_{i+1} is added to W_i or to Y_i, are analogous.

Suppose now, without restricting generality (the other two cases are analogous), that parents$(A_{i+1}) \subseteq X_i \cup Z_i$. Then we can derive from the conditional independence A.1 by applying the weak union axiom that

$$A_{i+1} \perp\!\!\!\perp_\delta W_i \cup Y_i \mid Z_i \cup X_i.$$

Applying the contraction axiom to this statement and $X_i \perp\!\!\!\perp_\delta W_i \cup Y_i \mid Z_i$ (which we know from the induction hypothesis), we arrive at

$$A_{i+1} \cup X_i \perp\!\!\!\perp_\delta W_i \cup Y_i \mid Z_i.$$

From (1.) and (2.) and the assumption that parents$(A_{i+1}) \subseteq X_i \cup Z_i$ it follows that $A_{i+1} \in X_{i+1} \cup Z_{i+1}$, i.e., that A_{i+1} is added either to X_i or to Z_i. If A_{i+1} is added to X_i, then the above statement is equivalent to

$$X_{i+1} \perp\!\!\!\perp_\delta W_{i+1} \cup Y_{i+1} \mid Z_{i+1},$$

and if it is added to Z_i, then

$$X_{i+1} \perp\!\!\!\perp_\delta W_{i+1} \cup Y_{i+1} \mid Z_{i+1} \;\equiv\; X_i \perp\!\!\!\perp_\delta W_i \cup Y_i \mid Z_i \cup A_{i+1},$$

i.e., the desired statement results from applying the weak union axiom.

The other two statements for $(i+1)$ are derived as follows: Applying the weak union axiom to $A_{i+1} \cup X_i \perp\!\!\!\perp_\delta W_i \cup Y_i \mid Z_i$ yields

$$\begin{aligned} &A_{i+1} \cup X_i \perp\!\!\!\perp_\delta W_i \mid Z_i \cup Y_i \qquad \text{and} \\ &A_{i+1} \cup X_i \perp\!\!\!\perp_\delta Y_i \;\mid Z_i \cup W_i. \end{aligned}$$

By applying the contraction axiom to these statements and the statement $Y_i \perp\!\!\!\perp_\delta W_i \mid Z_i$, which can be derived from the statements of the induction hypothesis by applying the decomposition axiom, we arrive at

$$\begin{aligned} &A_{i+1} \cup X_i \cup Y_i \;\perp\!\!\!\perp_\delta W_i \mid Z_i \qquad \text{and} \\ &A_{i+1} \cup X_i \cup W_i \perp\!\!\!\perp_\delta Y_i \;\mid Z_i. \end{aligned}$$

If A_{i+1} is added to X_i these statements are already equivalent to the desired ones. If A_{i+1} is added to Z_i, then the desired statements can be derived, as above, by applying the weak union axiom. This completes the induction, because, as already mentioned above, the other two cases, i.e., parents$(A_{i+1}) \in W_i \cup Z_i$ and parents$(A_{i+1}) \in Y_i \cup Z_i$, are analogous.

Since, as indicated above, $X \subseteq X_{|U_k|}$, $Y \subseteq Y_{|U_k|}$ and $Z = Z_{|U_k|}$, we can finally derive $X \perp\!\!\!\perp_\delta Y \mid Z$ by (at most) two applications of the decomposition axiom to the statement $X_{|U_k|} \perp\!\!\!\perp_\delta W_{|U_k|} \cup Y_{|U_k|} \mid Z_{|U_k|}$, thus completing the proof of the first part of the theorem.

The second part of the theorem is much easier to prove than the first. Since we already know from the first part of the theorem that the local and the global Markov property are equivalent (recall that the semi-graphoid axioms are a subset of the graphoid axioms), it suffices to show that the pairwise and the local Markov property are equivalent. That the local Markov property implies the pairwise can be seen from the fact that the parents of a node are a subset of its non-descendants and thus we get any pairwise conditional independence statement for a node by applying the weak union axiom (cf. the observations made in the paragraph following Definition 4.1.19 on page 106).

To show the other direction, we start from an arbitrary pairwise conditional independence statement $A \perp\!\!\!\perp_\delta B \mid \mathrm{nondescs}(A) - \{B\}$. Then we apply the intersection axiom, drawing on other pairwise conditional independence statements involving A, in order to shift attributes out of the separating set. Eventually only the parents of A remain and thus we have the desired local conditional independence statement.

A.4 Proof of Theorem 4.1.22

Theorem 4.1.22 *Let p_U be a strictly positive probability distribution on a set U of (discrete) attributes. An undirected graph $G = (U, E)$ is a conditional independence graph w.r.t. p_U, iff p_U is factorizable w.r.t. G.*

To prove this theorem we need two lemmata. The first of them provides us with a more general characterization of conditional probabilistic independence, while the second is merely a technical feature needed in the proof.

Lemma A.4.1 *Let A, B, and C be three attributes with respective domains* $\mathrm{dom}(A)$, $\mathrm{dom}(B)$, *and* $\mathrm{dom}(C)$. *Furthermore, let $p = P|_{ABC}$ be a probability distribution on the joint domain of A, B, and C derived from some probability measure P. Then*

$$A \perp\!\!\!\perp_p C \mid B \quad \Leftrightarrow \quad \exists g, h : \forall a \in \mathrm{dom}(A) : \forall b \in \mathrm{dom}(B) : \forall c \in \mathrm{dom}(C) : \\ p(A = a, B = b, C = c) = g(a, b) \cdot h(b, c).$$

Proof. Although this lemma is often stated and used, it is rarely proven. Conditional probabilistic independence implies (cf. page 75)

$$\forall a \in \mathrm{dom}(A) : \forall b \in \mathrm{dom}(B), P(B = b) \neq 0 : \forall c \in \mathrm{dom}(C) \\ P(A = a, B = b, C = c) = P(A = a, B = b) \cdot P(C = c \mid B = b).$$

and thus we can choose, for instance (cf. page 78),

$$\begin{aligned} g(a,b) &= P(A=a, B=b) \qquad \text{and} \\ h(b,c) &= \begin{cases} P(C=c \mid B=b), & \text{if } P(B=b) \neq 0, \\ 0, & \text{otherwise.} \end{cases} \end{aligned}$$

To prove the other direction, we sum

$$P(A=a, B=b, C=c) = g(a,b) \cdot h(b,c)$$

over all values $c \in \text{dom}(C)$, which yields

$$P(A=a, B=b) = g(a,b) \cdot \sum_{c \in \text{dom}(C)} h(b,c).$$

Since the left hand side does not depend on c, the right hand side must not depend on c either. Therefore we can infer

$$\exists f_h : \forall b \in \text{dom}(B) : \quad \sum_{c \in \text{dom}(C)} h(b,c) = f_h(b).$$

Analogously, by summing over all values $a \in \text{dom}(A)$, we get

$$\exists f_g : \forall b \in \text{dom}(B) : \quad \sum_{a \in \text{dom}(A)} g(a,b) = f_g(b).$$

Therefore we have

$$\begin{aligned} \forall a \in \text{dom}(A) : \forall b \in \text{dom}(B) : \quad & P(A=a, B=b) = g(a,b) \cdot f_h(b) \quad \text{and} \\ \forall b \in \text{dom}(B) : \forall c \in \text{dom}(C) : \quad & P(B=b, C=c) = f_g(b) \cdot h(b,c). \end{aligned}$$

Summing either the first of these equations over all values $a \in \text{dom}(A)$ or the second over all values $c \in \text{dom}(C)$ yields, in analogy to the above,

$$\forall b \in \text{dom}(B) : \quad P(B=b) = f_g(b) \cdot f_h(b).$$

Combining these results we finally arrive at

$$\begin{aligned} &\forall a \in \text{dom}(A) : \forall b \in \text{dom}(B) : \forall c \in \text{dom}(C) : \\ &\quad P(A=a, B=b) \cdot P(B=b, C=c) \\ &\quad = g(a,b) \cdot f_g(b) \cdot h(b,c) \cdot f_h(c) \\ &\quad = P(A=a, B=b, C=c) \cdot P(B=b), \end{aligned}$$

which shows that $A \perp\!\!\!\perp_P C \mid B$. It is clear that the lemma and its proof carry over directly to sets of attributes.

Lemma A.4.2 (Möbius inversion) *Let ξ and ψ be functions defined on the set of all subsets of a finite set V whose range of values is an Abelian group. Then the following two statements are equivalent:*

$$\text{(1)}\quad \forall X \subseteq V: \quad \xi(X) = \sum_{Y: Y \subseteq X} \psi(Y);$$

$$\text{(2)}\quad \forall Y \subseteq V: \quad \psi(Y) = \sum_{Z: Z \subseteq Y} (-1)^{|Y-Z|}\, \xi(Z).$$

Proof. (Basically as given in [Lauritzen 1996], only with modified notation.) We prove (2) $\Rightarrow$ (1), i.e., we show that the sum of the terms $\psi(Y)$ as defined in (2) over all subsets of some set X is equal to $\xi(X)$:

$$\begin{aligned}
\sum_{Y: Y \subseteq X} \psi(Y) &= \sum_{Y: Y \subseteq X} \sum_{Z: Z \subseteq Y} (-1)^{|Y-Z|}\, \xi(Z) \\
&= \sum_{Z: Z \subseteq Y} \xi(Z) \Bigg(\sum_{Y: Z \subseteq Y \subseteq X} (-1)^{|Y-Z|} \Bigg) \\
&= \sum_{Z: Z \subseteq Y} \xi(Z) \Bigg(\sum_{W: W \subseteq X-Z} (-1)^{|W|} \Bigg)
\end{aligned}$$

The latter sum is equal to zero unless $X - Z = \emptyset$, i.e., unless $Z = X$, because any finite, non-empty set has the same number of subsets of even as of odd cardinality. The proof (1) $\Rightarrow$ (2) is performed analogously.

Proof of Theorem 4.1.22. (The proof follows mainly [Lauritzen 1996].) The proof consists of two parts. In the first part it is shown that, if the distribution p_U is factorizable w.r.t. an undirected graph G, then G satisfies the global Markov property and thus is a conditional independence graph. This part of the proof, which uses Lemma A.4.1, is rather simple. In the second part, which is more complicated, a factorization of p_U is derived from p_U and the pairwise Markov property of an undirected graph G that is a conditional independence graph w.r.t. p_U. For this second part we need Lemma A.4.2.

For the first part of the proof let $\mathcal{M} \subseteq 2^U$ be a family of sets of attributes such that the subgraphs of G induced by the sets $M \in \mathcal{M}$ are the maximal cliques of G and let ϕ_M be the functions of the factorization of p_U w.r.t. G. Furthermore, let X, Y, and Z be three arbitrary disjoint subsets of attributes such that X and Y are u-separated by Z in G. We have to show that G has the global Markov property, i.e., that $X \perp\!\!\!\perp_{p_U} Y \mid Z$. Let

$$\hat{X} = \{A \in U - Z \mid \langle A \mid Z \mid Y \rangle_G\} \qquad \text{and} \qquad \hat{Y} = U - Z - \hat{X}.$$

Obviously, it is $X \subseteq \hat{X}$ (since X and Y are u-separated by Z), $Y \subseteq \hat{Y}$, and $\langle \hat{X} \mid Z \mid Y \rangle_G$. From the latter and the definition of $\hat{X}$ it follows $\langle \hat{X} \mid Z \mid \hat{Y} \rangle_G$

by the transitivity axiom.[6] (Intuitively: There cannot be an attribute A in $\hat{X}$ that is not u-separated from $\hat{Y}$, because, by construction, no attribute in $\hat{Y}$ is u-separated from Y and thus A could not be u-separated from Y, contradicting the definition of $\hat{X}$.)

Let $\mathcal{M}_{\hat{X}}$ be the family of all sets in $\mathcal{M}$ that contain an attribute in $\hat{X}$. No set in $\mathcal{M}_{\hat{X}}$ can contain an attribute in $\hat{Y}$, because otherwise—since the sets in $\mathcal{M}$ and thus the sets in $\mathcal{M}_{\hat{X}}$ induce cliques—we could infer that there is an edge connecting an attribute in $\hat{X}$ and an attribute in $\hat{Y}$, contradicting that $\hat{X}$ and $\hat{Y}$ are u-separated by Z. That is, we have

$$\forall M \in \mathcal{M}_{\hat{X}}: \quad M \subseteq \hat{X} \cup Z \qquad \text{and} \qquad \forall M \in \mathcal{M} - \mathcal{M}_{\hat{X}}: \quad M \subseteq \hat{Y} \cup Z.$$

Therefore, if we split the product of the factorization as follows:

$$\begin{aligned}
& p_U\Big(\bigwedge_{A_i \in U} A_i = a_i\Big) \\
&= \prod_{M \in \mathcal{M}} \phi_M\Big(\bigwedge_{A_i \in M} A_i = a_i\Big) \\
&= \left(\prod_{M \in \mathcal{M}_{\hat{X}}} \phi_M\Big(\bigwedge_{A_i \in M} A_i = a_i\Big)\right)\left(\prod_{M \in \mathcal{M} - \mathcal{M}_{\hat{X}}} \phi_M\Big(\bigwedge_{A_i \in M} A_i = a_i\Big)\right),
\end{aligned}$$

we can conclude that the first product depends only on the values of the attributes in $\hat{X} \cup Z$, while the second product depends only on the values of the attributes in $\hat{Y} \cup Z$. With Lemma A.4.1, extended to sets of attributes, we arrive at $\hat{X} \perp\!\!\!\perp_{p_U} \hat{Y} \mid Z$, from which we can infer $X \perp\!\!\!\perp_{p_U} Y \mid Z$ by (at most) two applications of the decomposition axiom.[7]

Note that the first part of the proof as it is given above does not exploit the presupposition that p_U is strictly positive and thus the implication proven holds also for more general probability distributions [Lauritzen 1996].

In the second part of the proof non-negative functions ϕ_M are constructed w.r.t. the maximal cliques M of G. Since it simplifies the proof, these functions are not constructed directly, but via their logarithms. This is possible, because, according to the presuppositions of the theorem, the probability distribution—and thus the factor potentials ϕ_M—must be strictly positive. (Note, by the way, that, in contrast to the first part of the proof, the second part cannot be strengthened to not strictly positive distributions. An example of a probability distribution that is not factorizable w.r.t. its conditional independence graph can be found in [Lauritzen 1996].) The idea of the proof

[6] The transitivity axiom is defined and shown to hold for u-separation on page 99.

[7] The decomposition axiom is part of the semi-graphoid axioms, defined in Definition 4.1.1 on page 92, which are satisfied by conditional probabilistic independence according to Theorem 4.1.2 on page 93.

is to define specific functions for all subsets of attributes and to show that they are identical to zero, if the subset is not a clique. Finally, the functions for the cliques are properly combined to determine functions for the maximal cliques only.

First we choose for each attribute $A_i \in U$ a fixed but arbitrary value $a_i^* \in \mathrm{dom}(A_i)$ and then we define for all $X \subseteq U$

$$\xi_X\Big(\bigwedge_{A_i \in X} A_i = a_i\Big) = \log P\Big(\bigwedge_{A_i \in X} A_i = a_i, \bigwedge_{A_j \in U-X} A_j = a_j^*\Big).$$

Since the values a_j^* are fixed, ξ_X depends only on the values of the attributes in X. Furthermore, we define for all $Y \subseteq U$

$$\psi_Y\Big(\bigwedge_{A_i \in U} A_i = a_i\Big) = \sum_{Z:Z\subseteq Y} (-1)^{|Y-Z|}\, \xi_Z\Big(\bigwedge_{A_i \in Y} A_i = a_i\Big).$$

From this equation it is clear that ψ_Y depends only on the values of the attributes in Y. Next we apply Lemma A.4.2 (Möbius inversion) to obtain that

$$\begin{aligned}\log P\Big(\bigwedge_{A_i \in U} A_i = a_i\Big) &= \xi_U\Big(\bigwedge_{A_i \in U} A_i = a_i\Big)\\ &= \sum_{X:X\subseteq U} \psi_X\Big(\bigwedge_{A_i \in X} A_i = a_i\Big).\end{aligned}$$

(Note that w.r.t. the application of the Möbius inversion the index counts as the function argument.)

In the next step we have to show that $\psi_X \equiv 0$ whenever the subgraph induced by X is not a clique of G. So let $X \subseteq U$ be a set of attributes that does not induce a clique of G. Then there must be two attributes $A, B \in X$ that are not adjacent in G. Since G is a conditional independence graph, it has the global Markov property, which implies the pairwise Markov property (cf. the paragraph following the definition of the Markov properties for undirected graphs on page 103). Therefore it is $A \perp\!\!\!\perp_{p_U} B \mid U - \{A, B\}$. To exploit this statement we first rewrite the function ψ_X as

$$\begin{aligned}&\psi_X\Big(\bigwedge_{A_i \in X} A_i = a_i\Big)\\ &= \sum_{Y:Y\subseteq X-\{A,B\}} (-1)^{|X-\{A,B\}-Y|}\Big(\xi_Y - \xi_{Y\cup\{A\}} - \xi_{Y\cup\{B\}} + \xi_{Y\cup\{A,B\}}\Big),\end{aligned}$$

where ξ_Z is an abbreviation of $\xi_Z\left(\bigwedge_{A_i \in Z} A_i = a_i\right)$. Consider a term of the sum, i.e., consider an arbitrary set $Y \subseteq X - \{A, B\}$. Let $Z = U - Y - \{A, B\}$ and let $Y = y$ and $Z = z$ (as in the proof of Theorem 4.1.2 in Section A.1) be abbreviations of $\bigwedge_{A_i \in Y} A_i = a_i$ and $\bigwedge_{A_i \in Z} A_i = a_i$, respectively. (This is somewhat sloppy, but simplifies the notation considerably.) Then

$$\begin{aligned}
&\xi_{Y \cup \{A,B\}} - \xi_{Y \cup \{A\}} \\
&= \log \frac{P(A = a, B = b,\ Y = y, Z = z^*)}{P(A = a, B = b^*, Y = y, Z = z^*)} \\
&\overset{(1)}{=} \log \frac{P(A = a \mid Y = y, Z = z^*) \cdot P(B = b,\ Y = y, Z = z^*)}{P(A = a \mid Y = y, Z = z^*) \cdot P(B = b^*, Y = y, Z = z^*)} \\
&\overset{(2)}{=} \log \frac{P(A = a^* \mid Y = y, Z = z^*) \cdot P(B = b,\ Y = y, Z = z^*)}{P(A = a^* \mid Y = y, Z = z^*) \cdot P(B = b^*, Y = y, Z = z^*)} \\
&\overset{(3)}{=} \log \frac{P(A = a^*, B = b,\ Y = y, Z = z^*)}{P(A = a^*, B = b^*, Y = y, Z = z^*)} \\
&= \xi_{Y \cup \{B\}} - \xi_Y.
\end{aligned}$$

(1) and (3) follow from the conditional independence of A and B given all other attributes. (2) follows, because the first factors in numerator and denominator cancel each other and therefore may be replaced by any other factors that cancel each other. We conclude that all terms in the sum defining a function ψ_X and thus the function itself must be zero if X is not a clique of G.

In the final step we have to get rid of the functions for non-maximal cliques. However, this is easily achieved. Any non-maximal clique is contained in a maximal clique. Therefore we can simply add the function of the non-maximal clique to the function of the maximal clique and replace the function of the non-maximal clique by the zero function. In this process we have to be careful, though, because a non-maximal clique may be contained in more than one maximal clique, but the function for it should, obviously, be used only once. Therefore, before adding functions, we assign each clique to exactly one maximal clique in which it is contained and then we add for each maximal clique the functions of the cliques assigned to it. The resulting functions are the logarithms of the factor potentials ϕ_M of a factorization of p_U w.r.t. G.

A.5 Proof of Theorem 4.1.24

Theorem 4.1.24 *Let π_U be a possibility distribution on a set U of (discrete) attributes and let $G = (U, E)$ be an undirected graph over U. If π_U is decomposable w.r.t. G, then G is a conditional independence graph w.r.t. π_U. If G is a conditional independence graph w.r.t. π_U and if it has hypertree structure, then π_U is decomposable w.r.t. G.*

Proof. The proof of the first part of the theorem is directly analogous to the proof of the probabilistic case as given in Section A.4, because we can exploit a possibilistic analog of Lemma A.4.1. This analog can be derived, together with its proof, from the probabilistic version by replacing any product by the minimum and any sum by the maximum (together with some other trivial modifications). Note that this part of the theorem does not require the graph G to have hypertree structure and thus is valid for arbitrary graphs.

The proof of the second part of the theorem follows mainly [Gebhardt 1997], although in our opinion the proof given in [Gebhardt 1997] is incomplete, since it does not make clear where it is necessary to exploit the running intersection property, especially where the requirement is needed that the intersection of a set with the union of all preceding sets must be contained in a single set. Here we try to amend this deficiency.

The proof is carried out by an induction on a construction sequence for the graph $G = (U, E)$, in which the cliques of G are added one by one until the full graph G is reached. This sequence is derived from the running intersection property of the family $\mathcal{M}$ of attribute sets that induce the maximal cliques of G. Let $m = |\mathcal{M}|$ and let $M_1, \ldots, M_m$ be the ordering underlying the definition of the running intersection property. Furthermore, let $G_1, \ldots, G_m = G$ be the sequence of graphs defined by

$$\forall i, 1 \leq i \leq m: \quad G_i = (U_i, U_i \times U_i \cap E), \qquad \text{where} \quad U_i = \bigcup_{j \leq i} M_j.$$

In the induction it is shown that for all $i, 1 \leq i \leq m$, the marginal distributions π_{U_i} are decomposable w.r.t. G_i. Clearly, π_{U_1} is decomposable w.r.t. G_1, because there is only one clique (induction anchor). So suppose that π_{U_i} is decomposable w.r.t. G_i for some i (induction hypothesis), i.e.

$$\begin{aligned} &\forall a_i \in \mathrm{dom}(A_1) : \ldots a_n \in \mathrm{dom}(A_n) : \\ &\qquad \pi_{U_i}\Big(\bigwedge_{A_k \in U_i} A_k = a_k \Big) = \min_{j \leq i} \pi_{M_j}\Big(\bigwedge_{A_k \in M_j} A_k = a_k \Big). \end{aligned}$$

In the step from i to $(i+1)$ the clique with attribute set M_{i+1} is added to G_i. If we can show that

$$M_{i+1} - S_i \perp\!\!\!\perp_{pU} U_i - S_i \mid S_i, \qquad \text{where} \quad S_i = M_{i+1} \cap U_i,$$

we have proven the theorem, because this conditional independence, translated to degrees of possibility, reads (note that $U_{i+1} = U_i \cup M_{i+1}$)

$$\begin{aligned} &\forall a_i \in \mathrm{dom}(A_1) : \ldots a_n \in \mathrm{dom}(A_n) : \\ &\quad \pi_{U_{i+1}}\Big(\bigwedge_{A_k \in U_{i+1}} A_k = a_k \Big) \\ &\qquad = \min\Big\{ \pi_{M_{i+1}}\Big(\bigwedge_{A_k \in M_{i+1}} A_k = a_k \Big), \pi_{U_i}\Big(\bigwedge_{A_k \in U_i} A_k = a_k \Big) \Big\} \end{aligned}$$

and combining this equation with the induction hypothesis yields

$$\forall a_i \in \mathrm{dom}(A_1) : \dots a_n \in \mathrm{dom}(A_n) :$$
$$\pi_{U_{i+1}}\Big(\bigwedge_{A_k \in U_{i+1}} A_k = a_k\Big) = \min_{j \leq i+1} \pi_{M_j}\Big(\bigwedge_{A_k \in M_j} A_k = a_k\Big).$$

To show that the above conditional independence statement holds, we exploit the global Markov property of G, that is, we show

$$\langle M_{i+1} - S_i \mid S_i \mid U_i - S_i \rangle_G,$$

from which the conditional independence follows. It is clear that the separation holds w.r.t. G_{i+1}, because any path from an attribute in $M_{i+1} - S_i$ to an attribute in $U_i - S_i$ must pass through an attribute in S_i, simply because in G_{i+1} all edges from an attribute in $M_{i+1} - S_i$ lead to an attribute in M_{i+1}. Therefore this separation can be invalidated only by a set M_j, $j > i + 1$, that still needs to be added to construct $G = G_m$. So suppose there is a set M_j, $j > i + 1$, that contains an attribute $A \in U_i - S_i$ and an attribute $B \in M_{i+1} - S_i$, thus bypassing the separating set S_i. Furthermore, if there are several such sets, let M_j be the set with the smallest index among these sets. From the running intersection property we know that there is a set M_k, $k < j$, so that $\{A, B\} \subseteq M_k$ (this and the following is what [Gebhardt 1997] fails to point out). It must be $k > i$, because A was added with M_{i+1} and was not present in U_i. It must also be $k \leq i + 1$, because we chose M_j to be the set with the smallest index $j > i + 1$ containing both A and B. Therefore $M_k = M_{i+1}$. However, this contradicts $A \in U_i - S_i$. It follows that there cannot be a set M_j through which the separating set S_i can be bypassed. In the same manner we can show that there cannot be a sequence of sets $M_{j_1}, \dots, M_{j_s}$, and a path $A_0, A_1, \dots, A_s, A_{s+1}$, such that $A_0 \in U_i - S_i$, $\{A_0, A_1\} \subseteq M_{j_1}, \dots, \{A_s, B\} \subseteq M_{j_s}$, $A_{s+1} \in M_{i+1} - S_i$ bypassing the separating set: With a similar argument as above, assuming again that the sets $M_{j_1}, \dots, M_{j_s}$ are the ones with the smallest indices having the desired property, we can show that there must be a set M_{j_k}, namely the one with the largest index, such that either $A_{k+1} \in M_{j_{k-1}}$ or $A_{k-1} \in M_{j_{k+1}}$, so that we can shorten the path by one element. Working recursively, we reduce this case to the case with only one set bypassing the separating set, which we already showed to be impossible. Therefore the above separation holds.

Note that the running intersection property is essential for the proof. Without it, we do not have the conditional independence statements that are exploited. Consider, for example, a structure like the one shown in Figure 4.8 on page 111, which was used to demonstrate that the theorem does not hold for arbitrary graphs. Since this graph does not have hypertree structure, there is no construction sequence. Indeed, independent of the edge we start with, and independent of the edge we add next, we do not have (in the general case) the necessary conditional independence statement that would allow us to extend the decomposition formula.

A.6 Proof of Theorem 4.1.26

Theorem 4.1.26 *Let p_U be a probability distribution on a set U of (discrete) attributes. A directed acyclic graph $\vec{G} = (U, \vec{E})$ is a conditional independence graph w.r.t. p_U, iff p_U is factorizable w.r.t. $\vec{G}$.*

Proof. The proof of the theorem consists of two parts. In the first part it is shown that, if the distribution p_U is factorizable w.r.t. a directed acyclic graph, then $\vec{G}$ satisfies the global Markov property and thus is a conditional independence graph. In the second part it is shown that the family of conditional probabilities of an attribute given its parents in the graph is a factorization of p_U. Both parts make use of Theorem 4.1.20 (see page 107), which states that the global and the local Markov property of a directed acyclic graph are equivalent if the relation representing the set of conditional independence statements that hold in p_U satisfies the semi-graphoid axioms (as conditional probabilistic independence does in general, cf. Theorem 4.1.2 on page 93).

For both parts of the proof let $n = |U|$ and $o : U \to \{1, \ldots, n\}$ be a topological order[8] of the attributes in $\vec{G}$. Let the attributes in U be numbered in such a way that $o(A_i) = i$. Due to the properties of a topological order all attributes preceding an attribute in the topological order are among its non-descendants and all parents of A_i precede it in the topological order. That is, we have

$$\forall 1 \leq i \leq n: \quad \text{parents}(A_i) \subseteq \{A_j \mid j < i\} \subseteq \text{nondescs}(A_i).$$

Apply the chain rule of probability (cf. page 79) to the attributes in U w.r.t. the topological order o. That is, write p_U as

$$\forall a_1 \in \text{dom}(A_1) : \ldots \forall a_n \in \text{dom}(A_n) :$$
$$p_u\Big(\bigwedge\nolimits_{i=1}^{n} A_i = a_i\Big) = \prod_{i=1}^{n} P\Big(A_i = a_i \,\Big|\, \bigwedge\nolimits_{j=1}^{i-1} A_j = a_j\Big).$$

For the first part of the proof[9] we know that p_U factorizes w.r.t. $\vec{G}$, i.e.

$$\forall a_1 \in \text{dom}(A_1) : \ldots \forall a_n \in \text{dom}(A_n) :$$
$$p_u\Big(\bigwedge\nolimits_{i=1}^{n} A_i = a_i\Big) = \prod_{i=1}^{n} P\Big(A_i = a_i \,\Big|\, \bigwedge\nolimits_{A_j \in \text{parents}(A_i)} A_j = a_j\Big).$$

[8]The notion of a topological order is defined in Definition 4.1.12 on page 98.

[9]Unfortunately, the first part of the proof is dealt with somewhat casually in [Lauritzen 1996] (which we follow for the proofs of some of the other theorems), so that we decided to give a different proof. It is based strongly on the proof of Theorem 4.1.20.

Therefore we have

$$\forall a_1 \in \mathrm{dom}(A_1) : \ldots \forall a_n \in \mathrm{dom}(A_n) :$$
$$\prod_{i=1}^{n} P\Big(A_i = a_i \,\Big|\, \bigwedge\nolimits_{j=1}^{i-1} A_j = a_j\Big)$$
$$= \prod_{i=1}^{n} P\Big(A_i = a_i \,\Big|\, \bigwedge\nolimits_{A_j \in \mathrm{parents}(A_i)} A_j = a_j\Big).$$

From this equation one can easily establish that corresponding factors must be equal by the following procedure: We sum the above equation over all values of A_n, i.e., the last attribute of the topological order. This attribute appears only in the last factor and thus all other factors can be moved out of the sum. Since the last factor is a conditional probability of the values of A_n, summing over all values of A_n yields 1 and therefore the term is simply canceled by the summation. We arrive at

$$\forall a_1 \in \mathrm{dom}(A_1) : \ldots \forall a_{n-1} \in \mathrm{dom}(A_{n-1}) :$$
$$\prod_{i=1}^{n-1} P\Big(A_i = a_i \,\Big|\, \bigwedge\nolimits_{j=1}^{i-1} A_j = a_j\Big)$$
$$= \prod_{i=1}^{n-1} P\Big(A_i = a_i \,\Big|\, \bigwedge\nolimits_{A_j \in \mathrm{parents}(A_i)} A_j = a_j\Big).$$

Comparing this result to the preceding equality, we conclude that

$$P\Big(A_n = a_n \,\Big|\, \bigwedge\nolimits_{j=1}^{n-1} A_j = a_j\Big) = P\Big(A_n = a_n \,\Big|\, \bigwedge\nolimits_{A_j \in \mathrm{parents}(A_n)} A_j = a_j\Big).$$

In the same manner, working recursively downward the topological order, we establish the equality for all attributes. Finally we have

$$\forall 1 \leq i \leq n : \quad A_i \perp\!\!\!\perp_{P_U} \{A_j \mid j < i\} - \mathrm{parents}(A_i) \mid \mathrm{parents}(A_i).$$

From these statements we can infer, working in the same way as in the proof of Theorem 4.1.20 (cf. Section A.3), that $\vec{G}$ has the global Markov property, since the set of statements above was all that was needed in that proof.

Another way to prove the first part of the theorem is the following: It is clear that for each attribute A of the graph a topological order can be constructed, so that all its non-descendants precede it in the topological order. For example, such a topological order can be found with the simple recursive algorithm stated on page 98 by preferring those childless attributes that are descendants of A. With respect to this specific topological order the conditional independence statement that can be derived for A by the method employed above is already identical to the desired local conditional independence statement. Note, though, that with this approach we may need a different topological order for each attribute of the graph and that we finally have to apply Theorem 4.1.20 in order to derive the global Markov property by which a conditional independence graph is defined.

For the second part of the proof we know that $\vec{G}$ has the global and thus the local Markov property (cf. Theorem 4.1.20). Therefore we have

$$\forall 1 \leq i \leq n: \quad A_i \perp\!\!\!\perp_{p_U} \text{nondescs}(A_i) - \text{parents}(A_i) \mid \text{parents}(A_i).$$

By applying the decomposition axiom[10] and exploiting the properties of a topological order (see above), we have

$$\forall 1 \leq i \leq n: \quad A_i \perp\!\!\!\perp_{p_U} \{A_j \mid j < i\} - \text{parents}(A_i) \mid \text{parents}(A_i).$$

Translating this to probabilities yields

$$\forall 1 \leq i \leq n: \\ P\Big(A_i = a_i \,\Big|\, \bigwedge\nolimits_{j=1}^{i-1} A_j = a_j\Big) = P\Big(A_i = a_i \,\Big|\, \bigwedge\nolimits_{A_j \in \text{parents}(A_i)} A_j = a_j\Big).$$

By combining this statement with the chain rule decomposition shown above we arrive at the desired factorization.

A.7 Proof of Theorem 4.1.27

Theorem 4.1.27 *Let π_U be a possibility distribution on a set U of (discrete) attributes. If a directed acyclic graph $\vec{G} = (U, \vec{E})$ is a conditional independence graph w.r.t. π_U, then π_U is decomposable w.r.t. $\vec{G}$.*

Proof. The proof is analogous to the corresponding part of the probabilistic case. We apply the possibilistic counterpart of the chain rule w.r.t. a topological order of the attributes and simplify the conditions by exploiting the conditional independences that can be derived from the local Markov property of the graph $\vec{G}$. There is only one minor difference, because, if conditional possibilistic independence is expressed in conditional possibilities, we have a relation that differs from the probabilistic case. However, how to deal with this difference was already explained on page 87 and therefore we do not repeat it here. It should be noted, though, that this difference is the main reason, why the converse of the theorem, i.e. that a decomposition w.r.t. $\vec{G}$ implies the global Markov property of $\vec{G}$, does not hold, as demonstrated by the example on page 113.

[10]The decomposition axiom is part of the semi-graphoid axioms, defined in Definition 4.1.1 on page 92, which are satisfied by conditional probabilistic independence according to Theorem 4.1.2 on page 93.

A.8 Proof of Theorem 5.4.8

Theorem 5.4.8 *Let* $D = (R, w_R)$ *be a database over a set* U *of attributes and let* $X \subseteq U$*. Furthermore, let* $\text{support}(D) = (\text{support}(R), w_{\text{support}(R)})$ *and* $\text{closure}(D) = (\text{closure}(R), w_{\text{closure}(R)})$ *as well as* $\pi_X^{(\text{support}(D))}$ *and* $\pi_X^{(\text{closure}(D))}$ *be defined as in Definition 5.4.3 on page 146, in Definition 5.4.7 on page 148, and in the paragraphs following these definitions, respectively. Then*

$$\forall t \in T_X^{(\text{precise})}: \quad \pi_X^{(\text{closure}(D))}(t) = \pi_X^{(\text{support}(D))}(t),$$

i.e., computing the maximum projection of the possibility distribution $\pi_U^{(D)}$ *induced by* D *to the attributes in* X *via the closure of* D *is equivalent to computing it via the support of* D.

Proof. [Borgelt and Kruse 1998c] The assertion of the theorem is proven in two steps. In the first, it is shown that, for an arbitrary tuple $t \in T_X^{(\text{precise})}$,

$$\pi_X^{(\text{closure}(D))}(t) \geq \pi_X^{(\text{support}(D))}(t),$$

and in the second that

$$\pi_X^{(\text{closure}(D))}(t) \leq \pi_X^{(\text{support}(D))}(t).$$

Both parts together obviously prove the theorem. So let $t \in T_X^{(\text{precise})}$ be an arbitrary precise tuple and let $w_0 = \sum_{u \in R} w_R(u)$. Furthermore, let

$$S = \{s \in \text{support}(R) \mid t \sqsubseteq s|_X\} \quad \text{and} \quad C = \{c \in \text{closure}(R) \mid t \sqsubseteq c|_X\}.$$

(1) $\pi_X^{(\text{closure}(D))}(t) \geq \pi_X^{(\text{support}(D))}(t)$:

We have to distinguish two cases, namely $S = \emptyset$ and $S \neq \emptyset$, the first of which is obviously trivial.

(a) $S = \emptyset$: $\pi_X^{(\text{support}(D))}(t) = 0 \leq \pi_X^{(\text{closure}(D))}(t) \in [0, 1]$.

(b) $S \neq \emptyset$: Due to the definitions of $\pi_X^{(\text{support}(D))}$ and $w_{\text{support}(R)}$

$$\begin{aligned}\pi_X^{(\text{support}(D))}(t) &= \frac{1}{w_0} \max_{s \in S} w_{\text{support}(R)}(s) \\ &= \frac{1}{w_0} \max_{s \in S} \sum_{u \in R, s \sqsubseteq u} w_R(u).\end{aligned}$$

Let $\hat{s} \in S$ be (one of) the tuple(s) $s \in S$ for which $w_{\text{support}(R)}(s)$ is maximal. Let $V = \{v \in R \mid \hat{s} \sqsubseteq v\}$, i.e., let V be the set of tuples from which the weight of $\hat{s}$ is computed. Then

$$\pi_X^{(\text{support}(D))}(t) = \frac{1}{w_0} w_{\text{support}(R)}(\hat{s}) = \frac{1}{w_0} \sum_{v \in V} w_R(v).$$

Since $V \subseteq R$, we have $v^* = \bigsqcap_{v \in V} v \in \text{closure}(R)$, because of the definition of the closure of a relation (cf. Definition 5.4.6 on page 147). Since $\hat{s} \in S$, we have $t \sqsubseteq \hat{s}|_X$ (because of the definition of S), and since $\forall v \in V : \hat{s} \sqsubseteq v$, we have $\hat{s} \sqsubseteq v^*$ (because the intersection of a set of tuples is the least specific tuple that is at least as specific as all tuples in the set), hence $t \sqsubseteq v^*|_X$. It follows that $v^* \in C$.
Let $W = \{w \in R \mid v^* \sqsubseteq w\}$, i.e. let W be the set of tuples from which the weight of v^* is computed. Since $v^* = \bigsqcap_{v \in V} v$ (due to the definition of v^*), we have $\forall v \in V : v^* \sqsubseteq v$ (due to the fact that the intersection of a set of tuples is at least as specific as all tuples in the set), and hence $V \subseteq W$. Putting everything together we arrive at

$$\begin{aligned} \pi_X^{(\text{closure}(D))}(t) &= \frac{1}{w_0} \max_{c \in C} w_{\text{closure}(R)}(c) \\ &\geq \frac{1}{w_0} w_{\text{closure}(R)}(v^*) \\ &= \frac{1}{w_0} \sum_{w \in W} w_R(w) \\ &\geq \frac{1}{w_0} \sum_{v \in V} w_R(v) \\ &= \pi_X^{(\text{support}(D))}(t). \end{aligned}$$

From what we have considered, the first inequality need not be an equality, since there may be another tuple in closure(R) to which a higher weight was assigned. The second inequality need not be an equality, because W may contain more tuples than V.

(2) $\pi_X^{(\text{closure}(D))}(t) \leq \pi_X^{(\text{support}(D))}(t)$:

Again we have to distinguish two cases, namely $C = \emptyset$ and $C \neq \emptyset$, the first of which is obviously trivial.

(a) $C = \emptyset$: $\pi_X^{(\text{closure}(D))}(t) = 0 \leq \pi_X^{(\text{support}(D))}(t) \in [0, 1]$.

(b) $C \neq \emptyset$: Due to the definitions of $\pi_X^{(\text{closure}(D))}$ and $w_{\text{closure}(R)}$

$$\begin{aligned} \pi_X^{(\text{closure}(D))}(t) &= \frac{1}{w_0} \max_{c \in C} w_{\text{closure}(R)}(c) \\ &= \frac{1}{w_0} \max_{c \in C} \sum_{u \in R, c \sqsubseteq u} w_R(u). \end{aligned}$$

Let $\hat{c} \in C$ be (one of) the tuple(s) $c \in C$ for which $w_{\text{closure}(R)}(c)$ is maximal. Let $W = \{w \in R \mid \hat{c} \sqsubseteq w\}$, i.e. let W be the set of tuples from which the weight of $\hat{c}$ is computed. Then

$$\pi_X^{(\text{closure}(D))}(t) = \frac{1}{w_0} w_{\text{closure}(R)}(\hat{c}) = \frac{1}{w_0} \sum_{w \in W} w_R(w).$$

Let $Q = \{q \in T_X^{(\text{precise})} \mid q \sqsubseteq \hat{c}\}$, i.e. let Q be the set of tuples "supporting" $\hat{c}$. Since $t \in T_X^{(\text{precise})}$ and $t \sqsubseteq \hat{c}|_X$ (due to $\hat{c} \in C$), there must be a tuple $s^* \in Q$, for which $t \sqsubseteq s^*|_X$. Since $s^* \in Q$, we have $s^* \sqsubseteq \hat{c} \in \text{closure}(R)$ (due to the definition of Q), and since $\forall c \in \text{closure}(R) : \exists u \in R : c \sqsubseteq u$ (due to the definition of the closure of a relation, cf. Definition 5.4.6 on page 147), it follows that $\exists u \in R : s^* \sqsubseteq u$ and hence we have $s^* \in \text{support}(R)$.

Let $V = \{v \in R \mid s^* \sqsubseteq v\}$, i.e. let V be the set of tuples from which the weight of s^* is computed. Since $s^* \sqsubseteq \hat{c}$ (see above), we have $\forall w \in W : s^* \sqsubseteq w$ and hence $W \subseteq V$. Thus we arrive at

$$
\begin{aligned}
\pi_X^{(\text{support}(D))}(r) &= \frac{1}{w_0} \max_{s \in S} w_{\text{support}(R)}(s) \\
&\geq \frac{1}{w_0} w_{\text{support}(R)}(s^*) \\
&= \frac{1}{w_0} \sum_{v \in V} w_R(v) \\
&\geq \frac{1}{w_0} \sum_{w \in W} w_R(w) \\
&= \pi_X^{(\text{closure}(D))}(t).
\end{aligned}
$$

The reasons underlying the inequalities are similar to those in (1). From (1) and (2) it follows that, since t is arbitrary,

$$
\forall t \in T_X^{(\text{precise})} : \quad \pi_X^{(\text{closure}(D))}(t) = \pi_X^{(\text{support}(D))}(t).
$$

This completes the proof.

A.9 Proof of Theorem 7.3.2

Theorem 7.3.2 *If a family $\mathcal{M}$ of subsets of objects of a given set U is constructed observing the two conditions stated on page 232, then this family $\mathcal{M}$ has the running intersection property.*

Proof. Adding a node set to a given family $\mathcal{M}$ either adds isolated nodes, i.e., nodes not contained in any subfamily, to a subfamily, or connects two or more subfamilies, or both. Hence one can show that the method referred to indeed results in a family $\mathcal{M}$ having the running intersection property by a simple inductive argument, which proves that all subfamilies that are created during the construction have the running intersection property:

A subfamily with a single node set trivially has the running intersection property (induction anchor). So assume that all subfamilies up to a certain size, i.e., with a certain number of node sets, have the running intersection property (induction hypothesis). If a new node set only adds isolated nodes to a subfamily, the enlarged family has the running intersection property, because in this case the second condition on page 232 is equivalent to the last part of the defining condition of a construction sequence (cf. Definition 4.1.23 on page 110). Hence the construction sequence of the subfamily (which must exist due to the induction hypothesis) is simply extended by one set.

So assume that a new node set connects two or more subfamilies (and maybe adds some isolated nodes, too). In order to show that there is a construction sequence for the resulting subfamily of node sets, we show first that any set of a family of sets having the running intersection property can be made the first set in a construction sequence for this family: Reconsider the join tree illustration of Graham reduction (cf. page 231). Obviously, the reduction can be carried out w.r.t. a join tree even if a given set (i.e., a given node of the join tree) is chosen in advance to be the last to be removed, simply because we can work from the leaves of the join tree towards the corresponding node. Since the reverse of the order in which the node sets are removed by Graham reduction is a construction sequence, there is a construction sequence starting with the chosen set, and since the set can be chosen arbitrarily, any set can be made the first of a construction sequence.

With this established, the remainder of the proof is simple: For each of the subfamilies that are connected by the new node set, we determine a construction sequence starting with the set M_k mentioned in the second condition stated on page 232. Then we form the construction sequence of the resulting enlarged subfamily as follows: The new node set is the first set in this sequence. To it we append the construction sequences we determined for the subfamilies, one after the other. This sequence is a construction sequence, because the sets M_k obviously satisfy the defining condition w.r.t. the first set due to the way in which they where chosen. Within the appended sequences the condition holds, because these sequences are construction sequences for the subfamilies. There is no interaction between these sequences, because the subfamilies are node-disjoint. Hence the new subfamily has the running intersection property.

A.10 Proof of Lemma 7.2.2

Lemma 7.2.2 *Let A, B, and C be three attributes with finite domains and let their joint probability distribution be strictly positive, i.e., $\forall a \in \mathrm{dom}(A) : \forall b \in \mathrm{dom}(B) : \forall c \in \mathrm{dom}(C) : P(A = a, B = b, C = c) > 0$. Then*

$$I_{\mathrm{gain}}^{(\mathrm{Shannon})}(C, AB) \;\geq\; I_{\mathrm{gain}}^{(\mathrm{Shannon})}(C, B),$$

with equality obtaining only if the attributes C and A are conditionally independent given B.

Proof. Let the domains of A, B, and C be $\text{dom}(A) = \{a_1, \ldots, a_{n_A}\}$, $\text{dom}(B) = \{b_1, \ldots, b_{n_B}\}$, and $\text{dom}(C) = \{c_1, \ldots, c_{n_C}\}$, respectively. In order to make the formulae easier to read, we use the following abbreviations, which are consistent with the abbreviations introduced in Section 7.2.2:

$$\begin{aligned}
p_{i..} &= P(C = c_i), & p_{ij.} &= P(C = c_i, A = a_j),\\
p_{.j.} &= P(A = a_j), & p_{i.k} &= P(C = c_i, B = b_k),\\
p_{..k} &= P(B = b_k), & p_{.jk} &= P(A = a_j, B = b_k), \quad \text{and}\\
& & p_{ijk} &= P(C = c_i, A = a_j, B = b_k),
\end{aligned}$$

i.e., the index i always refers to the attribute C, the index j always refers to the attribute A, and the index k always refers to the attribute B.

Since it makes the proof much simpler, we show that

$$I_{\text{gain}}^{(\text{Shannon})}(C, B) - I_{\text{gain}}^{(\text{Shannon})}(C, AB) \leq 0,$$

from which the original statement follows trivially. In addition, we drop the upper index "(Shannon)" in the following, since no confusion with Hartley information gain or Hartley entropy is to be expected. (For the definition of Shannon information gain in terms of Shannon entropies cf. Section 7.2.4.)

$$\begin{aligned}
&I_{\text{gain}}(C, B) - I_{\text{gain}}(C, AB)\\
&= H(C) + H(B) - H(CB) - (H(C) + H(AB) - H(CAB))\\
&= -H(CB) - H(AB) + H(CAB) + H(B)\\
&= \sum_{i=1}^{n_C} \sum_{k=1}^{n_B} p_{i.k} \log_2 p_{i.k} + \sum_{j=1}^{n_A} \sum_{k=1}^{n_B} p_{.jk} \log_2 p_{.jk}\\
&\quad - \sum_{i=1}^{n_C} \sum_{j=1}^{n_A} \sum_{k=1}^{n_B} p_{ijk} \log_2 p_{ijk} - \sum_{k=1}^{n_B} p_{..k} \log_2 p_{..k}\\
&= \sum_{i=1}^{n_C} \sum_{j=1}^{n_A} \sum_{k=1}^{n_B} p_{ijk} \log_2 \frac{p_{i.k} p_{.jk}}{p_{ijk} p_{..k}}\\
&= \frac{1}{\ln 2} \sum_{i=1}^{n_C} \sum_{j=1}^{n_A} \sum_{k=1}^{n_B} p_{ijk} \ln \frac{p_{i.k} p_{.jk}}{p_{ijk} p_{..k}}\\
&\leq \frac{1}{\ln 2} \sum_{i=1}^{n_C} \sum_{j=1}^{n_A} \sum_{k=1}^{n_B} p_{ijk} \left(\frac{p_{i.k} p_{.jk}}{p_{ijk} p_{..k}} - 1 \right)\\
&= \frac{1}{\ln 2} \left[\sum_{i=1}^{n_C} \sum_{j=1}^{n_A} \sum_{k=1}^{n_B} \frac{p_{i.k} p_{.jk}}{p_{..k}} - \underbrace{\sum_{i=1}^{n_C} \sum_{j=1}^{n_A} \sum_{k=1}^{n_B} p_{ijk}}_{=1} \right]\\
&= \frac{1}{\ln 2} \left[\left(\sum_{k=1}^{n_B} \frac{1}{p_{..k}} \sum_{i=1}^{n_C} \sum_{j=1}^{n_A} p_{i.k} p_{.jk} \right) - 1 \right]
\end{aligned}$$

$$= \frac{1}{\ln 2}\left[\left(\sum_{k=1}^{n_B}\frac{1}{p_{..k}}\underbrace{\left(\sum_{i=1}^{n_C}p_{i.k}\right)}_{=p_{..k}}\underbrace{\left(\sum_{j=1}^{n_A}p_{.jk}\right)}_{=p_{..k}}\right)-1\right]$$

$$= \frac{1}{\ln 2}\left(\underbrace{\left(\sum_{k=1}^{n_B}\frac{p_{..k}^2}{p_{..k}}\right)}_{=1}-1\right)$$

$$= \frac{1}{\ln 2}(1-1) \;=\; 0,$$

where the inequality follows from the fact that

$$\ln x \leq x - 1,$$

with equality obtaining only for $x = 1$. (This can most easily be seen from the graph of $\ln x$.) As a consequence, $I_{\text{gain}}(C, AB) = I_{\text{gain}}(C, B)$ only if

$$\forall i, j, k : \frac{p_{i.k}p_{.jk}}{p_{ijk}p_{..k}} = 1 \quad \Leftrightarrow \quad \forall i, j, k : p_{ij|k} = p_{i.|k}p_{.j|k},$$

where $p_{ij|k} = P(C = c_i, A = a_j \mid B = b_k)$ and $p_{i.|k}$ and $p_{.j|k}$ likewise. That is, $I_{\text{gain}}(C, AB) = I_{\text{gain}}(C, B)$ only holds if the attributes C and A are conditionally independent given attribute B.

A.11 Proof of Lemma 7.2.4

Lemma 7.2.4 *Let A, B, and C be attributes with finite domains. Then*

$$I^2_{\text{gain}}(C, AB) \;\geq\; I^2_{\text{gain}}(C, B).$$

Proof. We use the same notation and the same abbreviations as in the proof of Lemma 7.2.2 in Section A.10. Since it makes the proof much simpler, we show

$$I^2_{\text{gain}}(C, AB) - I^2_{\text{gain}}(C, B) \;\geq\; 0,$$

from which the original statement follows trivially. (For the definition of quadratic information gain in terms of quadratic entropies cf. Section 7.2.4.)

$$\begin{aligned}
&I^2_{\text{gain}}(C, AB) - I^2_{\text{gain}}(C, B)\\
&\quad= H^2(C) + H^2(AB) - H^2(CAB) - (H^2(C) + H^2(B) - H^2(CB))\\
&\quad= H^2(AB) - H^2(CAB) + H^2(CB) - H^2(B)\\
&\quad= 1 - \sum_{j=1}^{n_A}\sum_{k=1}^{n_B} p_{.jk}^2 - 1 + \sum_{i=1}^{n_C}\sum_{j=1}^{n_A}\sum_{k=1}^{n_B} p_{ijk}^2 + 1 - \sum_{i=1}^{n_C}\sum_{k=1}^{n_B} p_{i.k}^2 - 1 + \sum_{k=1}^{n_B} p_{..k}^2
\end{aligned}$$

$$\begin{aligned}
= \quad & -\sum_{j=1}^{n_A}\sum_{k=1}^{n_B} p_{.jk} \sum_{i=1}^{n_C} p_{ijk} + \sum_{k=1}^{n_B}\sum_{i=1}^{n_C}\sum_{j=1}^{n_A} p_{ijk}^2 \\
& -\sum_{i=1}^{n_C}\sum_{k=1}^{n_B} p_{i.k} \sum_{j=1}^{n_A} p_{ijk} + \sum_{k=1}^{n_B} p_{..k} \sum_{i=1}^{n_C}\sum_{j=1}^{n_A} p_{ijk} \\
= \quad & \sum_{i=1}^{n_C}\sum_{j=1}^{n_A}\sum_{k=1}^{n_B} p_{ijk}(p_{..k} - p_{i.k} - p_{.jk} + p_{ijk}) \;\geq\; 0
\end{aligned}$$

That the last sum is nonnegative results from the fact that each term of the sum is nonnegative. This can be seen as follows: From probability theory it is well known that $\forall E_1, E_2, E_2 \subseteq \Omega$:

$$\begin{aligned}
P(E_1 \cup E_2) &= P(E_1) + P(E_2) - P(E_1 \cap E_2) \qquad \text{and} \\
P(E_1 \cup E_2 \cup E_3) &= P(E_1) + P(E_2) + P(E_1) \\
&\quad -P(E_1 \cap E_2) - P(E_1 \cap E_3) - P(E_2 \cap E_3) \\
&\quad +P(E_1 \cap E_2 \cap E_3).
\end{aligned}$$

It follows that $\forall E_1, E_2, E_2 \subseteq \Omega$:

$$\begin{aligned}
& P(E_3) - P(E_1 \cap E_3) - P(E_2 \cap E_3) + P(E_1 \cap E_2 \cap E_3) \\
&\quad = P(E_1 \cup E_2 \cup E_3) - P(E_1 \cup E_2) \;\geq\; 0.
\end{aligned}$$

Hence, if we identify E_1 with $C = c_i$, E_2 with $A = a_j$, and E_3 with $B = b_k$, we have that

$$\forall i, j, k: \quad p_{..k} - p_{i.k} - p_{.jk} + p_{ijk} \;\geq\; 0.$$

Therefore each term of the sum must be nonnegative, since the other factor p_{ijk} is clearly nonnegative, as it is a probability.

Since the sum can be zero only if all terms are zero, we obtain that it is $I^2_{\text{gain}}(C, AB) = I^2_{\text{gain}}(C, B)$ only if

$$\forall i, j, k: \quad p_{ijk} = 0 \;\vee\; p_{..k} + p_{ijk} = p_{i.k} + p_{.jk}.$$

Suppose there are two different values i_1 and i_2 for which $p_{ijk} > 0$. In this case the second equation yields

$$\forall j, k: \quad p_{i_1jk} - p_{i_2jk} = p_{i_1.k} - p_{i_2.k}.$$

Clearly, this equation can hold only if there is at most one value j for which $p_{ijk} > 0$. A symmetric argument shows that there can be at most one value i for which $p_{ijk} > 0$ if there are two values j for which $p_{ijk} > 0$. Therefore the sum can be zero only if at most one of the attributes A and C has more than one value with a non-vanishing probability. Consequently, it is, in general, $I^2_{\text{gain}}(C, AB) \neq I^2_{\text{gain}}(C, B)$ even if the attributes C and A are conditionally independent given the attribute B.

A.12 Proof of Lemma 7.2.6

Lemma 7.2.6 *Let A, B, and C be three attributes with finite domains and let their joint probability distribution be strictly positive, i.e.,* $\forall a \in \text{dom}(A) : \forall b \in \text{dom}(B) : \forall c \in \text{dom}(C) : P(A = a, B = b, C = c) > 0$. *Then*

$$\chi^2(C, AB) \geq \chi^2(C, B),$$

with equality obtaining only if the attributes C and A are conditionally independent given B.

Proof. We use the same notation and the same abbreviations as in the proof of Lemma 7.2.2 in Section A.10. Since it makes the proof much simpler, we show

$$\frac{1}{N}\left(\chi^2(C, AB) - \chi^2(C, B)\right) \;\geq\; 0,$$

from which the original statement follows trivially. (For the definition of the χ^2 measure cf. Section 7.2.4.)

$$\begin{aligned}
&\frac{1}{N}\left(\chi^2(C, AB) - \chi^2(C, B)\right) \\
&= \sum_{i=1}^{n_C}\sum_{j=1}^{n_A}\sum_{k=1}^{n_B} \frac{(p_{ijk} - p_{i..}p_{.jk})^2}{p_{i..}p_{.jk}} - \sum_{i=1}^{n_C}\sum_{k=1}^{n_B} \frac{(p_{i.k} - p_{i..}p_{..k})^2}{p_{i..}p_{..k}} \\
&= \sum_{i=1}^{n_C}\sum_{k=1}^{n_B}\left(\sum_{j=1}^{n_A} \frac{p_{ijk}^2 - 2p_{ijk}p_{i..}p_{.jk} + p_{i..}^2 p_{.jk}^2}{p_{i..}p_{.jk}} \right. \\
&\qquad\qquad \left. - \frac{p_{i.k}^2 - 2p_{i.k}p_{i..}p_{..k} + p_{i..}^2 p_{..k}^2}{p_{i..}p_{..k}}\right) \\
&= \sum_{i=1}^{n_C}\sum_{k=1}^{n_B}\left(\sum_{j=1}^{n_A}\left(\frac{p_{ijk}^2}{p_{i..}p_{.jk}} - 2p_{ijk} + p_{i..}p_{.jk}\right) - \frac{p_{i.k}^2}{p_{i..}p_{..k}} + 2p_{i.k} - p_{i..}p_{..k}\right) \\
&= \sum_{i=1}^{n_C}\sum_{k=1}^{n_B}\left(\sum_{j=1}^{n_A} \frac{p_{ijk}^2}{p_{i..}p_{.jk}} - 2p_{i.k} + p_{i..}p_{..k} - \frac{p_{i.k}^2}{p_{i..}p_{..k}} + 2p_{i.k} - p_{i..}p_{..k}\right) \\
&= \sum_{i=1}^{n_C}\sum_{k=1}^{n_B} \frac{1}{p_{i..}p_{..k}}\left(p_{..k}\sum_{j=1}^{n_A}\frac{p_{ijk}^2}{p_{.jk}} - p_{i.k}\sum_{j=1}^{n_A} p_{ijk}\right) \\
&= \sum_{i=1}^{n_C}\sum_{k=1}^{n_B} \frac{1}{p_{i..}p_{..k}}\left[\left(\sum_{j_1=1}^{n_A} p_{.j_1k}\right)\left(\sum_{j_2=1}^{n_A}\frac{p_{ij_2k}^2}{p_{.j_2k}}\right) - \left(\sum_{j_1=1}^{n_A} p_{ij_1k}\right)\left(\sum_{j_2=1}^{n_A} p_{ij_2k}\right)\right] \\
&= \sum_{i=1}^{n_C}\sum_{k=1}^{n_B} \frac{1}{p_{i..}p_{..k}}\left(\sum_{j_1=1}^{n_A}\sum_{j_2=1}^{n_A} \frac{p_{.j_1k}p_{ij_2k}^2}{p_{.j_2k}} - \sum_{j_1=1}^{n_A}\sum_{j_2=1}^{n_A} p_{ij_1k}p_{ij_2k}\right)
\end{aligned}$$

$$= \sum_{i=1}^{n_C} \sum_{k=1}^{n_B} \frac{1}{p_{i..}p_{..k}} \left(\sum_{j_1=1}^{n_A} \sum_{j_2=1}^{n_A} \frac{p_{.j_1k}^2 p_{ij_2k}^2 - p_{ij_1k}p_{ij_2k}p_{.j_1k}p_{.j_2k}}{p_{.j_1k}p_{.j_2k}} \right)$$

$$= \sum_{i=1}^{n_C} \sum_{k=1}^{n_B} \frac{1}{2p_{i..}p_{..k}} \sum_{j_1=1}^{n_A} \sum_{j_2=1}^{n_A} \frac{(p_{.j_1k}p_{ij_2k} - p_{ij_1k}p_{.j_2k})^2}{p_{.j_1k}p_{.j_2k}}$$

$$= \sum_{i=1}^{n_C} \sum_{k=1}^{n_B} \sum_{j_1=1}^{n_A} \sum_{j_2=1}^{n_A} \frac{(p_{.j_1k}p_{ij_2k} - p_{ij_1k}p_{.j_2k})^2}{2p_{i..}p_{..k}p_{.j_1k}p_{.j_2k}} \geq 0,$$

where the semi-last step follows by duplicating the term in parentheses and then interchanging the indices j_1 and j_2 in the second instance (which is possible, because they have the same range). From the result it is immediately clear that $\chi^2(C, AB) \geq \chi^2(C, B)$: Since each term of the sum is a square divided by a product of (positive) probabilities, each term and thus the sum must be non-negative. It also follows that the sum can be zero only if all of its terms are zero, which requires their numerators to be zero:

$$\forall i, j_1, j_2, k : p_{.j_1k}p_{ij_2k} - p_{ij_1k}p_{.j_2k} = 0 \quad \Leftrightarrow \quad \forall i, j_1, j_2, k : \frac{p_{ij_2k}}{p_{.j_2k}} = \frac{p_{ij_1k}}{p_{.j_1k}}$$
$$\Leftrightarrow \quad \forall i, j_1, j_2, k : p_{i|j_2k} = p_{i|j_1k},$$

where $p_{i|j_\alpha k} = P(C = c_i \mid A = a_{j_\alpha}, B = b_k)$ with $\alpha \in \{1, 2\}$. That is, $\chi^2(C, AB) = \chi^2(C, B)$ only holds if the attributes C and A are conditionally independent given attribute B.

A.13 Proof of Theorem 7.3.4

Theorem 7.3.4 *Let m be a symmetric evaluation measure satisfying*

$$\forall A, B, C: \quad m(C, AB) \geq m(C, B)$$

with equality obtaining only if the attributes A and C are conditionally independent given B. Let G be a singly connected undirected perfect map of a probability distribution p over a set U of attributes. Then constructing a maximum weight spanning tree for the attributes in U with m (computed from p) providing the edge weights uniquely identifies G.

In order to prove this theorem, it is convenient to prove first the following lemma, by which an important property of the measure m is established:

Lemma A.13.1 *Let m be a symmetric evaluation measure satisfying*

$$\forall A, B, C: \quad m(C, AB) \geq m(C, B)$$

with equality obtaining only if the attributes C and A are conditionally independent given B. If A, B, and C are three attributes satisfying $A \perp\!\!\!\perp C \mid B$, but neither $A \perp\!\!\!\perp B \mid C$ nor $C \perp\!\!\!\perp B \mid A$, then

$$m(A, C) \ < \ \min\{m(A, B), m(B, C)\}.$$

Proof. From the fact that $A \perp\!\!\!\perp C \mid B$ we know that

$$m(C, AB) = m(C, B) \qquad \text{and} \qquad m(A, CB) = m(A, B).$$

Since it is $A \not\perp\!\!\!\perp B \mid C$ and $C \not\perp\!\!\!\perp B \mid A$, we have

$$m(C, AB) > m(C, A) \qquad \text{and} \qquad m(A, CB) > m(A, C).$$

Consequently, $m(C, A) < m(C, B)$ and $m(C, A) < m(A, B)$.

Proof of Theorem 7.3.4. Let C and A be two arbitrary attributes in U that are not adjacent in G. Since G is singly connected there is a unique path connecting C and A in G. We show that any edge connecting two consecutive nodes on this path has a higher weight than the edge (C, A).

Let B be the successor of C on the path connecting C and A in G. Then it is $C \perp\!\!\!\perp_p A \mid B$, but neither $C \perp\!\!\!\perp_p B \mid A$ nor $A \perp\!\!\!\perp_p B \mid C$. Consequently, it is $m(C, A) < m(C, B)$ and $m(C, A) < m(B, A)$. If B is the predecessor of A on the path, we already have that all edges on the path have a higher weight than the edge (C, A). Otherwise we have that the edge (C, B) has a higher weight than the edge (C, A). For the remaining path, i.e., the path that connects B and A, the above argument is applied recursively.

Hence any edge between two consecutive nodes on the path connecting any two attributes C and A has a higher weight than the edge (C, A). From this it is immediately clear, for example by considering how the Kruskal algorithm [Kruskal 1956] works, that constructing the optimum weight spanning tree with m providing the edge weights uniquely identifies G.

A.14 Proof of Theorem 7.3.5

Theorem 7.3.5 *Let m be a symmetric evaluation measure satisfying*

$$\forall A, B, C : \quad m(C, AB) \ \geq \ m(C, B)$$

with equality obtaining only if the attributes A and C are conditionally independent given B and

$$\forall A, C : \quad m(C, A) \ \geq \ 0$$

with equality obtaining only if the attributes A and C are (marginally) independent. Let $\vec{G}$ be a singly connected directed acyclic graph of a probability *distribution p over a set U of attributes. Then constructing a maximum weight spanning tree for the attributes in U with m (computed from p) providing the edge weights uniquely identifies the* skeleton *of* $\vec{G}$, *i.e., the undirected graph than results if all edge directions are discarded.*

Proof. Let C and A be two arbitrary attributes in U that are not adjacent in $\vec{G}$. Since $\vec{G}$ is singly connected, there is a unique path connecting C and A. Suppose first that this path does not contain a node with converging edges (from its predecessor and its successor on the path). In this case the proof of Theorem 7.3.4 can be transferred, because, according to d-separation,[11] we have $C \perp\!\!\!\perp_p A \mid B$, but neither $C \perp\!\!\!\perp_p B \mid A$ nor $A \perp\!\!\!\perp_p B \mid C$ (because $\vec{G}$ is a perfect map). Therefore the value of m must be less for the edge (C, A) than for any pair of consecutive nodes on the path connecting C and A.

Suppose next that the path connecting C and A in $\vec{G}$ contains at least one node with converging edges (from its predecessor and its successor on the path). According to the d-separation criterion C and A must be marginally independent and hence it is $m(C, A) = 0$. However, no pair (B_i, B_j) of consecutive nodes on the path is marginally independent (since $\vec{G}$ is a perfect map) and thus $m(B_i, B_j) > 0$.

Hence any edge between two nodes on a path connecting two nonadjacent nodes in the perfect map $\vec{G}$ has a higher weight than the edge connecting them directly. From this it is immediately clear, for example by considering how the Kruskal algorithm [Kruskal 1956] works, that constructing the optimum weight spanning tree with m providing the edge weights uniquely identifies the skeleton of $\vec{G}$.

A.15 Proof of Theorem 7.3.6

Theorem 7.3.6 [Chow and Liu 1968] *Let p be a strictly positive probability distribution over a set U of attributes. Then a best tree-structured approximation*[12] *of p w.r.t. the Kullback–Leibler information divergence*[13] *is obtained by constructing an undirected maximum weight spanning tree of U with mutual information*[14] *providing the edge weights, then directing the edges away from an arbitrarily chosen root node, and finally computing the (conditional) probability distributions associated with the edges of the tree from the given distribution p.*

In order to prove this theorem, it is convenient to prove first the following lemma, by which an important property of the Kullback–Leibler information divergence is established.

[11] d-separation was defined in Definition 4.1.14 on page 98.
[12] The notion of a tree-structured approximation was introduced on page 242.
[13] Kullback–Leibler information divergence was defined in Definition 7.1.3 on page 171.
[14] Mutual (Shannon) information was defined in Definition 7.1.4 on page 174.

Lemma A.15.1 *Let p_1 and p_2 be two strictly positive probability distributions on the same set $\mathcal{E}$ of events. The Kullback–Leibler information divergence is non-negative and zero only if $p_1 \equiv p_2$.*

Proof. The proof is carried out by roughly the same means as the proof of Lemma 7.2.2 in Section A.10.

$$\begin{aligned}
\sum_{E\in\mathcal{E}} p_1(E)\log_2\frac{p_2(E)}{p_1(E)} &= \frac{1}{\ln 2}\sum_{E\in\mathcal{E}} p_1(E)\ln\frac{p_2(E)}{p_1(E)} \\
&\leq \frac{1}{\ln 2}\sum_{E\in\mathcal{E}} p_1(E)\left(\frac{p_2(E)}{p_1(E)} - 1\right) \\
&= \frac{1}{\ln 2}\left(\sum_{E\in\mathcal{E}} p_2(E) - \sum_{E\in\mathcal{E}} p_1(E)\right) \\
&= \frac{1}{\ln 2}(1-1) \;=\; 0
\end{aligned}$$

where the inequality follows from the fact that $\ln x \leq x - 1$, with equality obtaining only for $x = 1$. (This can most easily be seen from the graph of $\ln x$.) Consequently,

$$\sum_{E\in\mathcal{E}} p_1(E)\log_2\frac{p_1(E)}{p_2(E)} = -\sum_{E\in\mathcal{E}} p_1(E)\log_2\frac{p_2(E)}{p_1(E)} \geq 0.$$

In addition,

$$\sum_{E\in\mathcal{E}} p_1(E)\log_2\frac{p_1(E)}{p_2(E)} = 0$$

only if $\forall E \in \mathcal{E} : \frac{p_1(E)}{p_2(E)} = 1$, i.e., if $\forall E \in \mathcal{E} : p_1(E) = p_2(E)$. Note that from this result we also have that the expression $\sum_{E\in\mathcal{E}} p_1(E)\log_2 p_2(E)$ is maximized for a fixed p_1 by choosing $p_2 \equiv p_1$

Proof of Theorem 7.3.6. The proof follows mainly [Pearl 1988]. In a first step it is shown that the Kullback–Leibler divergence of the distribution p and the distribution p_t represented by a given tree is minimized by associating with the edges of the tree the (conditional) probability distributions of the child attribute given the parent attribute that can be computed from p. (This is also called the *projection* of p onto the tree.) Let $U = \{A_1, \dots, A_n\}$ and let, without restricting generality, attribute A_1 be the root of the tree. Furthermore, let $A_{k(j)}$ be the parent of attribute A_j in the tree. For convenience, we denote all probabilities derived from p with P, all derived from p_t by P_t. Finally, we use $P(u)$ as an abbreviation for $P(\bigwedge_{A_i\in U} A_i = a_i)$, $P(a_1)$ as an abbreviation for $P(A_1 = a_1)$ and analogous abbreviations for other expressions. Then

$$\begin{aligned}
& I_{\mathrm{KLdiv}}(p, p_t) \\
&= \sum_{\forall A_i \in U:\ a_i \in \mathrm{dom}(A_i)} P(u) \log_2 \frac{P(u)}{P_t(a_1) \prod_{j=2}^{n} P_t(a_j \mid a_{k(j)})} \\
&= -H^{(\mathrm{Shannon})}(A_1 \ldots A_n) - \sum_{\substack{\forall A_i \in U: \\ a_i \in \mathrm{dom}(A_i)}} P(u) \log_2 P_t(a_1) \prod_{j=2}^{n} P_t(a_j \mid a_{k(j)}) \\
&= -H^{(\mathrm{Shannon})}(A_1 \ldots A_n) - \sum_{a_1 \in \mathrm{dom}(A_1)} P(a_1) \log_2 P_t(a_1) \\
&\quad - \sum_{j=2}^{n} \sum_{\substack{a_j \in \mathrm{dom}(A_j) \\ a_{k(j)} \in \mathrm{dom}(A_{k(j)})}} P(a_j, a_{k(j)}) \log_2 P_t(a_j \mid a_{k(j)})
\end{aligned}$$

From Lemma A.15.1 we know that $\sum_{E \in \mathcal{E}} p_1(E) \log_2 p_2(E)$ is maximized for a fixed distribution p_1 by choosing $p_2 \equiv p_1$. Consequently, the above expression is minimized by choosing

$$\begin{aligned}
\forall a_1 \in \mathrm{dom}(A_1): \quad & P_t(a_1) = P(a_1) \qquad \text{and} \\
\forall a_j \in \mathrm{dom}(A_j): \forall a_{k(j)} \in \mathrm{dom}(A_{k(j)}): \quad & P_t(a_j \mid a_{k(j)}) = P(a_j \mid a_{k(j)}).
\end{aligned}$$

This proves one part of the theorem, namely that the (conditional) probabilities must be those computed from the distribution p.

To show that the best tree is obtained by constructing a maximum weight spanning tree with Shannon information gain providing the edge weights, we simply substitute these equations into the Kullback–Leibler information divergence. This yields

$$\begin{aligned}
& I_{\mathrm{KLdiv}}(p, p_t) \\
&= -H^{(\mathrm{Shannon})}(A_1 \ldots A_n) - \sum_{a_1 \in \mathrm{dom}(A_1)} P(a_1) \log_2 P(a_1) \\
&\quad - \sum_{j=2}^{n} \sum_{\substack{a_j \in \mathrm{dom}(A_j) \\ a_{k(j)} \in \mathrm{dom}(A_{k(j)})}} P(a_j, a_{k(j)}) \log_2 P(a_j \mid a_{k(j)}) \\
&= -H^{(\mathrm{Shannon})}(A_1 \ldots A_n) + H^{(\mathrm{Shannon})}(A_1) \\
&\quad - \sum_{j=2}^{n} \sum_{\substack{a_j \in \mathrm{dom}(A_j) \\ a_{k(j)} \in \mathrm{dom}(A_{k(j)})}} P(a_j, a_{k(j)}) \left(\log_2 \frac{P(a_j, a_{k(j)})}{P(a_j) P(a_{k(j)})} + \log_2 P(a_j) \right) \\
&= -H^{(\mathrm{Shannon})}(A_1 \ldots A_n) - \sum_{i=1}^{n} H^{(\mathrm{Shannon})}(A_i) - \sum_{j=2}^{n} I_{\mathrm{gain}}^{(\mathrm{Shannon})}(A_j, A_{k(j)}).
\end{aligned}$$

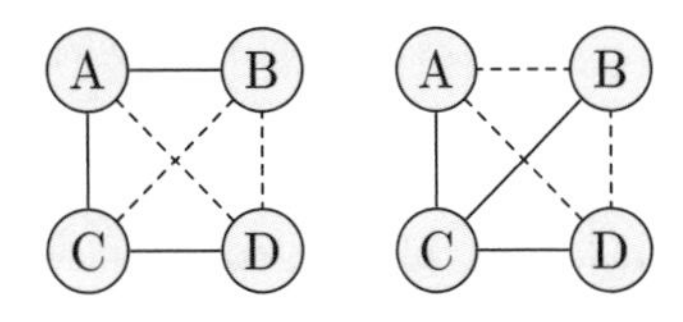

Figure A.1 Maximal cliques with four or more nodes cannot be created without breaking the rules for adding edges.

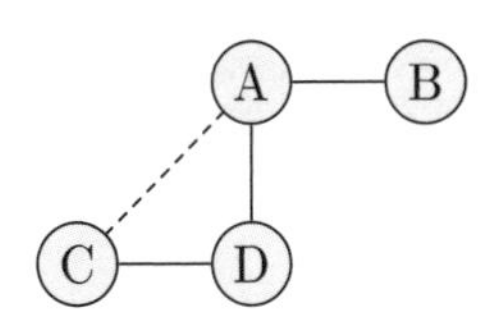

Figure A.2 The node A can be bypassed only by an edge connecting the node D to a neighbor of A.

Since the first two terms are independent of the structure of the tree, the Kullback–Leibler information divergence is minimized by choosing the tree that maximizes the third term, which is the sum of the edge weights of the tree. Hence the tree underlying the best tree-structured approximation is obtained by constructing the maximum weight spanning tree with Shannon information gain providing the edge weights. This proves the theorem.

A.16 Proof of Theorem 7.3.8

Theorem 7.3.8 *If an undirected tree is extended by adding edges only between nodes with a common neighbor in the tree and if the added edges alone do not form a cycle, then the resulting graph has hypertree structure and its maximal cliques contain at most three nodes.*

Proof. Consider first the size of the maximal cliques. Figure A.1 shows, with solid edges, the two possible structurally different spanning trees for four nodes. In order to turn these into cliques the dotted edges have to be added. However, in the graph on the left the edge (B, D) connects two nodes not having a common neighbor in the original tree and in the graph on the right the additional edges form a cycle. Therefore it is impossible to get a clique with a size greater than three without breaking the rules for adding edges.

In order to show that the resulting graph has hypertree structure, it is sufficient to show that all cycles with a length greater than three have a chord (cf. Section 4.2.2). This is easily verified with the following argument. Neither the original tree nor the graph without the edges of this tree contain a cycle. Therefore in all cycles there must be a node A at which an edge from the original tree meets an added edge. Let the former edge connect the nodes B and A and the latter connect the nodes C and A. Since edges may only be added between nodes that have a common neighbor in the tree, there must be a node D that is adjacent to A as well as to C in the original tree.

This node may or may not be identical to B. If it is identical to B and the cycle has a length greater than three, then the edge (B, C) clearly is a chord. Otherwise the edge (A, D) is a chord, because D must also be in the cycle. To see this, consider Figure A.2, which depicts the situation referred to. To close the cycle we are studying there must be a path connecting B and C that does not contain A. However, from the figure it is immediately clear that any such path must contain D, because A can only be bypassed via an edge that has been added between D and a neighbor of A (note that this neighbor may or may not be B).

Appendix B

Software Tools

In this appendix we briefly describe some software tools that support reasoning with graphical models and/or inducing them from a database of sample cases. The following list is also available on our WWW pages:

`http://fuzzy.cs.uni-magdeburg.de/books/gm/tools.html`.

Of course, we do not claim this list to be complete (definitely it is not). Nor does it represent a ranking of the tools, since they are ordered alphabetically. More extensive lists of probabilistic network tools have been compiled by Russel Almond (an old list, which is not maintained anymore):

`http://www.stat.washington.edu/almond/belief.html`,

Kevin Patrick Murphy:

`http://www.cs.berkeley.edu/~murphyk/Bayes/bnsoft.html`

and Google:

`http://directory.google.com/Top/Computers/`
`Artificial_Intelligence/Belief_Networks/Software/`.

The Bayesian Network Repository is also a valuable resource. It lists examples of Bayesian networks and datasets, from which they can be learned:

`http://www.cs.huji.ac.il/labs/compbio/Repository/`.

The software we developed in connection with this book is available at

`http://fuzzy.cs.uni-magdeburg.de/books/gm/software.html`.

Tools for troubleshooting Microsoft products, which are based on Bayesian networks (but do not allow you to access them directly), can be found at

`http://support.microsoft.com/support/tshoot/default.asp`.

BayesBuilder

SNN, University of Nijmegen
PO Box 9101, 6500 HB Nijmegen, The Netherlands
`http://www.mbfys.kun.nl/snn/Research/bayesbuilder/`

BayesBuilder is a tool for (manually) constructing Bayesian networks and drawing inferences with them. It supports neither parameter nor structure learning of Bayesian networks. The graphical user interface of this program is written in Java and is easy to use. However, the program is available only for Windows, because the underlying inference engine is written in C++ and has only been compiled for Windows yet. BayesBuilder is free software.

Bayesian Knowledge Discoverer / Bayesware Discoverer

Knowledge Media Institute / Department of Statistics
The Open University
Walton Hall, Milton Keynes MK7 6AA, United Kingdom
`http://kmi.open.ac.uk/projects/bkd/`

Bayesware Ltd.
`http://bayesware.com/`

The Bayesian Knowledge Discoverer is a software tool that can learn Bayesian networks from data (structure as well as parameters). The dataset to learn from may contain missing values, which are handled by an approach called "bound and collapse" that is based on probability intervals. The Bayesian Knowledge Discoverer is free software, but it has been succeeded by a commercial version, the Bayesware Discoverer. This program has a nice graphical user interface with some powerful visualization options. A 30 days trial version may be retrieved free of charge. Bayesware Discoverer is available for Windows, Unix, and Macintosh.

Bayes Net Toolbox

Kevin Patrick Murphy
Department of Computer Science, UC Berkeley
387 Soda Hall, Berkeley, CA 94720-1776, USA
`http://www.cs.berkeley.edu/~murphyk/Bayes/bnt.html`

The Bayes Net Toolbox is an extension for Matlab, a well-known and widely used mathematical software package. It supports several different algorithms for drawing inferences in Bayesian networks as well as several algorithms for learning the parameters and the structure of Bayesian networks from a dataset of sample cases. It does not have a graphical user interface of its own, but profits from the visualization capabilities of Matlab. The Bayes Net Toolbox is distributed under the Gnu Library General Public License and is available for all systems that can run Matlab, an installation of which is required.

Belief Network Power Constructor

Jie Cheng
Dept. of Computing Science, University of Alberta
155 Athabasca Hall, Edmonton, Alberta, Canada T6G 2E1
`http://www.cs.ualberta.ca/~jcheng/bnpc.htm`

The Bayesian Network Power Constructor uses a three-phase algorithm that is based on conditional independence tests to learn a Bayesian network from data. The conditional independence tests rely on mutual information, which is used to determine whether a (set of) node(s) can reduce or even block the information flow from one node to another. The program comes with a graphical user interface, though a much less advanced one than those of, for instance, HUGIN and Netica (see below). It does not support drawing inferences, but has the advantage that it is free software. It is available only for Windows.

GeNIe / SMILE

Decision Systems Laboratory, University of Pittsburgh
B212 SLIS Building, 135 North Bellefield Avenue, Pittsburgh, PA 15260, USA
`http://www2.sis.pitt.edu/~genie/`

SMILE (Structural Modeling, Inference and Learning Engine) is a library of functions for building Bayesian networks and drawing inferences with them. It does support neither parameter nor structural learning of Bayesian networks. GeNIe (Graphical Network Interface) is a graphical user interface for SMILE, that makes the functions of SMILE easily accessible. While SMILE is platform-independent, GeNIe is available only for Windows, since it relies heavily on the Microsoft Foundation classes. Both packages are distributed free of charge.

Hugin

Hugin Expert A/S
Niels Jernes Vej 10, 9220 Aalborg, Denmark
`http://www.hugin.com`

Hugin is one of the oldest and best-known tools for Bayesian network construction and inference. It comes with an easy to use graphical user interface, but also has an API (application programmers interface) for several programming languages, so that the inference engine can be used in other programs. It supports estimating the parameters of a Bayesian network from a dataset of sample cases. In a recent version it has also been extended by a learning algorithm for the structure of a Bayesian network, which is based on conditional independence tests. Hugin is a commercial tool, but a demo version with restricted capabilities may be retrieved free of charge. Hugin is available for Windows and Solaris (Sun Unix).

Netica

Norsys Software Corp.
2315 Dunbar Street, Vancouver, BC, Canada V6R 3N1
`http://www.norsys.com`

Like Hugin, Netica is a commercial tool with an advanced graphical user interface. It supports Bayesian network construction and inference and also comprises an API (application programmers interface) for C++, so that the inference engine may be used in other programs. Netica offers quantitative network learning (known structure, parameter estimation) from a dataset of sample cases, which may contain missing values. It does not support structural learning. A version of Netica with restricted capabilities may be retrieved free of charge, but the price of a full version is also moderate. Netica is available for Windows and Macintosh.

Pulcinella

IRIDA, Université Libre de Bruxelles
50, Av. F. Roosevelt, CP 194/6, B-1050 Brussels, Belgium
`http://iridia.ulb.ac.be/pulcinella/Welcome.html`

Pulcinella is more general than the other programs listed in this appendix, as it is based on the framework of valuation systems [Shenoy 1992a]. Pulcinella supports reasoning by propagating uncertainty with local computations w.r.t. different uncertainty calculi, but does not support learning graphical models from a dataset of sample cases in any way. The current version of Pulcinella does not have a graphical user interface, but an outdated version of such an interface may be retrieved for Solaris (Sun Unix). Pulcinella is available for Solaris (Sun Unix) and Macintosh, but requires a Common Lisp system.

Tetrad

Tetrad Project, Department of Philosophy
Carnegie Mellon University, Pittsburgh, PA, USA
`http://hss.cmu.edu/html/departments/philosophy/TETRAD/tetrad.htm`

Tetrad is based on the algorithms developed in [Spirtes *et al.* 1993], i.e. on conditional independence test approaches to learn Bayesian networks from data, and, of course, subsequent research in this direction. It can learn the structure as well as the parameters of a Bayesian network from a dataset of sample cases, but does not support drawing inferences. Currently the program is being ported to Java (Tetrad IV). Older versions are available for MSDOS (Tetrad II) and Windows (Tetrad III). Tetrad II is commercial, but available at a moderate fee. Free beta versions are available of Tetrad III and Tetrad IV.

WinMine / MSBN

Machine Learning and Statistics Group
Microsoft Research, One Microsoft Way, Redmond, WA 98052-6399, USA
`http://research.microsoft.com/~dmax/WinMine/tooldoc.htm`

WinMine is a toolkit, i.e. a set of programs for different tasks, rather than an integrated program. Most programs in this toolkit are command line driven, but there is a graphical user interface for the data converter and a network visualization program. WinMine learns the structure and the parameters of Bayesian networks from data and uses decision trees to represent the conditional distributions. It does not support drawing inferences. However, Microsoft Research also offers MSBN (Microsoft Bayesian Networks), a tool for (manually) building Bayesian networks and drawing inferences with them, MSBN comes with a graphical user interface. Both programs, WinMine as well as MSBN, are available for Windows only.

Bibliography

[Agrawal *et al.* 1993] R. Agrawal, T. Imielienski, and A. Swami. Mining Association Rules between Sets of Items in Large Databases. *Proc. Conf. on Management of Data (Washington, DC, USA)*, 207–216. ACM Press, New York, NY, USA 1993

[Agrawal *et al.* 1996] R. Agrawal, H. Mannila, R. Srikant, H. Toivonen, and A. Verkamo. Fast Discovery of Association Rules. In: [Fayyad *et al.* 1996], 307–328

[Aha 1992] D.W. Aha. Tolerating Noisy, Irrelevant and Novel Attributes in Instance-based Learning Algorithms. *Int. Journal of Man-Machine Studies* 36(2):267–287. Academic Press, San Diego, CA, USA 1992

[Akaike 1974] H. Akaike. A New Look at the Statistical Model Identification. *IEEE Trans. on Automatic Control* 19:716–723. IEEE Press, Piscataway, NJ, USA 1974

[Andersen *et al.* 1989] S.K. Andersen, K.G. Olesen, F.V. Jensen, and F. Jensen. HUGIN — A Shell for Building Bayesian Belief Universes for Expert Systems. *Proc. 11th Int. J. Conf. on Artificial Intelligence (IJCAI'89, Detroit, MI, USA)*, 1080–1085. Morgan Kaufmann, San Mateo, CA, USA 1989

[Anderson 1935] E. Anderson. The Irises of the Gaspe Peninsula. *Bulletin of the American Iris Society* 59:2–5. Philadelphia, PA, USA 1935

[Anderson 1995] J.A. Anderson. *An Introduction to Neural Networks.* MIT Press, Cambridge, MA, USA 1995

[Andreassen *et al.* 1987] S. Andreassen, M. Woldbye, B. Falck, and S.K. Andersen. MUNIN — A Causal Probabilistic Network for Interpretation of Electromyographic Findings. *Proc. 10th Int. J. Conf. on Artificial Intelligence (IJCAI'87, Milan, Italy)*, 366–372. Morgan Kaufmann, San Mateo, CA, USA 1987

[Azvine *et al.* 2000] B. Azvine and N. Azarmi and D. Nauck, eds. *Soft Computing and Intelligent Systems: Prospects, Tools and Applications.* LNAI 1804, Springer-Verlag, Berlin, Germany 2000

[Baim 1988] P.W. Baim. A Method for Attribute Selection in Inductive Learning Systems. *IEEE Trans. on Pattern Analysis and Machine Intelligence* 10:888-896. IEEE Press, Piscataway, NJ, USA 1988

[Baldwin *et al.* 1995] J.F. Baldwin, T.P. Martin, and B.W. Pilsworth. *FRIL — Fuzzy and Evidential Reasoning in Artificial Intelligence.* Research Studies Press/J. Wiley & Sons, Taunton/Chichester, United Kingdom 1995

[Bartlett 1935] M.S. Bartlett. Contingency Table Interactions. *Journal of the Royal Statistical Society, Supplement* 2:248–252. Blackwell, Oxford, United Kingdom 1935

[Bauer *et al.* 1997] E. Bauer, D. Koller, and Y. Singer. Update Rules for Parameter Estimation in Bayesian Networks. *Proc. 13th Conf. on Uncertainty in Artificial Intelligence (UAI'97, Providence, RI, USA)*, 3–13. Morgan Kaufmann, San Mateo, CA, USA 1997

[Bernardo *et al.* 1992] J. Bernardo, J. Berger, A. Dawid, and A. Smith, eds. *Bayesian Statistics 4.* Oxford University Press, New York, NY, USA 1992

[Bezdek 1981] J.C. Bezdek. *Pattern Recognition with Fuzzy Objective Function Algorithms.* Plenum Press, New York, NY, USA 1981

[Bezdek and Pal 1992] J.C. Bezdek and N. Pal. *Fuzzy Models for Pattern Recognition.* IEEE Press, New York, NY, USA 1992

[Bezdek *et al.* 1999] J.C. Bezdek, J. Keller, R. Krishnapuram, and N. Pal. *Fuzzy Models and Algorithms for Pattern Recognition and Image Processing.* Kluwer, Dordrecht, Netherlands 1999

[Bock 1974] H.H. Bock. *Automatische Klassifikation (Cluster-Analyse).* Vandenhoek & Ruprecht, Göttingen, Germany 1974

[Bodendiek and Lang 1995] R. Bodendiek and R. Lang. *Lehrbuch der Graphentheorie.* Spektrum Akademischer Verlag, Heidelberg, Germany 1995

[Borgelt 1995] C. Borgelt. *Diskussion verschiedener Ansätze zur Modellierung von Unsicherheit in relationalen Datenbanken.* Studienarbeit, TU Braunschweig, Germany 1995

[Borgelt 1998] C. Borgelt. A Decision Tree Plug-In for DataEngine. *Proc. 2nd Int. Data Analysis Symposium.* MIT GmbH, Aachen, Germany 1998. Reprinted in: *Proc. 6th European Congress on Intelligent Techniques and Soft Computing (EUFIT'98, Aachen, Germany)*, Vol. 2:1299–1303. Verlag Mainz, Aachen, Germany 1998

[Borgelt 1999] C. Borgelt. A Naive Bayes Classifier Plug-In for DataEngine. *Proc. 3rd Int. Data Analysis Symposium*, 87–90. MIT GmbH, Aachen, Germany 1999

[Borgelt *et al.* 1996] C. Borgelt, J. Gebhardt, and R. Kruse. Concepts for Probabilistic and Possibilistic Induction of Decision Trees on Real World

Data. *Proc. 4th European Congress on Intelligent Techniques and Soft Computing (EUFIT'96, Aachen, Germany)*, Vol. 3:1556–1560. Verlag Mainz, Aachen, Germany 1996

[Borgelt and Gebhardt 1997] C. Borgelt and J. Gebhardt. Learning Possibilistic Networks with a Global Evaluation Method. *Proc. 5th European Congress on Intelligent Techniques and Soft Computing (EUFIT'97, Aachen, Germany)*, Vol. 2:1034–1038. Verlag Mainz, Aachen, Germany 1997

[Borgelt *et al.* 1998a] C. Borgelt, J. Gebhardt, and R. Kruse. Chapter F1.2: Inference Methods. In: [Ruspini *et al.* 1998], F1.2:1–14

[Borgelt and Gebhardt 1999] C. Borgelt and J. Gebhardt. A Naive Bayes Style Possibilistic Classifier. *Proc. 7th European Congress on Intelligent Techniques and Soft Computing (EUFIT'99, Aachen, Germany)*, CD-ROM. Verlag Mainz, Aachen, Germany 1999

[Borgelt and Kruse 1997a] C. Borgelt and R. Kruse. Evaluation Measures for Learning Probabilistic and Possibilistic Networks. *Proc. 6th IEEE Int. Conf. on Fuzzy Systems (FUZZ-IEEE'97, Barcelona, Spain)*, Vol. 2:1034–1038. IEEE Press, Piscataway, NJ, USA 1997

[Borgelt and Kruse 1997b] C. Borgelt and R. Kruse. Some Experimental Results on Learning Probabilistic and Possibilistic Networks with Different Evaluation Measures. *Proc. 1st Int. J. Conf. on Qualitative and Quantitative Practical Reasoning (ECSQARU/FAPR'97, Bad Honnef, Germany)*, LNAI 1244, 71–85. Springer-Verlag, Berlin, Germany 1997

[Borgelt and Kruse 1998a] C. Borgelt and R. Kruse. Probabilistic and Possibilistic Networks and How to Learn Them from Data. In: [Kaynak *et al.* 1998], 403–426

[Borgelt and Kruse 1998b] C. Borgelt and R. Kruse. Attributauswahlmaße für die Induktion von Entscheidungsbäumen: Ein Überblick. In: [Nakhaeizadeh 1998a], 77–98

[Borgelt and Kruse 1998c] C. Borgelt and R. Kruse. Efficient Maximum Projection of Database-Induced Multivariate Possibility Distributions. *Proc. 7th IEEE Int. Conf. on Fuzzy Systems (FUZZ-IEEE'98, Anchorage, AK, USA)*, CD-ROM. IEEE Press, Piscataway, NJ, USA 1998

[Borgelt and Kruse 1998d] C. Borgelt and R. Kruse. Possibilistic Networks with Local Structure. *Proc. 6th European Congress on Intelligent Techniques and Soft Computing (EUFIT'98, Aachen, Germany)*, Vol. 1:634–638. Verlag Mainz, Aachen, Germany 1998

[Borgelt *et al.* 1998b] C. Borgelt, R. Kruse, and G. Lindner. Lernen probabilistischer and possibilistischer Netze aus Daten: Theorie und Anwendung. *Künstliche Intelligenz* (Themenheft Data Mining) 12(2):11–17. ScienTec, Bad Ems, Germany 1998

[Borgelt and Kruse 1999] C. Borgelt and R. Kruse. A Critique of Inductive Causation. *Proc. 5th European Conf. on Symbolic and Quantitative Approaches to Reasoning and Uncertainty (ECSQARU'99, London, United Kingdom)*, LNAI 1638, 68–79. Springer-Verlag, Heidelberg, Germany 1999

[Borgelt and Timm 2000] C. Borgelt and H. Timm. Advanced Fuzzy Clustering and Decision Tree Plug-Ins for DataEngine. In: [Azvine *et al.* 2000], 188–212.

[Borgelt and Kruse 2001] C. Borgelt and R. Kruse. An Empirical Investigation of the K2 Metric. *Proc. 6th European Conf. on Symbolic and Quantitative Approaches to Reasoning and Uncertainty (ECSQARU'01, Toulouse, France)*. Springer-Verlag, Heidelberg, Germany 2001

[Borgelt *et al.* 2001] C. Borgelt, H. Timm, and R. Kruse. Probabilistic Networks and Fuzzy Clustering as Generalizations of Naive Bayes Classifiers. In: [Reusch and Temme 2001], 121–138.

[Boulay *et al.* 1987] B.D. Boulay, D. Hogg, and L. Steels, eds. *Advances in Artificial Intelligence 2*. North-Holland, Amsterdam, Netherlands 1987

[Boutilier *et al.* 1996] C. Boutilier, N. Friedman, M. Goldszmidt, and D. Koller. Context Specific Independence in Bayesian Networks. *Proc. 12th Conf. on Uncertainty in Artificial Intelligence (UAI'96, Portland, OR, USA)*, 115–123. Morgan Kaufmann, San Mateo, CA, USA 1996

[Breiman *et al.* 1984] L. Breiman, J.H. Friedman, R.A. Olshen, and C.J. Stone. *Classification and Regression Trees*. Wadsworth International Group, Belmont, CA, USA 1984

[Buntine 1991] W. Buntine. Theory Refinement on Bayesian Networks. *Proc. 7th Conf. on Uncertainty in Artificial Intelligence (UAI'91, Los Angeles, CA, USA)*, 52–60. Morgan Kaufmann, San Mateo, CA, USA 1991

[Buntine 1994] W. Buntine. Operations for Learning with Graphical Models. *Journal of Artificial Intelligence Research* 2:159–225. Morgan Kaufman, San Mateo, CA, USA 1994

[de Campos 1996] L.M. de Campos. *Independence Relationships and Learning Algorithms for Singly Connected Networks*. DECSAI Technical Report 96-02-04, Universidad de Granada, Spain 1996

[de Campos *et al.* 1995] L.M. de Campos, J. Gebhardt, and R. Kruse. Axiomatic Treatment of Possibilistic Independence. In: [Froidevaux and Kohlas 1995], 77–88

[de Campos *et al.* 2000] L.M. de Campos, J.F. Huete, and S. Moral. Independence in Uncertainty Theories and Its Application to Learning Belief Networks. In: [Gabbay and Kruse 2000], 391–434.

[Carnap 1958] R. Carnap. *Introduction to Symbolic Logic and Its Applications*. Dover, New York, NY, USA 1958

[Castillo *et al.* 1997] E. Castillo, J.M. Gutierrez, and A.S. Hadi. *Expert Systems and Probabilistic Network Models.* Springer-Verlag, New York, NY, USA 1997

[Chapman *et al.* 1999] P. Chapman, J. Clinton, T. Khabaza, T. Reinartz, and R. Wirth. *The CRISP-DM Process Model.* NCR, Denmark 1999. `http://www.ncr.dk/CRISP/`.

[Cheeseman *et al.* 1988] P. Cheeseman, J. Kelly, M. Self, J. Stutz, W. Taylor, and D. Freeman. AutoClass: A Bayesian Classification System. *Proc. 5th Int. Workshop on Machine Learning*, 54–64. Morgan Kaufmann, San Mateo, CA, USA 1988

[Chickering 1995] D.M. Chickering. Learning Bayesian Networks is NP-Complete. *Proc. 5th Int. Workshop on Artificial Intelligence and Statistics (Fort Lauderdale, FL, USA)*, 121–130. Springer-Verlag, New York, NY, USA 1995

[Chickering *et al.* 1994] D.M. Chickering, D. Geiger, and D. Heckerman. *Learning Bayesian Networks is NP-Hard (Technical Report MSR-TR-94-17).* Microsoft Research, Advanced Technology Division, Redmond, WA, USA 1994

[Chickering *et al.* 1997] D.M. Chickering, D. Heckerman, and C. Meek. A Bayesian Approach to Learning Bayesian Networks with Local Structure. *Proc. 13th Conf. on Uncertainty in Artificial Intelligence (UAI'97, Providence, RI, USA)*, 80–89. Morgan Kaufmann, San Mateo, CA, USA 1997

[Chow and Liu 1968] C.K. Chow and C.N. Liu. Approximating Discrete Probability Distributions with Dependence Trees. *IEEE Trans. on Information Theory* 14(3):462–467. IEEE Press, Piscataway, NJ, USA 1968

[Clarke *et al.* 1993] M. Clarke, R. Kruse, and S. Moral, eds. *Symbolic and Quantitative Approaches to Reasoning and Uncertainty (LNCS 747).* Springer-Verlag, Berlin, Germany 1993

[Cooper and Herskovits 1992] G.F. Cooper and E. Herskovits. A Bayesian Method for the Induction of Probabilistic Networks from Data. *Machine Learning* 9:309–347. Kluwer, Dordrecht, Netherlands 1992

[Cowell 1992] R. Cowell. BAIES — A Probabilistic Expert System Shell with Qualitative and Quantitative Learning. In: [Bernardo *et al.* 1992], 595–600

[Daróczy 1970] Z. Daróczy. Generalized Information Functions. *Information and Control* 16(1):36–51. Academic Press, San Diego, CA, USA 1970

[Dasarathy 1990] B.V. Dasarathy. *Nearest Neighbor (NN) Norms: NN Pattern Classifcation Techniques.* IEEE Computer Science Press, Los Alamitos, CA, USA 1990

[Dasgupta 1999] S. Dasgupta. Learning Polytrees. *Proc. 15th Conf. on Uncertainty in Artificial Intelligence.* Morgan Kaufmann, San Mateo, CA, USA 1999

[Date 1986] C.J. Date. *An Introduction to Database Systems, Vol. 1.* Addison-Wesley, Reading, MA, USA 1986

[Darwin 1859] C. Darwin. *The Origin of Species.* Penguin, London, United Kingdom 1980 (first published in 1859)

[Dawid 1979] A. Dawid. Conditional Independence in Statistical Theory. *SIAM Journal on Computing* 41:1–31. Society of Industrial and Applied Mathematics, Philadelphia, PA, USA 1979

[Dawkins 1976] R. Dawkins. *The Selfish Gene.* Oxford University Press, Oxford, United Kingdom 1976

[Dawkins 1986] R. Dawkins. *The Blind Watchmaker.* Longman, Harlow, United Kingdom 1986

[Dechter 1990] R. Dechter. Decomposing a Relation into a Tree of Binary Relations. *Journal of Computer and Systems Sciences* 41:2–24. Academic Press, San Diego, CA, USA 1990

[Dechter 1996] R. Dechter. Bucket Elimination: A Unifying Framework for Probabilistic Inference. *Proc. 12th Conf. on Uncertainty in Artificial Intelligence (UAI'96, Portland, OR, USA)*, 211–219. Morgan Kaufmann, San Mateo, CA, USA 1996

[Dechter and Pearl 1992] R. Dechter and J. Pearl. Structure Identification in Relational Data. *Artificial Intelligence* 58:237–270. North-Holland, Amsterdam, Netherlands 1992

[Dempster 1967] A.P. Dempster. Upper and Lower Probabilities Induced by a Multivalued Mapping. *Annals of Mathematical Statistics* 38:325–339. Institute of Mathematical Statistics, Hayward, CA, USA 1967

[Dempster 1968] A.P. Dempster. Upper and Lower Probabilities Generated by a Random Closed Interval. *Annals of Mathematical Statistics* 39:957–966. Institute of Mathematical Statistics, Hayward, CA, USA 1968

[Dempster *et al.* 1977] A.P. Dempster, N. Laird, and D. Rubin. Maximum Likelihood from Incomplete Data via the EM Algorithm. *Journal of the Royal Statistical Society (Series B)* 39:1–38. Blackwell, Oxford, United Kingdom 1977

[Detmer and Gebhardt 2001] H. Detmer and J. Gebhardt. Markov-Netze für die Eigenschaftsplanung und Bedarfsvorschau in der Automobilindustrie. *Künstliche Intelligenz* (Themenheft Unsicherheit und Vagheit) 3/01:16–22. arendtap, Bremen, Germany 2001

[Dirichlet 1839] P.G.L. Dirichlet. Sur un nouvelle methode pour la determination des integrales multiples. *C. R. Acad. Sci. Paris* 8:156–160. Paris, France 1839

[Dougherty *et al.* 1995] J. Dougherty, R. Kohavi, and M. Sahami. Supervised and Unsupervised Discretization of Continuous Features. *Proc. 12th Int. Conf. on Machine Learning (ICML'95, Lake Tahoe, CA, USA)*, 194–202. Morgan Kaufmann, San Mateo, CA, USA 1995

[Dubois and Prade 1988] D. Dubois and H. Prade. *Possibility Theory.* Plenum Press, New York, NY, USA 1988

[Dubois and Prade 1992] D. Dubois and H. Prade. When Upper Probabilities are Possibility Measures. *Fuzzy Sets and Systems* 49:65–74. North-Holland, Amsterdam, Netherlands 1992

[Dubois *et al.* 1993] D. Dubois, S. Moral, and H. Prade. *A Semantics for Possibility Theory Based on Likelihoods.* Annual report, CEC-ESPRIT III BRA 6156 DRUMS II, 1993

[Dubois *et al.* 1996] D. Dubois, H. Prade, and R. Yager, eds. *Fuzzy Set Methods in Information Engineering: A Guided Tour of Applications.* J. Wiley & Sons, New York, NY, USA 1996

[Duda and Hart 1973] R.O. Duda and P.E. Hart. *Pattern Classification and Scene Analysis.* J. Wiley & Sons, New York, NY, USA 1973

[Everitt 1981] B.S. Everitt. *Cluster Analysis.* Heinemann, London, United Kingdom 1981

[Everitt 1998] B.S. Everitt. *The Cambridge Dictionary of Statistics.* Cambridge University Press, Cambridge, United Kingdom 1998

[Ezawa and Norton 1995] K.J. Ezawa and S.W. Norton. Knowledge Discovery in Telecommunication Services Data Using Bayesian Network Models. *Proc. 1st Int. Conf. on Knowledge Discovery and Data Mining (KDD'95, Montreal, Canada)*, 100–105. AAAI Press, Menlo Park, CA, USA 1995

[Ezawa *et al.* 1996] K.J. Ezawa, M. Singh, and S.W. Norton. Learning Goal Oriented Bayesian Networks for Telecommunications Risk Management. *Proc. 13th Int. Conf. on Machine Learning (ICML'96, Bari, Italy)*, 194–202. Morgan Kaufmann, San Mateo, CA, USA 1995

[Farinas del Cerro and Herzig 1994] L. Farinas del Cerro and A. Herzig. Possibility Theory and Independence. *Proc. 5th Int. Conf. on Information Processing and Management of Uncertainty in Knowledge-based Systems (IPMU'94, Paris, France)*, LNCS 945, 820–825. Springer-Verlag, Heidelberg, Germany 1994

[Fayyad *et al.* 1996] U.M. Fayyad, G. Piatetsky-Shapiro, P. Smyth, and R. Uthurusamy, eds. *Advances in Knowledge Discovery and Data Mining.* AAAI Press and MIT Press, Menlo Park and Cambridge, MA, USA 1996

[Feller 1968] W. Feller. *An Introduction to Probability Theory and Its Applications, Vol. 1 (3rd edition).* J. Wiley & Sons, New York, NY, USA 1968

[Feynman *et al.* 1963] R.P. Feynman, R.B. Leighton, and M. Sands. *The Feynman Lectures on Physics, Vol. 1: Mechanics, Radiation, and Heat.* Addison-Wesley, Reading, MA, USA 1963

[Feynman *et al.* 1965] R.P. Feynman, R.B. Leighton, and M. Sands. *The Feynman Lectures on Physics, Vol. 3: Quantum Mechanics.* Addison-Wesley, Reading, MA, USA 1965

[Fisher 1936] R.A. Fisher. The use of multiple measurements in taxonomic problems. *Annals of Eugenics* 7(2):179–188. Cambridge University Press, Cambridge, United Kingdom 1936

[Fisher 1987] D.H. Fisher. Knowledge Acquisition via Incremental Conceptual Clustering. *Machine Learning* 2:139-172. Kluwer, Dordrecht, Netherlands 1987

[Fonck 1994] P. Fonck. Conditional Independence in Possibility Theory. *Proc. 10th Conf. on Uncertainty in Artificial Intelligence (UAI'94, Seattle, WA, USA)*, 221–226. Morgan Kaufmann, San Mateo, CA, USA 1994

[Fredkin and Toffoli 1982] E. Fredkin and T. Toffoli. Conservative Logic. *Int. Journal of Theoretical Physics* 21(3/4):219–253. Plenum Press, New York, NY, USA 1982

[Friedman and Goldszmidt 1996] N. Friedman and M. Goldszmidt. Building Classifiers using Bayesian Networks. *Proc. 13th Nat. Conf. on Artificial Intelligence (AAAI'96, Portland, OR, USA)*, 1277–1284. AAAI Press, Menlo Park, CA, USA 1996

[Froidevaux and Kohlas 1995] C. Froidevaux and J. Kohlas, eds. *Symbolic and Quantitative Approaches to Reasoning and Uncertainty (LNCS 946)*. Springer-Verlag, Berlin, Germany 1995

[Frydenberg 1990] M. Frydenberg. The Chain Graph Markov Property. *Scandinavian Journal of Statistics* 17:333–353. Swets & Zeitlinger, Amsterdam, Netherlands 1990

[Gabbay and Kruse 2000] D. Gabbay and R. Kruse, eds. *DRUMS Handbook on Abduction and Learning.* Kluwer, Dordrecht, Netherlands 2000

[Gath and Geva 1989] I. Gath and A.B. Geva. Unsupervised Optimal Fuzzy Clustering. *IEEE Trans. Pattern Anal. Mach. Intelligence* 11:773–781. IEEE Press, Piscataway, NJ, USA 1989

[Gebhardt 1997] J. Gebhardt. *Learning from Data: Possibilistic Graphical Models.* Habilitation Thesis, University of Braunschweig, Germany 1997

[Gebhardt and Kruse 1992] J. Gebhardt and R. Kruse. A Possibilistic Interpretation of Fuzzy Sets in the Context Model. *Proc. 1st IEEE Int. Conf. on Fuzzy Systems (FUZZ-IEEE'92, San Diego, CA, USA)*, 1089-1096. IEEE Press, Piscataway, NJ, USA 1992

[Gebhardt and Kruse 1993a] J. Gebhardt and R. Kruse. A New Approach to Semantic Aspects of Possibilistic Reasoning. In: [Clarke *et al.* 1993], 151–160

[Gebhardt and Kruse 1993b] J. Gebhardt and R. Kruse. The Context Model — An Integrating View of Vagueness and Uncertainty. *Int. Journal of Approximate Reasoning* 9:283–314. North-Holland, Amsterdam, Netherlands 1993

[Gebhardt and Kruse 1995] J. Gebhardt and R. Kruse. Learning Possibilistic Networks from Data. *Proc. 5th Int. Workshop on Artificial Intelligence and Statistics (Fort Lauderdale, FL, USA)*, 233–244. Springer-Verlag, New York, NY, USA 1995

[Gebhardt and Kruse 1996a] J. Gebhardt and R. Kruse. POSSINFER — A Software Tool for Possibilistic Inference. In: [Dubois *et al.* 1996], 407–418

[Gebhardt and Kruse 1996b] J. Gebhardt and R. Kruse. Tightest Hypertree Decompositions of Multivariate Possibility Distributions. *Proc. 7th Int. Conf. on Information Processing and Management of Uncertainty in Knowledge-based Systems (IPMU'96, Granada, Spain)*, 923–927. Universidad de Granada, Spain 1996

[Gebhardt and Kruse 1996c] J. Gebhardt and R. Kruse. Automated Construction of Possibilistic Networks from Data. *Journal of Applied Mathematics and Computer Science*, 6(3):101–136. University of Zielona Góra, Zielona Góra, Poland 1996

[Gebhardt and Kruse 1998] J. Gebhardt and R. Kruse. Information Source Modeling for Consistent Data Fusion. *Proc. Int. Conf. on Multisource-Multisensor Information Fusion (FUSION'98, Las Vegas, Nevada, USA)*, 27–34. CSREA Press, USA 1996

[Gebhardt 1999] J. Gebhardt. *EPL (Planung und Steuerung von Fahrzeugen, Eigenschaften und Teilen) — Detailmodellierung, Methoden und Verfahren.* ISC Gebhardt and Volkswagen AG, K-DOB-11 (internal report), Wolfsburg, Germany 1999

[Geiger *et al.* 1990] D. Geiger, T.S. Verma, and J. Pearl. Identifying Independence in Bayesian Networks. *Networks* 20:507–534. J. Wiley & Sons, Chichester, United Kingdom 1990

[Geiger and Heckerman 1991] D. Geiger and D. Heckerman. Advances in Probabilistic Reasoning. *Proc. 7th Conf. on Uncertainty in Artificial Intelligence (UAI'91, Los Angeles, CA, USA)*, 118–126. Morgan Kaufmann, San Mateo, CA, USA 1997

[Geiger 1990] D. Geiger. *Graphoids — A Qualitative Framework for Probabilistic Inference.* PhD thesis, University of California at Los Angeles, CA, USA 1990

[Geiger 1992] D. Geiger. An entropy-based learning algorithm of Bayesian conditional trees. *Proc. 8th Conf. on Uncertainty in Artificial Intelligence (UAI'92, Stanford, CA, USA)*, 92–97. Morgan Kaufmann, San Mateo, CA, USA 1992

[Gentsch 1999] P. Gentsch. *Data Mining Tools: Vergleich marktgängiger Tools.* WHU Koblenz, Germany 1999

[Gibbs 1902] W. Gibbs. *Elementary Principles of Statistical Mechanics.* Yale University Press, New Haven, Connecticut, USA 1902

[Good 1965] I.J. Good. *The Estimation of Probabilities: An Essay on Modern Bayesian Methods.* MIT Press, Cambridge, MA, USA 1965

[Goodman *et al.* 1991] I.R. Goodman, M.M. Gupta, H.T. Nguyen, and G.S. Rogers, eds. *Conditional Logic in Expert Systems.* North-Holland, Amsterdam, Netherlands 1991

[Gould 1981] S.J. Gould. *The Mismeasure of Man.* W.W. Norton, New York, NY, USA 1981. Reprinted by Penguin Books, New York, NY, USA 1992

[Greiner 1989] W. Greiner. *Mechanik, Teil 1 (Series: Theoretische Physik).* Verlag Harri Deutsch, Thun/Frankfurt am Main, Germany 1989. English edition: *Classical Mechanics.* Springer-Verlag, Berlin, Germany 2002

[Greiner *et al.* 1987] W. Greiner, L. Neise, and H. Stöcker. *Thermodynamik und Statistische Mechanik (Series: Theoretische Physik).* Verlag Harri Deutsch, Thun/Frankfurt am Main, Germany 1987. English edition: *Thermodynamics and Statistical Physics.* Springer-Verlag, Berlin, Germany 2000

[Hammersley and Clifford 1971] J.M. Hammersley and P.E. Clifford. *Markov Fields on Finite Graphs and Lattices.* Unpublished manuscript, 1971. Cited in: [Isham 1981]

[Hanson and Bauer 1989] S.J. Hanson and M. Bauer. Conceptual Clustering, Categorization, and Polymorphy. *Machine Learning* 3:343-372. Kluwer, Dordrecht, Netherlands 1989

[Hartley 1928] R.V.L. Hartley. Transmission of Information. *The Bell System Technical Journal* 7:535–563. Bell Laboratories, Murray Hill, NJ, USA 1928

[von Hasseln 1998] H. von Hasseln. IPF für bedingte Wahrscheinlichkeiten. In: [Nakhaeizadeh 1998a], 183–192.

[Haykin 1994] S. Haykin. *Neural Networks — A Comprehensive Foundation.* Prentice-Hall, Upper Saddle River, NJ, USA 1994

[Heckerman 1991] D. Heckerman. *Probabilistic Similarity Networks.* MIT Press, Cambridge, MA, USA 1991

[Heckerman 1998] D. Heckerman. *A Tutorial on Learning with Bayesian Networks.* In: [Jordan 1998], 301–354.

[Heckerman *et al.* 1994] D. Heckerman, J.S. Breese, and K. Rommelse. Troubleshooting Under Uncertainty. *Proc. 5th Int. Workshop on Principles of Diagnosis*, 121–130. AAAI Press, Menlo Park, CA, USA 1994

[Heckerman *et al.* 1995] D. Heckerman, D. Geiger, and D.M. Chickering. Learning Bayesian Networks: The Combination of Knowledge and Statistical Data. *Machine Learning* 20:197–243. Kluwer, Dordrecht, Netherlands 1995

[Herskovits and Cooper 1990] E. Herskovits and G.F. Cooper. Kutato: An entropy-driven System for the Construction of Probabilistic Expert Systems from Databases. *Proc. 6th Conf. on Uncertainty in Artificial Intelligence (UAI'90, Cambridge, MA, USA)*, 117–128. North-Holland, New York, NY, USA 1990

[Hestir *et al.* 1991] K. Hestir, H.T. Nguyen, and G.S. Rogers. A Random Set Formalism for Evidential Reasoning. In: [Goodman *et al.* 1991], 209–344

[Higashi and Klir 1982] M. Higashi and G.J. Klir. Measures of Uncertainty and Information based on Possibility Distributions. *Int. Journal of General Systems* 9:43–58. Gordon and Breach, Newark, NJ, USA 1982

[Hisdal 1978] E. Hisdal. Conditional Possibilities, Independence, and Noninteraction. *Fuzzy Sets and Systems* 1:283–297. North-Holland, Amsterdam, Netherlands 1978

[Höppner *et al.* 1999] F. Höppner, F. Klawonn, R. Kruse, and T. Runkler. *Fuzzy Cluster Analysis.* J. Wiley & Sons, Chichester, United Kingdom 1999

[Huete and de Campos 1993] J.F. Huete and L.M. de Campos. Learning Causal Polytrees. In: [Clarke *et al.* 1993], 180–185.

[Huffman 1952] D.A. Huffman. A Method for the Construction of Minimum Redundancy Codes. *Proc. Institute of Radio Engineers* 40(9):1098–1101. Institute of Radio Engineers, Menasha, WI, USA 1952

[Isham 1981] V. Isham. An Introduction to Spatial Point Processes and Markov Random Fields. *Int. Statistical Review* 49:21–43. Int. Statistical Institute, Voorburg, Netherlands 1981

[Jamshidian and Jennrich 1993] M. Jamshidian and R.I. Jennrich. Conjugate Gradient Acceleration of the EM Algorithm. *Journal of the American Statistical Society* 88(412):221–228. American Statistical Society, Providence, RI, USA 1993

[Jensen 1996] F.V. Jensen. *An Introduction to Bayesian Networks.* UCL Press, London, United Kingdom 1996

[Jordan 1998] M.I. Jordan, ed. *Learning in Graphical Models.* MIT Press, Cambridge, MA, USA 1998

[Kanal and Lemmer 1986] L.N. Kanal and J.F. Lemmer, eds. *Uncertainty in Artificial Intelligence.* North-Holland, Amsterdam, Netherlands 1986

[Kanal *et al.* 1989] L.N. Kanal, T.S. Levitt, and J.F. Lemmer, eds. *Uncertainty in Artificial Intelligence 3.* North-Holland, Amsterdam, Netherlands 1989

[Kaynak *et al.* 1998] O. Kaynak, L. Zadeh, B. Türksen, and I. Rudas, eds. *Computational Intelligence: Soft Computing and Fuzzy-Neuro Integration with Applications (NATO ASI Series F).* Springer-Verlag, New York, NY, USA 1998

[Kira and Rendell 1992] K. Kira and L. Rendell. A Practical Approach to Feature Selection. *Proc. 9th Int. Conf. on Machine Learning (ICML'92, Aberdeen, United Kingdom)*, 250–256. Morgan Kaufmann, San Mateo, CA, USA 1992

[Kirkpatrick *et al.* 1983] S. Kirkpatrick, C.D. Gelatt, and M.P. Vercchi. Optimization by Simulated Annealing. *Science* 220:671–680. High Wire Press, Stanford, CA, USA 1983

[Kline 1980] M. Kline. *Mathematics — The Loss of Certainty.* Oxford University Press, New York, NY, USA 1980

[Klir and Folger 1988] G.J. Klir and T.A. Folger. *Fuzzy Sets, Uncertainty and Information.* Prentice-Hall, Englewood Cliffs, NJ, USA 1988

[Klir and Mariano 1987] G.J. Klir and M. Mariano. On the Uniqueness of a Possibility Measure of Uncertainty and Information. *Fuzzy Sets and Systems* 24:141–160. North-Holland, Amsterdam, Netherlands 1987

[Kolmogorov 1933] A.N. Kolmogorov. *Grundbegriffe der Wahrscheinlichkeitsrechnung.* Springer-Verlag, Heidelberg, 1933. English edition: *Foundations of the Theory of Probability.* Chelsea, New York, NY, USA 1956

[Kolodner 1993] J. Kolodner. *Case-Based Reasoning.* Morgan Kaufmann, San Mateo, CA, USA 1993

[Kononenko 1994] I. Kononenko. Estimating Attributes: Analysis and Extensions of RELIEF. *Proc. 7th European Conf. on Machine Learning (ECML'94, Catania, Italy)*, 171–182. Springer-Verlag, New York, NY, USA 1994

[Kononenko 1995] I. Kononenko. On Biases in Estimating Multi-Valued Attributes. *Proc. 1st Int. Conf. on Knowledge Discovery and Data Mining (KDD'95, Montreal, Canada)*, 1034–1040. AAAI Press, Menlo Park, CA, USA 1995

[Koza 1992] J.R. Koza. *Genetic Programming 1 & 2.* MIT Press, Cambridge, CA, USA 1992/1994

[Krippendorf 1986] K. Krippendorf. *Information Theory and Statistics (Quantitative Applications in the Social Sciences 62).* Sage Publications, London, United Kingdom 1986

[Kruse and Schwecke 1990] R. Kruse and E. Schwecke. Fuzzy Reasoning in a Multidimensional Space of Hypotheses. *Int. Journal of Approximate Reasoning* 4:47–68. North-Holland, Amsterdam, Netherlands 1990

[Kruse *et al.* 1991] R. Kruse, E. Schwecke, and J. Heinsohn. *Uncertainty and Vagueness in Knowledge-based Systems: Numerical Methods (Series: Artificial Intelligence).* Springer-Verlag, Berlin, Germany 1991

[Kruse *et al.* 1994] R. Kruse, J. Gebhardt, and F. Klawonn. *Foundations of Fuzzy Systems*, J. Wiley & Sons, Chichester, United Kingdom 1994. Translation of the book: *Fuzzy Systeme (Series: Leitfäden und Monographien der Informatik).* Teubner, Stuttgart, Germany 1993

[Kruskal 1956] J.B. Kruskal. On the Shortest Spanning Subtree of a Graph and the Traveling Salesman Problem. *Proc. American Mathematical Society* 7(1):48–50. American Mathematical Society, Providence, RI, USA 1956

[Kullback and Leibler 1951] S. Kullback and R.A. Leibler. On Information and Sufficiency. *Annals of Mathematical Statistics* 22:79–86. Institute of Mathematical Statistics, Hayward, CA, USA 1951

[Langley *et al.* 1992] P. Langley, W. Iba, and K. Thompson. An Analysis of Bayesian Classifiers. *Proc. 10th Nat. Conf. on Artificial Intelligence (AAAI'92, San Jose, CA, USA)*, 223–228. AAAI Press/MIT Press, Menlo Park/Cambridge, CA, USA 1992

[Langley and Sage 1994] P. Langley and S. Sage. Induction of Selective Bayesian Classifiers. *Proc. 10th Conf. on Uncertainty in Artificial Intelligence (UAI'94, Seattle, WA, USA)*, 399–406. Morgan Kaufmann, San Mateo, CA, USA 1994

[Larsen and Marx 1986] R.J. Larsen and M.L. Marx. *An Introduction to Mathematical Statistics and Its Applications*. Prentice-Hall, Englewood Cliffs, NJ, USA 1986

[Lauritzen and Spiegelhalter 1988] S.L. Lauritzen and D.J. Spiegelhalter. Local Computations with Probabilities on Graphical Structures and Their Application to Expert Systems. *Journal of the Royal Statistical Society, Series B*, 2(50):157–224. Blackwell, Oxford, United Kingdom 1988

[Lauritzen *et al.* 1990] S.L. Lauritzen, A.P. Dawid, B.N. Larsen, and H.G. Leimer. Independence Properties of Directed Markov Fields. *Networks* 20:491–505. J. Wiley & Sons, Chichester, United Kingdom 1990

[Lauritzen 1996] S.L. Lauritzen. *Graphical Models*. Oxford University Press, Oxford, United Kingdom 1996

[Lemmer 1996] J.F. Lemmer. The Causal Markov Condition, Fact or Artifact? *SIGART Bulletin* 7(3):3–16. Association for Computing Machinery, New York, NY, USA 1996

[Little 1977] C.H.C. Little, ed. *Combinatorial Mathematics V (LNM 622)*. Springer-Verlag, New York, NY, USA 1977

[Lopez de Mantaras 1991] R. Lopez de Mantaras. A Distance-based Attribute Selection Measure for Decision Tree Induction. *Machine Learning* 6:81–92. Kluwer, Dordrecht, Netherlands 1991

[Maier 1983] D. Maier. *The Theory of Relational Databases*. Computer Science Press, Rockville, MD, USA 1983

[Metropolis *et al.* 1953] N. Metropolis, N. Rosenblut, A. Teller, and E. Teller. Equation of State Calculations for Fast Computing Machines. *Journal of Chemical Physics* 21:1087–1092. American Institute of Physics, Melville, NY, USA 1953

[Michalski and Stepp 1983] R.S. Michalski and R.E. Stepp. Learning from Observation: Conceptual Clustering. In: [Michalski *et al.* 1983], 331–363

[Michalski *et al.* 1983] R.S. Michalski, J.G. Carbonell, and T.M. Mitchell, ed. *Machine Learning: An Artificial Intelligence Approach.* Morgan Kaufmann, San Mateo, CA, USA 1983

[Michalewicz 1996] Z. Michalewicz. *Genetic Algorithms + Data Structures = Evolution Programs.* Springer-Verlag, Berlin, Germany 1996

[Michie 1989] D. Michie. Personal Models of Rationality. *Journal of Statistical Planning and Inference* 21, Special Issue on Foundations and Philosophy of Probability and Statistics. Swets & Zeitlinger, Amsterdam, Netherlands 1989

[Mucha 1992] H.-J. Mucha. *Clusteranalyse mit Mikrocomputern.* Akademie-Verlag, Berlin, Germany 1992

[Muggleton 1992] S. Muggleton, ed. *Inductive Logic Programming.* Academic Press, San Diego, CA, USA 1992

[Murphy and Aha 1994] P.M. Murphy and D. Aha. *UCI Repository of Machine Learning Databases.* University of California at Irvine, CA, USA 1994. `ftp://ics.uci.edu/pub/machine-learning-databases`

[Nakhaeizadeh 1998a] G. Nakhaeizadeh, ed. *Data Mining: Theoretische Aspekte und Anwendungen.* Physica-Verlag, Heidelberg, Germany 1998

[Nakhaeizadeh 1998b] G. Nakhaeizadeh. Wissensentdeckung in Datenbanken und Data Mining: Ein Überblick. In: [Nakhaeizadeh 1998a], 1–33

[Nauck and Kruse 1997] D. Nauck and R. Kruse. A Neuro-Fuzzy Method to Learn Fuzzy Classification Rules from Data. *Fuzzy Sets and Systems* 89:277–288. North-Holland, Amsterdam, Netherlands 1997

[Nauck *et al.* 1997] D. Nauck, F. Klawonn, and R. Kruse. *Foundations of Neuro-Fuzzy Systems.* J. Wiley & Sons, Chichester, United Kingdom 1997

[Neapolitan 1990] E. Neapolitan. *Probabilistic Reasoning in Expert Systems.* J. Wiley & Sons, New York, NY, USA 1990

[von Neumann 1932] J. von Neumann. *Mathematische Grundlagen der Quantenmechanik.* Springer-Verlag, Berlin, Germany 1932. English edition: *Mathematical Foundations of Quantum Mechanics.* Princeton University Press, Princeton, NJ, USA 1955

[Nguyen 1978] H.T. Nguyen. On Random Sets and Belief Functions. *Journal of Mathematical Analysis and Applications* 65:531–542. Academic Press, San Diego, CA, USA 1978

[Nguyen 1984] H.T. Nguyen. Using Random Sets. *Information Science* 34:265–274. Institute of Information Scientists, London, United Kindom 1984

[Nilsson 1998] N.J. Nilsson. *Artificial Intelligence: A New Synthesis.* Morgan Kaufman, San Francisco, CA, USA 1998

[Nürnberger *et al.* 1999] A. Nürnberger, C. Borgelt, and A. Klose. Improving Naive Bayes Classifiers Using Neuro-Fuzzy Learning. *Proc. Int. Conf. on Neural Information Processing (ICONIP'99, Perth, Australia).* 154–159, IEEE Press, Piscataway, NJ, USA 1999

[Pearl 1986] J. Pearl. Fusion, Propagation, and Structuring in Belief Networks. *Artificial Intelligence* 29:241–288. North-Holland, Amsterdam, Netherlands 1986

[Pearl 1988] J. Pearl. *Probabilistic Reasoning in Intelligent Systems: Networks of Plausible Inference.* Morgan Kaufmann, San Mateo, CA, USA 1988 (2nd edition 1992)

[Pearl and Paz 1987] J. Pearl and A. Paz. Graphoids: A Graph Based Logic for Reasoning about Relevance Relations. In: [Boulay *et al.* 1987], 357–363

[Pearl and Verma 1991a] J. Pearl and T.S. Verma. A Theory of Inferred Causation. *Proc. 2nd Int. Conf. on Principles of Knowledge Representation and Reasoning.* Morgan Kaufmann, San Mateo, CA, USA 1991

[Pearl and Verma 1991b] J. Pearl and T.S. Verma. A Statistical Semantics for Causation. *Proc. of the 3rd Int. Workshop on AI and Statistics (Fort Lauderdale, FL, USA).* Reprinted in: *Statistics and Computing 2*, 91–95. Chapman & Hall, New York, NY, USA 1992

[Poole 1993] D. Poole. Probabilistic Horn Abduction and Bayesian Networks. *Artificial Intelligence*, 64(1):81-129. North-Holland, Amsterdam, Netherlands 1993

[Popper 1934] K.R. Popper. *Logik der Forschung.* 1st edition: Springer-Verlag, Vienna, Austria 1934. 9th edition: J.C.B. Mohr, Tübingen, Germany 1989. English edition: *The Logic of Scientific Discovery.* Hutchinson, London, United Kingdom 1959

[Press *et al.* 1992] W.H. Press, S.A. Teukolsky, W.T. Vetterling, and B.P. Flannery. *Numerical Recipes in C — The Art of Scientific Computing (2nd edition).* Cambridge University Press, Cambridge, United Kingdom 1992

[Prim 1957] R.C. Prim. Shortest Connection Networks and Some Generalizations. *The Bell System Technical Journal* 36:1389-1401. Bell Laboratories, Murray Hill, NJ, USA 1957

[Quinlan 1986] J.R. Quinlan. Induction of Decision Trees. *Machine Learning* 1:81–106. Kluwer, Dordrecht, Netherlands 1986

[Quinlan 1993] J.R. Quinlan. *C4.5: Programs for Machine Learning.* Morgan Kaufmann, San Mateo, CA, USA 1993

[de Raedt and Bruynooghe 1993] L. de Raedt and M. Bruynooghe. A Theory of Clausal Discovery. *Proc. 13th Int. J. Conf. on Artificial Intelligence.* Morgan Kaufmann, San Mateo, CA, USA 1993

[Rasmussen 1992] L.K. Rasmussen. *Blood Group Determination of Danish Jersey Cattle in the F-blood Group System (Dina Research Report 8).* Dina Foulum, Tjele, Denmark 1992

[Rebane and Pearl 1987] G. Rebane and J. Pearl. The Recovery of Causal Polytrees from Statistical Data. *Proc. 3rd Workshop on Uncertainty in Artificial Intelligence (Seattle, WA, USA)*, 222–228. USA 1987. Reprinted in: [Kanal *et al.* 1989], 175–182

[Reichenbach 1944] H. Reichenbach. *Philosophic Foundations of Quantum Mechanics.* University of California Press, Berkeley, CA, USA 1944. Reprinted by Dover Publications, New York, NY, USA 1998

[Reichenbach 1947] H. Reichenbach. *Elements of Symbolic Logic.* Macmillan, New York, NY, USA 1947

[Reichenbach 1956] H. Reichenbach. *The Direction of Time.* University of California Press, Berkeley, CA, USA 1956

[Reusch and Temme 2001] B. Reusch and K.-H. Temme, eds. *Computational Intelligence in Theory and Practice.* Series: Advances in Soft Computing. Physica-Verlag, Heidelberg, Germany 2001

[Rissanen 1983] J. Rissanen. A Universal Prior for Integers and Estimation by Minimum Description Length. *Annals of Statistics* 11:416–431. Institute of Mathematical Statistics, Hayward, CA, USA 1983

[Rissanen 1987] J. Rissanen. Stochastic Complexity. *Journal of the Royal Statistical Society (Series B)*, 49:223-239. Blackwell, Oxford, United Kingdom 1987

[Robinson 1977] R.W. Robinson. Counting Unlabeled Acyclic Digraphs. In: [Little 1977], 28–43

[Rojas 1993] R. Rojas. *Theorie der Neuronalen Netze: Eine systematische Einführung.* Springer-Verlag, Berlin, Germany 1993

[Ruelle 1993] D. Ruelle. *Zufall und Chaos.* Springer-Verlag, Heidelberg, Germany 1993

[Ruspini *et al.* 1998] E. Ruspini, P. Bonissone, and W. Pedrycz, eds. *Handbook of Fuzzy Computation.* Institute of Physics Publishing, Bristol, United Kingdom 1998

[Russel *et al.* 1995] K. Russel, J. Binder, D. Koller, and K. Kanazawa. Local Learning in Probabilistic Networks with Hidden Variables. *Proc. 1st Int. Conf. on Knowledge Discovery and Data Mining (KDD'95, Montreal, Canada)*, 1146–1152. AAAI Press, Menlo Park, CA, USA 1995

[Saffiotti and Umkehrer 1991] A. Saffiotti and E. Umkehrer. PULCINELLA: A General Tool for Propagating Uncertainty in Valuation Networks. *Proc. 7th Conf. on Uncertainty in Artificial Intelligence (UAI'91, Los Angeles, CA, USA)*, 323–331. Morgan Kaufmann, San Mateo, CA, USA 1991

[Sahami 1996] M. Sahami. Learning Limited Dependence Bayesian Classifiers. *Proc. 2nd Int. Conf. on Knowledge Discovery and Data Mining (KDD'96, Portland, OR, USA)*, 335–338. AAAI Press, Menlo Park, CA, USA 1996

[Salmon 1963] W.C. Salmon. *Logic.* Prentice-Hall, Englewood Cliffs, NJ, USA 1963

[Salmon 1984] W.C. Salmon. *Scientific Explanation and the Causal Structure of the World.* Princeton University Press, Princeton, NJ, USA 1984

[Savage 1954] L.J. Savage. *The Foundations of Statistics.* J. Wiley & Sons, New York, NY, USA 1954. Reprinted by Dover Publications, New York, NY, USA 1972

[Schwarz 1978] G. Schwarz. Estimating the Dimension of a Model. *Annals of Statistics* 6:461–464. Institute of Mathematical Statistics, Hayward, CA, USA 1978

[Shachter *et al.* 1990] R.D. Shachter, T.S. Levitt, L.N. Kanal, and J.F. Lemmer, eds. *Uncertainty in Artificial Intelligence 4.* North-Holland, Amsterdam, Netherlands 1990

[Shafer 1976] G. Shafer. *A Mathematical Theory of Evidence.* Princeton University Press, Princeton, NJ, USA 1976

[Shafer and Pearl 1990] G. Shafer and J. Pearl. *Readings in Uncertain Reasoning.* Morgan Kaufmann, San Mateo, CA, USA 1990

[Shafer and Shenoy 1988] G. Shafer and P.P. Shenoy. *Local Computations in Hypertrees (Working Paper 201).* School of Business, University of Kansas, Lawrence, KS, USA 1988

[Shannon 1948] C.E. Shannon. The Mathematical Theory of Communication. *The Bell System Technical Journal* 27:379–423. Bell Laboratories, Murray Hill, NJ, USA 1948

[Shenoy 1991b] P.P. Shenoy. *Conditional Independence in Valuation-based Systems (Working Paper 236).* School of Business, University of Kansas, Lawrence, KS, USA 1991

[Shenoy 1992a] P.P. Shenoy. Valuation-based Systems: A Framework for Managing Uncertainty in Expert Systems. In: [Zadeh and Kacprzyk 1992], 83–104

[Shenoy 1992b] P.P. Shenoy. Conditional Independence in Uncertainty Theories. *Proc. 8th Conf. on Uncertainty in Artificial Intelligence (UAI'92, Stanford, CA, USA)*, 284–291. Morgan Kaufmann, San Mateo, CA, USA 1992

[Shenoy 1993] P.P. Shenoy. Valuation Networks and Conditional Independence. *Proc. 9th Conf. on Uncertainty in AI (UAI'93, Washington, DC, USA)*, 191–199. Morgan Kaufmann, San Mateo, CA, USA 1993

[Singh and Valtorta 1993] M. Singh and M. Valtorta. An Algorithm for the Construction of Bayesian Network Structures from Data. *Proc. 9th Conf. on Uncertainty in Artificial Intelligence (UAI'93, Washington, DC, USA)*, 259–265. Morgan Kaufmann, San Mateo, CA, USA 1993

[Smith *et al.* 1993] J.E. Smith, S. Holtzman, and J.E. Matheson. Structuring Conditional Relationships in Influence Diagrams. *Operations Research* 41(2):280–297. INFORMS, Linthicum, MD, USA 1993

[Spirtes *et al.* 1989] P. Spirtes, C. Glymour, and R. Scheines. *Causality from Probability (Technical Report CMU-LCL-89-4)*. Department of Philosophy, Carnegie-Mellon University, Pittsburgh, PA, USA 1989

[Spirtes *et al.* 1993] P. Spirtes, C. Glymour, and R. Scheines. *Causation, Prediction, and Search (Lecture Notes in Statistics 81)*. Springer-Verlag, New York, NY, USA 1993

[Spohn 1990] W. Spohn. A General Non-Probabilistic Theory of Inductive Reasoning. In: [Shachter *et al.* 1990], 149–158

[Srikant and Agrawal 1996] R. Srikant and R. Agrawal. Mining Quantitative Association Rules in Large Relational Tables. *Proc. Int. Conf. Management of Data (Montreal, Quebec, Canada)*, 1–12. ACM Press, New York, NY, USA 1996

[Steck and Tresp 1999] H. Steck and V. Tresp. Bayesian Belief Networks for Data Mining. *Lernen, Wissensentdeckung und Adaptivität (LWA'99)*. Otto-von-Guericke-Universität Magdeburg, Germany 1999

[Studený 1992] M. Studený. Conditional Independence Relations have no Finite Complete Characterization. *Trans. 11th Prague Conf. on Information Theory, Statistical Decision Functions, and Random Processes*, 377–396. Academia, Prague, Czechoslovakia 1992

[Tarjan and Yannakakis 1984] R.E. Tarjan and M. Yannakakis. Simple linear-time algorithms to test chordality of graphs, test acyclicity of hypergraphs and selectively reduce acyclic hypergraphs. *SIAM Journal on Computing* 13:566–579. Society of Industrial and Applied Mathematics, Philadelphia, PA, USA 1984

[Ullman 1988] J.D. Ullman. *Principles of Database and Knowledge-Base Systems, Vol. 1 & 2.* Computer Science Press, Rockville, MD, USA 1988

[Verma and Pearl 1990] T.S. Verma and J. Pearl. Causal Networks: Semantics and Expressiveness. In: [Shachter *et al.* 1990], 69–76

[Vollmer 1981] G. Vollmer. Ein neuer dogmatischer Schlummer? — Kausalität trotz Hume und Kant. *Akten des 5. Int. Kant-Kongresses (Mainz, Germany)*, 1125–1138. Bouvier, Bonn, Germany 1981

[Walley 1991] P. Walley. *Statistical Reasoning with Imprecise Probabilities.* Chapman & Hall, New York, NY, USA 1991

[Wang 1983a] P.Z. Wang. From the Fuzzy Statistics to the Falling Random Subsets. In: [Wang 1983b], 81–96

[Wang 1983b] P.P. Wang, ed. *Advances in Fuzzy Sets, Possibility and Applications.* Plenum Press, New York, NY, USA 1983

[Wang and Mendel 1992] L.-X. Wang and J.M. Mendel. Generating fuzzy rules by learning from examples. *IEEE Trans. on Systems, Man, & Cybernetics* 22:1414–1227. IEEE Press, Piscataway, NJ, USA 1992

[Wehenkel 1996] L. Wehenkel. On Uncertainty Measures Used for Decision Tree Induction. *Proc. 7th Int. Conf. on Information Processing and Management of Uncertainty in Knowledge-based Systems (IPMU'96, Granada, Spain)*, 413–417. Universidad de Granada, Spain 1996

[von Weizsäcker 1992] C.F. Weizsäcker. *Zeit und Wissen.* Hanser, München, Germany 1992

[Wettschereck 1994] D. Wettschereck. *A Study of Distance-Based Machine Learning Algorithms.* PhD Thesis, Oregon State University, OR, USA 1994

[Whittaker 1990] J. Whittaker. *Graphical Models in Applied Multivariate Statistics.* J. Wiley & Sons, Chichester, United Kingdom 1990

[Wright 1921] S. Wright. Correlation and Causation. *Journal of Agricultural Research* 20(7):557-585. US Dept. of Agriculture, Beltsville, MD, USA 1921.

[Zadeh 1975] L.A. Zadeh. The Concept of a Linguistic Variable and Its Application to Approximate Reasoning. *Information Sciences* 9:43–80. Elsevier Science, New York, NY, USA 1975

[Zadeh 1978] L.A. Zadeh. Fuzzy Sets as a Basis for a Theory of Possibility. *Fuzzy Sets and Systems* 1:3–28. North-Holland, Amsterdam, Netherlands 1978

[Zadeh and Kacprzyk 1992] L.A. Zadeh and J. Kacprzyk. *Fuzzy Logic for the Management of Uncertainty.* J. Wiley & Sons, New York, NY, USA 1992

[Zell 1994] A. Zell. *Simulation Neuronaler Netze.* Addison-Wesley, Bonn, Germany 1994

[Zey 1997] R. Zey, ed. *Lexikon der Forscher und Erfinder.* Rowohlt, Reinbek/Hamburg, Germany 1997

[Zhang and Poole 1996] N.L. Zhang and D. Poole. Exploiting Causal Independence in Bayesian Network Inference. *Journal of Artificial Intelligence Research* 5:301–328. Morgan Kaufman, San Mateo, CA, USA 1996

[Zhou and Dillon 1991] X. Zhou and T.S. Dillon. A statistical-heuristic Feature Selection Criterion for Decision Tree Induction. *IEEE Trans. on Pattern Analysis and Machine Intelligence (PAMI)*, 13:834–841. IEEE Press, Piscataway, NJ, USA 1991

Index